AN INTRODUCTION to
NONLINEAR CHEMICAL DYNAMICS

TOPICS IN PHYSICAL CHEMISTRY
A Series of Advanced Textbooks and Monographs

Series Editor
Donald G. Truhlar, University of Minnesota

AN INTRODUCTION to NONLINEAR CHEMICAL DYNAMICS

Oscillations, Waves, Patterns, and Chaos

IRVING R. EPSTEIN
JOHN A. POJMAN

New York   Oxford
Oxford University Press
1998

Oxford University Press

Oxford   New York
Athens   Auckland   Bangkok   Bogotá
Buenos Aires   Calcutta   Cape Town   Chennai   Dar es Salaam
Delhi   Florence   Hong Kong   Istanbul   Karachi
Kuala Lumpur   Madrid   Melbourne
Mexico City   Mumbai   Nairobi   Paris   São Paulo   Singapore
Taipei   Tokyo   Toronto   Warsaw

and associated companies in
Berlin   Ibadan

Published by Oxford University Press, Inc.,
198 Madison Avenue, New York, New York 10016

Oxford is a registered trademark of Oxford University Press

Library of Congress Cataloging-in-Publication Data
Epstein, Irving R. (Irving Robert), 1945–
An introduction to nonlinear chemical dyamics: oscillations,
waves, patterns, and chaos / Irving R. Epstein, John A. Pojman.
p.   cm. — (Topics in physical chemistry)
Includes bibliographical references and index.
ISBN 0–19–509670–3
1. Nonlinear chemical kinetics.   I. Pojman, John A.
(John Anthony). 1962–   .   II. Title.   III. Series: Topics in physical
chemistry series.
QD502.2.E67   1998
541.3'94—dc21   97-48850

1 3 5 7 9 8 6 4 2

Printed in the United States of America
on acid-free paper

To Our Families and Students

Preface

What led us to write this book? A fair question. Is the world clamoring for a lucid introduction to nonlinear chemical dynamics? Did our wives and children implore us to share our insights into this arcane subject with an eager but ignorant world?

We would be the last to suggest that the subject matter treated in this volume is the stuff that bestsellers are made of. Nonetheless, these topics are of interest to a growing number of scientists and engineers, not only chemists, but also physicists, biologists, and others in a variety of obviously and not so obviously related fields. Three decades ago, a book devoted largely to chemical oscillations would have been inconceivable. Most chemists then viewed such behavior as a form of perpetual motion, rendered impossible by the Second Law of Thermodynamics. Fifteen years ago, one might have imagined writing a book of this sort, but it would have been a thin work indeed, since only two chemical oscillators were known, both discovered by serendipity, and only one understood to any extent at a mechanistic level.

Times change, and today chemical oscillations and the more encompassing field of nonlinear chemical dynamics are among the most rapidly growing areas of chemical research. Both of us, teaching at very different universities, have observed that disproportionate numbers of graduate and undergraduate students flock to do research in this area. The visually compelling phenomena and their potential significance to a wide range of problems make nonlinear chemical dynamics a subject about which colleagues in such disparate fields as neurobiology, polymer science, and combustion engineering seek to become better informed.

One of the greatest handicaps facing potential entrants into, or consumers of the fruits of, this field of research is the lack of an introductory text. This gap cannot be accounted for by the inherent difficulty of the subject. The mathematical and chemical tools required are almost all possessed by any well-trained undergraduate in the sciences. Most of the necessary experimental apparatus is, by modern standards, nearly primitive and certainly inexpensive. This last feature accounts in large measure for the many significant contributions that have been made to this field by scientists from Eastern Europe and the developing world.

There are, to be sure, some excellent books of which the reader should be aware and to which we owe a real debt. Limitations of space, both on this page and in our immediately accessible memory banks, confine us to mentioning but a few here. More will be found in the references. Nicolis and Prigogine (1977) discuss the foundations of nonequilibrium thermodynamics, laying to rest the misguided notion that chemical oscillation and its cousin, *dissipative structures*, violate some physical principle or other. They also show, largely by using the classic Brusselator model, that periodic oscillation and spatial wave propagation can arise from very simple mathematical models. They do not, however, at a time when experimental observations of nonlinear dynamical phenomena in chemistry were extremely limited, devote much attention to experimental aspects. Field and Burger (1985) edited a comprehensive collection of essays, nearly all of which focus on the behavior of the archetypal Belousov–Zhabotinsky reaction. Gray and Scott (1990) and then Scott (1991) produced lucid treatments, first of simple models of nonlinear oscillations and waves in chemical systems, and then of chemical chaos. Much of the most exciting work in this rapidly moving field has appeared in collections of conference papers, beginning with the results of the 1968 Prague conference on biological and biochemical oscillators (Chance et al., 1973) that may be said to have launched nonlinear chemical dynamics as a serious field of inquiry.

However, in our view, none of these volumes is satisfactory as a text for a course at the advanced undergraduate or introductory graduate level, or as a means for a newcomer to the field to obtain an overview and a relatively painless means of access to a basic competence in his or her area of interest. We believe strongly that the subject can be taught and learned at this level!

When teaching a course or writing a book, it is always tempting to focus on theory; it lends itself more easily to the blackboard or to the printed page. Chemistry, though, and nonlinear chemical dynamics in particular, is an experimental science. When chemical oscillations existed primarily in Lotka's models (Lotka, 1925), there *was* no subject of nonlinear chemical dynamics. When Turing structures could be found only in the papers of mathematical biologists, they played only a tiny role in this field. We have tried in this book to convey both the experimental and the theoretical background of the subject. We describe how to build a flow reactor, for example. We provide appendices that contain recipes for lecture demonstrations and guidelines for laboratory experiments. We recommend that the reader try at least some of the demonstrations. They are just the sort of thing that hooked many chemists at an early age; solutions suddenly switch from one color to another—not just once, but repeatedly. The demonstrations have provoked gaping mouths and perceptive questions from audiences ranging from elementary schoolchildren to university boards of trustees.

We have chosen to divide the book into two parts. In Part I, we present an overview of the subject. We start with a brief history and then move on to review some of the basic mathematics and chemistry. We next discuss the flow reactor, or CSTR, an experimental tool borrowed from chemical engineers, which led to the rapid expansion of nonlinear chemical dynamics in the 1970s and 1980s. The CSTR allows one to design new chemical oscillators, avoiding the earlier procedure of stumbling upon them. Having outlined how to build a chemical oscillator, we proceed to look at the task of dissecting them—that is, of constructing molecular-level descriptions or mechanisms. A realistic view of most chemical systems takes into account their behavior in space as well as in time. In the systems of interest here, this means that one must consider diffusion and how it can lead to pattern formation and wave propagation, a subject we consider in Chapter 6. We close our overview with a brief discussion of some of the computational tools that have furthered the understanding of these complex chemical systems.

A one-semester course in nonlinear chemical dynamics or an introduction for someone intent on entering the field as a researcher might consist of Part I (some or even much of which will be review, depending on the reader's background) supplemented by one or two of the chapters in Part II. Some of these "special" topics are treated in other books, but others (e.g., delays, polymers, convection) are treated in a context that is, for the first time here, both chemical and pedagogical. Each chapter in Part II can be read independently of the others, though readers may (and we hope will) find that there will be symbiotic effects among certain combinations of chapters (e.g., Chapters 9 and 15 or 11, 12, and 13), some of which may not yet have occurred to the less than omniscient authors. A reasonable year's course in this subject could cover all of the special topics plus selected papers from the current literature and/or projects proposed by the instructor and the students. We have found that students are eager to get in there and do something on their own and that it is not unrealistic for them to do so. Indeed, some of the earliest and most significant discoveries in both of our laboratories were made by undergraduates.

Both of us "wandered" into this field after being trained in other areas. Having seen many others come this way and never look back, we are convinced that the direction we took is a natural one. We hope that this book will make the road just a little smoother for those who follow us. Remember, if you have any doubts, try the demonstrations!

Waltham   I.R.E.
Hattiesburg   J.A.P.
September 1997

Acknowledgments

We are indebted to many who helped in the preparation of this work. We would like to thank those at USM who spent many hours in the library tracking down references, specifically: Lydia Lee Lewis, Dionne Fortenberry, Victor Ilyashenko, Tim Meehan, Gauri Misra, Stanislav Solovyov, Randy Washington, and William West.

The students in the spring 1996 Nonlinear Chemical Dynamics class at USM tested a beta version of this book, and we thank them for their comments.

The following reviewed individual chapters: Arthur Winfree, Patrick De Kepper, Ken Kustin, Robert Olsen, Milos Dolnik, Albert Goldbeter, Dilip Kondepudi, Miklós Orbán, István Nagypál, Milos Marek, Kenneth Showalter, Steven Scott, Desederio Vasquez, and Vitaly Volpert.

We thank P. Camacho, Kristztina Kurin-Csörgei, Patrick De Kepper, J. Lechleiter, Miklós Orbán, Stefan Müller, Reuben Simoyi, Qui Tran-Cong, Tomohiko Yamaguchi, and Anatol M. Zhabotinsky for providing figures.

We thank Hans-Jürgen Krug who provided copies of nineteenth-century references.

We also thank Bob Rogers at Oxford University Press for his enthusiastic support for this project during its preparation.

We of course are responsible for any errors that remain.

Contents

PART I

OVERVIEW

I

Introduction—A Bit of History

Oscillations of chemical origin have been present as long as life itself. Every living system contains scores, perhaps hundreds, of chemical oscillators. The systematic study of oscillating chemical reactions and of the broader field of nonlinear chemical dynamics is of considerably more recent origin, however. In this chapter, we present a brief and extremely idiosyncratic overview of some of the history of nonlinear chemical dynamics.

In 1828, Fechner described an electrochemical cell that produced an oscillating current, this being the first published report of oscillations in a chemical system. Ostwald observed in 1899 that the rate of chromium dissolution in acid periodically increased and decreased. Because both systems were inhomogeneous, it was believed then, and through much of our own century, that homogeneous oscillating reactions were impossible. Degn wrote in 1972 (p. 302): "It is hard to think of any other question which already occupied chemists in the nineteenth century and still has not received a satisfactory answer." In that same year, though, answers were coming. How it took so long for the nature of oscillating chemical reactions to be understood and how that understanding eventually came about will be the major focus of this chapter.

Although oscillatory behavior can be seen in many chemical systems, we shall concentrate primarily on homogeneous, isothermal reactions in aqueous solution. In later chapters, we shall broaden our horizons a bit. While the study of oscillating reactions did not become well established until the mid-1970s, theoretical discussions go back to at least 1910. We consider here some of the early theoretical and experimental work that led up to the ideas of Prigogine on nonequilibrium thermodynamics and to the experimental and theoretical work of Belousov,

Zhabotinsky, Field, Körös, and Noyes, all of whom did much to persuade chemists that chemical oscillations, traveling fronts, and other phenomena that now comprise the repertoire of nonlinear chemical dynamics were deserving of serious study.

1.1 Lotka–Volterra Model

Alfred Lotka was one of the more interesting characters in the history of science. He wrote a handful of theoretical papers on chemical oscillation during the early decades of this century and authored a monograph (1925) on theoretical biology that is filled with insights that still seem fresh today. He then left science and spent the majority of his working life in the employ of an insurance company. In 1910, Lotka showed that a set of consecutive reactions can give rise to damped oscillations on the way to equilibrium. He continued his search for oscillatory behavior arising from mass action kinetics in a second paper published a decade later (Lotka, 1920a). This latter model, though it does not apply to any real chemical system, has provided considerable inspiration to ecologists. Vito Volterra used ideas similar to Lotka's to investigate a wide range of ecological problems, including the effects of migration and of several species simultaneously interacting (D'Ancona, 1954). The best known model of this type is today called the Lotka–Volterra model and is often used to characterize predator–prey interactions (Lotka, 1920a, 1920b; Volterra, 1926).

The model consists of three irreversible steps. X is the population of rabbits, which reproduce autocatalytically. A is the amount of grass, which we assume to be constant, or at least in great excess compared with its consumption by the rabbits. Y is the population of lynxes (bobcats), and P represents dead lynxes.

$$A + X \rightarrow 2X \tag{1.1}$$

$$X + Y \rightarrow 2Y \tag{1.2}$$

$$Y \rightarrow P \tag{1.3}$$

As indicated, each step is irreversible: rabbits will never turn back into grass, nor dead lynxes into live ones. We can write down a system of differential equations to describe the behavior of the predator and prey species:

$$dx/dt = k_x ax - k_y xy \tag{1.4}$$

$$dy/dt = k_y xy - k_d y \tag{1.5}$$

where k_x is a *rate constant* reflecting how fast rabbits reproduce; k_y specifies how fast lynxes reproduce given a number, x, of rabbits to eat; and k_d indicates the mortality rate of lynxes. For any set of these constants, the numbers of rabbits and lynxes will oscillate with a period that depends on k_x, k_y, k_d, and a. The "net reaction," which proceeds monotonically, is the conversion of grass, which is assumed to grow as fast as it is consumed, to dead lynxes. Figure 1.1 shows the oscillations in the rabbit and lynx populations.

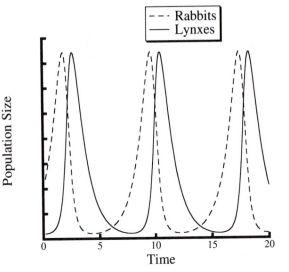

Figure 1.1 Numerical solution of the Lotka–Volterra model with $A = k_x = k_y = k_d = 1$.

A key feature of this system, and of most chemical systems that exhibit oscillations, is *autocatalysis*, which means that the rate of growth of a species, whether animal or chemical, increases with the population or concentration of that species. Even autocatalytic systems can reach a *steady state* in which the net rate of increase of all relevant species is zero—for example, the rate of reproduction of rabbits is exactly balanced by that species' consumption by lynxes, and lynxes die at the same rate that baby lynxes are born. Mathematically, we find such a state by setting all the time derivatives equal to zero and solving the resulting algebraic equations for the populations. As we shall see later, a steady state is not necessarily *stable*; that is, the small perturbations or fluctuations that always exist in a real system may grow, causing the system to evolve away from the steady state.

The oscillations in the two populations result from the difference in phases between rabbit reproduction and lynx reproduction. The rabbits reproduce because grass, A, is in constant supply. The lynx population will also increase, but only after the rabbit population has grown. Once the lynx population gets too high, since the grass supply is limited, rabbits will be eaten more rapidly than new rabbits are born, and their population will begin to decrease, which, in turn, will lead to a decrease in the lynx population. The rabbit population can then begin to rise again. Thus, there will be a time lag between changes in the two populations.

Figure 1.2 shows the number of lynx furs turned in to the Hudson Bay Company from 1820 to 1920. Distinct oscillations are seen with a period of about nine years. No data were available on the rabbit population, so we cannot be certain that the oscillations are due to a predator–prey interaction. However, controlled experiments have been performed in the laboratory with paramecia (*Paramecium aurelia*) that eat the yeast *Saccharomyces exiguns* (Figure 1.3). Notice how the predator population lags behind the population changes in the prey.

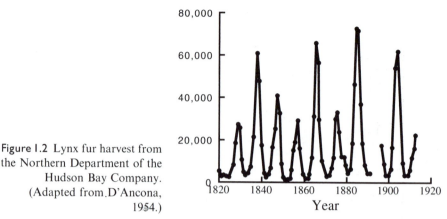

Figure 1.2 Lynx fur harvest from the Northern Department of the Hudson Bay Company. (Adapted from D'Ancona, 1954.)

1.2 Bray Reaction

The first homogeneous isothermal chemical oscillator to be described was the reaction of iodate, iodine and hydrogen peroxide, studied by William C. Bray at the University of California, Berkeley, and later by Bray's student Herman Liebhafsky (Bray, 1921; Bray and Liebhafsky, 1931). Hydrogen peroxide decomposes to oxygen and water. The rate of evolution of oxygen and the iodine (I_2) concentration were found to vary nearly periodically (Figure 1.4). Nonetheless, for the next fifty years, chemists would write that the reaction was not really homogeneous and that the oscillations were an artifact of dust or bubbles. Bray explicitly addressed these possible objections in 1921 by using carefully filtered

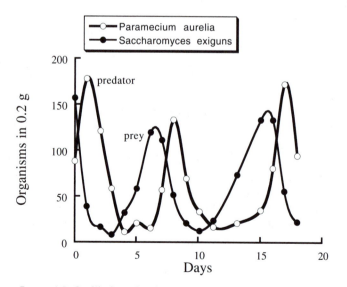

Figure 1.3 Oscillations in the populations of paramecia and yeast. (Adapted from D'Ancona, 1954.)

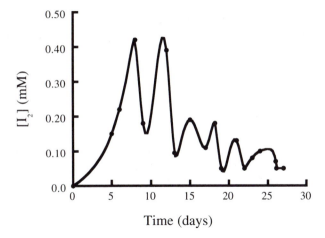

Figure 1.4 Oscillations in the iodine concentration in the Bray reaction at 25 °C. (Adapted from Bray, 1921.)

and stirred solutions at room temperature. The rate of reaction was so slow that oxygen diffused out before bubbles could form.

Noyes and coworkers (Sharma and Noyes, 1975) revived the work of Bray and Liebhafsky in the 1970s and, through careful experiments and mathematical modeling, building on the theoretical groundwork that had been laid by studies of non-equilibrium thermodynamics, succeeded in convincing the chemical community that the Bray reaction represented a genuine chemical oscillator.

1.3 The Belousov–Zhabotinsky Reaction

The beginning of modern nonlinear chemical dynamics can be traced to Boris Pavlovich Belousov (1893–1970), who was looking for an inorganic analog of the Krebs cycle, a key metabolic process in which citric acid is an intermediate. He began his studies in 1950 while head of the Laboratory of Biophysics in the former U.S.S.R. Ministry of Health. (Before the 1917 revolution, Belousov had studied chemistry in Zurich.) He investigated a solution of bromate, citric acid, and ceric ions (Ce^{4+}). He expected to see the monotonic conversion of yellow Ce^{4+} into colorless Ce^{3+}. Instead, the solution repeatedly cleared and then became yellow again! He studied the system carefully, including the effects of temperature and initial concentrations. Belousov also noted that, unstirred in a graduated cylinder, the solution exhibited traveling waves of yellow. He submitted a manuscript in 1951, but it was rejected. The editor of one of the unsympathetic journals informed him (Winfree, 1984) that his "supposed discovered discovery" was impossible! The paper could only be published if he furnished additional evidence; a simple recipe and photographs of the different phases of oscillation were deemed to be insufficient. He labored six more years and submitted a revised manuscript to another journal, but that editor insisted that the paper be shortened to a letter before further consideration. Belousov decided to give up on publishing his work, but he kept his manuscript, which circulated among colleagues in Moscow. His only publication on this reaction appears in the unrefereed abstracts of a conference on radiation biology (Belousov, 1958). A manuscript that Belousov wrote in 1951 describing his work was posthumously published in Russian (Belousov, 1981) and later in English translation (Belousov, 1985).

In 1961, Anatol Zhabotinsky, a graduate student in biophysics at Moscow State University, began looking at the same system at the suggestion of his professor, S. E. Schnoll. Although Zhabotinsky did not have Belousov's 1958 paper, he did have access to the original recipe: 0.2 g $KBrO_3$, 0.16 g $Ce(SO_4)_2$, 2 g citric acid, 2 mL H_2SO_4 (1:3), and enough water to make 10 mL of solution. He replaced citric acid with malonic acid and obtained a better formulation, which did not produce precipitate. (Information on demonstrations and experiments is presented in Appendices 1 and 2.)

Zhabotinsky wrote a manuscript that he sent to Belousov for his comments in 1962. Belousov replied by mail, but avoided any direct meeting, and the two never met face to face. At least ten papers on the Belousov–Zhabotinsky (BZ) reaction were published in Russian before the first one in English (Degn, 1967). A conference was held in Prague in 1968 on "Biological and Biochemical Oscillators," and Zhabotinsky presented some of his results. This meeting motivated many in the Eastern bloc to study the BZ reaction, and the publication of the proceedings in English (Chance et al., 1973) brought the BZ reaction to the attention of several Western chemists as well.

In several of his experiments, Belousov used the redox indicator ferroin to heighten the color change during oscillations. Ferroin is red in reduced solution and blue in oxidized form, providing a more easily visible variation than the pale yellow to colorless change of the ceric–cerous system. Zaikin and Zhabotinsky found that ferroin alone could catalyze the BZ reaction without cerium (Zaikin and Zhabotinskii, 1970). This advance allowed them to study unstirred solutions in thin layers, in which they discovered propagating chemical waves. Thus a homogeneous system was shown to exhibit not only temporal but also spatial self-organization.

In 1980, the Lenin prize was awarded to Belousov, Zhabotinsky, V. I. Krinsky, G. R. Ivanitsky, and A. Zaikin for their work on the BZ reaction. Belousov had died in 1970.

Before, and even during, the development of the BZ reaction, a number of papers were being written in the West on why true homogeneous oscillating reactions were impossible. Some claimed that the Bray reaction was not a homogeneous reaction, but, instead, that the oscillations resulted from the presence of small particles (Rice and Reiff, 1927). Many believed that the oscillations were an artifact of heterogeneous phenomena like bubble formation (Peard and Cullis, 1951; Shaw and Pritchard, 1968). Others argued that such spontaneous temporal self-organization (periodic changes in concentration of putative reactants and products) violated the Second Law of Thermodynamics (Swartz, 1969; and references in Degn, 1972; and Winfree, 1984). To understand why such objections were so influential and why they do not, in fact, preclude chemical oscillation, we need to digress into a brief discussion on thermodynamics.

1.4 The Second Law

Many of those who found themselves unable to accept chemical oscillation as a reality based their refusal on the Second Law of Thermodynamics. The power of the Second Law lies in its ability to predict the direction of spontaneous change from the deceptively simple condition that

$$\Delta S_{total} > 0 \qquad (1.6)$$

where ΔS_{total} is the total entropy change of an isolated system or of the universe for the change of state of interest. It is the Second Law that forbids those seemingly plausible perpetual motion machines foisted on patent offices by loony inventors or on unsuspecting students by fiendish teachers of physical chemistry.

In chemical reactions, it is difficult to keep track of the entropy of the universe. If the reaction is performed under the commonly encountered conditions of constant temperature and pressure, the condition given in eq. (1.6) becomes equivalent to the requirement that the change in the Gibbs free energy, G, be negative (Atkins, 1995). The essential result is that there exists a function of state—that is, a function like the entropy or the free energy—that depends only on the current condition (temperature, pressure, volume, concentrations) of the system and not on its past history, and which changes monotonically in any spontaneous process, bringing the system ever closer to its final equilibrium state. Those who objected to the notion of oscillating reactions understandably, but mistakenly, considered an oscillating reaction to be analogous to a pendulum, which passes through its equilibrium point during each cycle of oscillation. They thereby concluded that an oscillating reaction would require the free energy of the system to oscillate as the reactants were converted to products and then back to reactants, thus contradicting the Second Law. Figure 1.5 shows permissible and invalid temporal evolutions of free energy.[1]

As the figure indicates, a chemical oscillator is really quite different from a pendulum. When a chemical reaction oscillates, it never passes through its equilibrium point. Instead, chemical oscillation is a far-from-equilibrium phenomenon, governed by the laws of *non-equilibrium thermodynamics* (de Groot and Mazur, 1984). Beginning in the 1930s, Lars Onsager, Ilya Prigogine, and others realized that thermodynamic ideas could be applied to systems far from equilibrium, but that a new theory was required. Prigogine and his coworkers in Brussels focused on chemical systems, pointing out that a system could organize (decrease its entropy), so long as the net entropy change in the universe was positive (Nicolis and Prigogine, 1977, 1989; Prigogine, 1980). Thus, for example, the con-

---

[1] Note that in a system with several concentrations, the concentration of a species may oscillate through its equilibrium value, as shown in Figure 1.6a, so long as the free energy decreases and not all of the concentrations *simultaneously* pass through their equilibrium values.

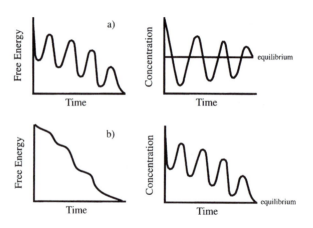

Figure 1.5 Two types of conceivable oscillations in closed systems. (a) Oscillations around equilibrium; this is not consistent with the Second Law because the free energy must monotonically decrease to the equilibrium value. (b) Oscillations on the way to equilibrium, consistent with the Second Law.

centrations of the intermediates in a reaction can increase and decrease with time while the free energy monotonically decreases as a result of the continuing conversion of high free energy reactants into low free energy products. Any decrease in entropy caused by the periodic concentration changes is more than compensated by an entropy increase from the other processes (Figure 1.6).

Prigogine pointed out in 1955 that *open systems* (i.e., systems open to exchange of matter and/or energy with their surroundings) kept far from equilibrium could exhibit spontaneous self-organization by dissipating energy to the surroundings to compensate for the entropy decrease in the system. He called the temporal or spatial structures that can arise in such a fashion *dissipative structures*. A closed system must reach equilibrium and so can exhibit only transitory oscillations as it approaches equilibrium. Sustained oscillations require an open system with a constant influx of new reagents and the removal of waste products.

Oscillating reactions no more violate the laws of nature than do living organisms. In our body, the spontaneous formation of proteins occurs even though the ΔG of peptide bond formation is positive. Protein synthesis occurs because it is coupled to other reactions whose ΔG is negative, yielding a net decrease in free energy (Lehninger, 1993). What oscillate in the BZ reaction in a beaker are not the concentrations of reactants (bromate and malonic acid) and products (carbon dioxide and bromomalonic acid), but the concentrations of intermediates such as bromide and bromous acid. The free energy of the reaction is always decreasing since high free energy reactants are continuously converted to low free energy products.

Figure 1.6 Schematic diagram showing how an open system can become ordered but increase the disorder of the surroundings. The overall entropy change of the universe is positive.

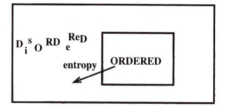

1.5 Nonlinear Oscillations

To understand the role of a system being far from equilibrium, consider a bottle of beer. If it is gently tipped on its side, the beer will pour out smoothly. However, if you tip the bottle over rapidly, then the beer will gush out in pulsations. A reaction near equilibrium is like the bottle slightly tipped—it will react smoothly and monotonically. A reaction far from equilibrium can undergo oscillations on the way to equilibrium. To complete our analogy, a beer bottle operating as an open system would have a pump that continuously flows with beer; the pulsating flow could continue indefinitely, much to the delight of the participants in the experiment.

1.6 The Brusselator

Despite the fact that it generates sustained oscillatory behavior from simple "chemical reactions" with mass action kinetics, the Lotka–Volterra model is not an appropriate description of any actual chemical, as opposed to ecological, system. Its fatal flaw is that, in a sense, it is *too* successful in generating oscillatory behavior. It is possible to prove that the model has an oscillatory solution for any values of the rate constants and "food supply" A and initial values of X and Y in eqs. (1.1)–(1.3), and that the amplitude and period of the oscillations obtained depend on all of these quantities; there is an infinite array of oscillatory solutions. If the system is perturbed, say, by adding a bit more A or X or Y, it continues to oscillate, but now with a new period and amplitude, until it is perturbed again. In the presence of any significant amount of noise, the behavior would hardly be recognizable as periodic, since it would constantly be jumping from one oscillatory behavior to another. Real chemical systems do not behave this way. They oscillate only within a finite range of parameters, and they have a single mode (amplitude and frequency) of oscillation, to which they return if the system is perturbed. Is it possible to construct a reasonable model that has these features? The first chemically "respectable" model was proposed by Prigogine and Lefever in 1968 and dubbed the "Brusselator" by Tyson in 1973.

$$A \rightarrow X \qquad k_1 \tag{1.7}$$
$$B + Y \rightarrow Y + D \qquad k_2 \tag{1.8}$$
$$2X + Y \rightarrow 3X \qquad k_3 \tag{1.9}$$
$$X \rightarrow E \qquad k_4 \tag{1.10}$$

By analysis and numerical simulation, Prigogine and Lefever demonstrated that their model shows homogeneous oscillations and propagating waves like those seen in the BZ system. The Brusselator was extremely important because it showed that a chemically reasonable mechanism could exhibit self-organization.

In 1977, Nicolis and Prigogine summarized the work of the Brussels school in a book entitled *Self-Organization in Nonequilibrium Systems*. For his contributions to the study of nonequilibrium systems, Ilya Prigogine was awarded the 1977 Nobel prize in chemistry.

1.7 Back to the BZ Reaction

Although chemical engineers had been considering oscillatory behavior in reactors since 1945 (Higgins, 1967), chemists remained skeptical of claims of oscillations. As Degn described the situation in 1972 (p. 302): "Although there now seems to be abundant evidence for the existence of homogeneous oscillating reactions, there are still theoreticians who resist the idea, and also a few experimentalists think that alleged homogeneous oscillations are caused by dust particles, although nobody has explained how." Despite the existence of the abstract Brusselator model, what was missing was a detailed *chemical mechanism* for an oscillating reaction to convince chemists that oscillating reactions were a legitimate area of study and not some form of a perpetual motion device.

A chemist wants to understand chemical reactions on the molecular level, and a mechanism is a map of those processes. A mechanism is a series of elementary reactions that involve actual molecular collisions that lead to transformation. To write that $2H_2 + O_2 \rightarrow 2H_2O$ is not to assert that two molecules of hydrogen collide with one molecule of oxygen and somehow rearrange themselves to fly apart into two molecules of water. This is only the net reaction, which shows the necessary stoichiometry. For a chemist to truly understand the reaction of hydrogen and oxygen, he or she needs to know all the individual steps that lead to the overall reaction. This is an extremely difficult task for all but the simplest of reactions.

Zhabotinsky and his coworkers had made significant progress in understanding the mechanism of the BZ reaction, but their work was largely unknown in the West. Few scientists outside the Eastern bloc could read Russian. Even though many Soviet journals were translated into English, a Cold War mentality, and perhaps a touch of arrogance, prevented most Western scientists from keeping up with that literature. Another, more pernicious, impediment to scientific exchange was the prohibition on travel. Except in rare cases, Soviet scientists could not attend Western conferences, and few Western scientists were permitted into the former Soviet Union. Today with our faxes and electronic mail, it is difficult to imagine how isolated the Soviet scientists were.

The first publications in English to recognize the work of Belousov and Zhabotinsky were by the Danish biochemist Hans Degn (1967). He measured the cerium(IV) concentration, platinum electrode potential (responding to the ratio [Ce(IV)]/[Ce(III)]), and the evolution of carbon dioxide in the BZ reaction, but he was unable to propose a complete mechanism for the system. The Prague conference in 1968 afforded an opportunity for Zhabotinsky and some of the other Russian scientists to present their work and to meet some of the more intrepid Westerners.

1.8 The BZ Reaction Comes to North America

The work that led to the "legitimizing" of the study of oscillating reactions is a fascinating story of serendipity and hard work. In 1969, Richard Field, a recent graduate of the University of Rhode Island, began a postdoctoral fellowship with the well-known kineticist Richard Noyes at the University of Oregon in Eugene. Robert Mazo, a theorist in the same department, had just returned from a sabbatical year in Brussels with the Prigogine group. In Brussels, Mazo had seen a demonstration of the BZ reaction and of the banding that occurs when the reaction is run in an unstirred graduated cylinder (Busse, 1969). Mazo presented a seminar in the Oregon chemistry department and demonstrated the banding. After his talk, Noyes wondered if the banding would reappear if the solution were stirred. When he and his coworkers tried the stirred reaction in the lab, they rediscovered the oscillatory behavior!

Field and Noyes were fascinated by the oscillations and set out to determine the mechanism of this remarkable behavior. Endre Körös had recently arrived in Eugene for a sabbatical visit from Eötvös University in Budapest. He had come to study iodine exchange reactions, a specialty of Noyes', but when he saw the BZ reaction, Körös decided to work on it, too. Körös suggested using a bromide-selective electrode to follow the reaction, which proved to be a major boon to understanding the importance of bromide in the reaction. During this period, Kasperek and Bruice carried out a study on the BZ reaction that included some significant additional observations but no mechanism (Kasperek and Bruice, 1971). This publication spurred the Eugene group, but Bruice discontinued the work, since he was skeptical that the oscillations arose from a purely homogeneous system.

The details of the mechanism that was developed in Eugene are discussed in later chapters. The important fact is that Field, Körös, and Noyes were able to explain the qualitative behavior of the BZ reaction using the same principles of chemical kinetics and thermodynamics that govern "ordinary" chemical reactions. They published their mechanism, now known as the FKN mechanism, in a classic paper in 1972 (Field et al., 1972; Noyes et al., 1972). A quantitative numerical simulation of the oscillatory behavior based on the FKN mechanism was published a few years later (Edelson et al., 1975).

Arthur Winfree, a biologist with an interest in spatial and temporal patterns, had attended the Prague conference and had decided to pursue the study of pattern formation in the BZ reaction. In 1972, the cover of *Science* magazine featured Winfree's photo of spiral wave patterns in the BZ reaction (Winfree, 1972). Figure 1.7 shows such patterns. Field and Noyes immediately saw how to understand the development of such patterns with the aid of the FKN mechanism, and they published an explanation the same year (Field and Noyes, 1972).

One final step solidified the hold that the BZ reaction and chemical oscillation were taking on the imaginations of chemists in the 1970s. Field and Noyes managed to simplify the FKN mechanism, which, with its twenty or so elementary steps and chemical species, was too large and complicated for numerical work with any but the most powerful computers and software then available, let alone analytical solutions. With penetrating chemical insight, they obtained a model that had only three variable concentrations yet maintained all the essential features of the full BZ reaction. The model was dubbed the "Oregonator" (Field and Noyes, 1974b).

Figure 1.7 Target patterns and spiral waves in the Belousov–Zhabotinsky reaction observed in a Petri dish. (Courtesy of T. Yamaguchi.)

Field continued his work as a professor at the University of Montana, Körös returned to Budapest, and Noyes continued at Eugene. To this day, Hungarian chemists continue to have a disproportionate impact on the study of oscillations and waves.

1.9 Systematic Design

By the late 1970s many chemists were aware of oscillating reactions and agreed that they were genuine and even interesting. However, only two fundamentally different chemical oscillators, the Bray and the BZ reactions, were known, and both had been discovered by serendipity. Several variants of the BZ reaction had been developed, using other metals in place of cerium to catalyze the reaction, or replacing the malonic acid with a similar organic species, or even a version in which, with an appropriate choice of the organic substrate, no catalyst was required (Orbán et al., 1978; Orbán and Körös, 1979). In 1973, two high school chemistry teachers in San Francisco (Briggs and Rauscher, 1973) developed a visually appealing lecture demonstration by combining the malonic acid and the metal ion catalyst of the BZ reaction with the ingredients of the Bray reaction. (A recipe is given in Appendix 1.) There were, of course, many biological oscillators known, but as far as chemical oscillators were concerned, the only techniques for finding them—accident and variation on a theme—did not inspire confidence in the ability of chemists to elucidate the nature of these systems. Efforts to specify a nontrivial set of necessary and sufficient conditions for a chemical reaction to oscillate proved fruitless.

In the mid-1970s, two efforts began that were to converge toward a systematic approach to building chemical oscillators. At Brandeis University, Irving Epstein and Kenneth Kustin had been working with a few talented undergraduates on some problems related to the BZ reaction (Jacobs and Epstein, 1976; Kaner and Epstein, 1978). They identified several autocatalytic inorganic reactions that they thought could be turned into new chemical oscillators, and they attempted to obtain support for the project. Their proposal was rejected three times by funding agencies on the grounds that it would never succeed, but they persisted and were eventually funded. Meanwhile, at the Paul Pascal Research Center in Bordeaux, a group of scientists led by Adolphe Pacault had pioneered the use of the continuous-flow stirred tank reactor (CSTR), a tool familiar to chemical engineers but essentially unknown to chemists, to provide an open system suited to the study of oscillating reactions (Pacault et al. 1976). Two members of the Bordeaux group, Patrick De Kepper and Jacques Boissonade, proposed an abstract model that suggested how oscillations might be obtained in a CSTR by perturbing a bistable chemical system (Boissonade and De Kepper, 1980). The funding obtained by the Brandeis group made it possible for De Kepper to join them in late 1979. Within a few months the team had developed the first systematically designed oscillating reaction, the arsenite–iodate–chlorite system (De Kepper et al., 1981a, 1981b). The technique, which is described in Chapter 4, was quickly refined and then exploited to develop literally dozens of new oscillators over the next decade.

1.10 The Plot Thickens

As the number of chemical oscillators grew in the 1980s, interest turned to more exotic phenomena, several of which are discussed in Part II of this book. Spatial pattern formation, especially the spiral waves seen in the BZ reaction, drew a great deal of interest, both for its own inherent beauty and because of the similar structures seen in a variety of biological systems. The stationary patterns posited by the mathematician Alan Turing in 1952 as a "chemical basis for morphogenesis" were seen for the first time in 1990 (Castets et al., 1990) in an experimental system derived from one of the systematically designed oscillators of the 1980s.

Like their scientific cousins, the physicists, chemists began to explore how coupled and forced oscillators behave. They also probed how their chemical oscillators and patterns were affected by diffusion, convection, and external fields. A topic that created great interest among physicists and mathematicians in the 1970s and 1980s is deterministic chaos. A chaotic system is unpredictable, but not random. Deterministic equations describe the system, but the system is so sensitive to its initial conditions that its future behavior is inherently unpredictable beyond some relatively short period of time. Chaos was found in many physical systems, from lasers to water faucets, in the 1970s. What then of chemistry? Schmitz and Hudson reported in 1977 that the BZ system behaves chaotically under certain conditions in a flow reactor (Schmitz et al., 1977). Further studies in the 1980s and 1990s have confirmed the existence of chaos in chemical systems and have shed new light on its origins.

1.11 Conclusions

The history we have given here has, of necessity, left out many important and fascinating details. What we have tried to convey is a sense that nonlinear chemical dynamics has not had an easy path into the hearts and minds of chemists. Skepticism and prejudice delayed for many years the serious study of the phenomena that we shall be discussing. In the end, the combination of persuasive experimental and theoretical work with persistent advocates and visually stimulating phenomena won out. Elements of nonlinear chemical dynamics are now found in "mainstream" textbooks and introductory courses. Research grants are awarded, international conferences are held, junior faculty are hired. In short, the field is now respectable. This book is an effort to make it not only respectable, but also accessible and comprehensible.

Many practitioners of nonlinear chemical dynamics have turned toward biological systems, toward geological phenomena, and toward materials science. There are likely to be additional exciting applications in the coming decade. Before one can jump into these new areas, it is necessary to acquire the basic tools, which is where we begin the next chapter.

2

Fundamentals

Before plunging into the meat of our discussions, we will review some basic but necessary ideas. Much of this material will be familiar to many readers, and we encourage you to move quickly through it or to skip it completely if appropriate. If you have not encountered these concepts before, you will find it worthwhile to invest some time here and perhaps to take a look at some of the more detailed references that we shall mention. We begin with a review of chemical kinetics. We then consider how to determine the stability of steady states in an open system using analytical and graphical techniques. Finally, we look at some of the methods used to represent data in nonlinear dynamics.

2.1 Chemical Kinetics

The problems that we are interested in involve the rates of chemical reactions, the study of which forms the basis of *chemical kinetics*. This is a rich and beautiful subject, worthy of whole volumes. For those interested in a less superficial view than we have room to present here, we recommend several excellent texts on kinetics (Jordan, 1979; Cox, 1994; Espenson, 1995). We review here a minimal set of fundamentals necessary for what comes later.

2.1.1 General Definitions and Terms

Consider a simple reaction,

$$A + B \rightarrow C \tag{2.1}$$

If we can follow the concentration of A in time, then a natural measure of the rate of reaction is the rate of consumption of A—the derivative of the concentration of A with respect to time—in units, for example, of moles per liter per second $(\mathrm{M\,s^{-1}})$:

$$\text{Rate of consumption of } A = v_A = -\frac{d[A]}{dt} \tag{2.2}$$

The minus sign makes the rate of consumption positive, because $d[A]/dt < 0$ for a reactant. The stoichiometry of eq. (2.1) tells us that for every molecule of A that is consumed, one molecule of B is simultaneously destroyed, and one molecule of C is produced. Therefore,

$$\frac{d[C]}{dt} = -\frac{d[A]}{dt} = -\frac{d[B]}{dt} \tag{2.3}$$

This observation suggests that we can define a more general *rate of reaction* for any chemical reaction with any stoichiometry in terms of the rate of consumption or production of any of the reactants or products:

$$v = \frac{1}{n_j}\frac{d[J]}{dt} \tag{2.4}$$

where n_j is the stoichiometric coefficient of species J with the convention that n_j is positive if J is a product and negative if J is a reactant. If we make reaction (2.1) a bit more complicated, say,

$$A + 2B \rightarrow 3C \tag{2.5}$$

then eq. (2.4) gives

$$v = -\frac{d[A]}{dt} = \frac{-1}{2}\frac{d[B]}{dt} = \frac{1}{3}\frac{d[C]}{dt} \tag{2.6}$$

The *molecularity* of a reaction is defined as the number of reactant molecules appearing in the stoichiometric equation. Thus, reaction (2.1) is bimolecular, while eq. (2.5) is termolecular. Often, the rate of a reaction is experimentally found to be proportional to the concentrations of the reactants (and, less frequently, the products or some other, *catalyst*, species, Q) to some power:

$$\text{Rate} = k[A]^a[B]^b[C]^c[Q]^q \tag{2.7}$$

The equation that specifies the rate of a reaction is known as the *rate law* for the reaction. The proportionality constant k, which is independent of the reactant and product concentrations but depends on the temperature of the system, is called the *rate constant*. Since the left-hand side of eq. (2.7) has units of $\mathrm{M\,s^{-1}}$, the units of k must be $\mathrm{M^{(1-a-b-c-q)}\,s^{-1}}$. Rate constants for certain reactions can be predicted a priori from statistical mechanical considerations, but this is an extremely difficult and error-prone task, especially if the molecules involved have any significant degree of complexity. Typically, rate constants are measured experimentally from a determination of concentration vs. time in a single experiment or from analysis of rate data for a range of initial concentrations. Determining the rate laws and rate constants of a large number of chemical reactions has been one of the major accomplishments of chemical kinetics in the past century.

The exponents a, b, c, and q specify the *order* of the reaction with respect to the respective species. Thus, the reaction described by the rate law (2.7) would be said to be ath order in species A, bth order in B, etc. The order of the reaction as a whole is defined as the sum of the orders of all the species, in this case $a + b + c + q$. Typically, c and q would be zero. Note that a reaction need not have an order with respect to all of its species, nor a total order. We could have a rate law like

$$-\frac{d[A]}{dt} = k[A][B]/(k' + [B])$$ (2.8)

so that the reaction is first order in A, but of undefined order with respect to B and overall. The order of a reaction need not be either integral or positive.

Note that the stoichiometric coefficients in a balanced chemical equation like eq. (2.5) bear no necessary relationship to the orders that appear in the empirical (i.e., experimentally derived) rate law. This statement becomes obvious if one considers that the chemical equation can be multiplied on both sides by any arbitrary number and remain an accurate representation of the stoichiometry even though all the coefficients will change. However, the orders for the "new" reaction will remain the same as they were for the old one. There are cases in which the rate law depends only on the reactant concentrations and in which the orders of the reactants equal their molecularity. A reaction in which the order of each reactant is equal to its molecularity is said to obey the *Law of Mass Action* or to behave according to *mass action kinetics*.

The rate laws that we have considered so far are *differential rate laws*; they specify the time derivative of a concentration in terms of other concentrations. It is often more useful to work with an *integrated rate law*, in which the concentration is given as a function of the initial concentrations and the time. If the form of the differential rate law is not too complex, one can derive the integrated rate law from it by using the stoichiometric equation to express all of the concentrations in terms of a single concentration (say, $[A]$) and then integrating the resulting one-variable differential equation subject to the known initial conditions.

For example, suppose that the differential rate law for reaction (2.5) is

$$-\frac{d[A]}{dt} = k[A][B]$$ (2.9)

and we start with initial concentrations $[A]_0$ and $[B]_0$. If we represent $[A]$ and $[B]$ at time t by $x(t)$ and $y(t)$, respectively, then the stoichiometry requires that

$$\Delta A = x - [A]_0 = \frac{1}{2}\Delta B = \frac{1}{2}(y - [B]_0)$$ (2.10)

$$y = [B]_0 + 2(x - [A]_0)$$ (2.11)

Defining the initial stoichiometric excess as $u_0 = [B]_0 - 2[A]_0$, with a little algebra the rate law of eq. (2.9) becomes

$$dx/dt = -kxy = -kx(u_0 + 2x)$$ (2.12)

Integrating eq. (2.12) from time 0 to time t yields[1]

$$(1/u_0)\{\ln(x/x_0) - \ln[(u_0 + 2x)/(u_0 + 2x_0)]\} = -kt \qquad (2.13)$$

Some more algebra enables us to solve for $x(t)$ explicitly:

$$x(t) = x_0 u_0 \exp(-u_0 kt)/\{u_0 + 2x_0[1 - \exp(-u_0 kt)]\} \qquad (2.14)$$

As a check on our work, note that if u_0 is positive, that is, B is in excess initially, then $x(t)$, or $[A]$ will approach zero, while if u_0 is negative, then $[A]$ will approach $-u_0/2 = [A]_0 - [B]_0/2$ as $t \to \infty$, in in accordance with the stoichiometry. We can determine k by measuring $[A]$ as a function of time: a plot of $\ln(1/x + 2/u_0)$ vs. t should yield a straight line with slope $u_0 k$ and intercept $\ln(1/x_0 + 2/u_0)$.

2.1.2 Mechanisms

As discussed in the previous chapter, there is an important distinction between the stoichiometric equations that describe the overall arithmetic of a chemical reaction—the relative numbers of molecules that are consumed and produced—and the set of equations that constitute a mechanism for the reaction. A mechanism is composed of a set of reactions called *elementary steps*, each of which is taken to represent an actual molecular event that leads to the overall reaction. The complete set of elementary steps must yield the correct stoichiometry and the experimentally observed rate law of the reaction.

Thus, if eq. (2.1) represented an elementary step, we would be saying that a molecule of A and a molecule of B actually collide and rearrange their atoms to form a molecule of C. Because collisions involving more than two molecules simultaneously are extremely improbable, nearly all elementary steps, except those involving species like solvent molecules that are present in very high concentration, are either unimolecular or bimolecular. For elementary steps, and *only* for elementary steps, the molecularity and the order are necessarily equal; an elementary step has mass action kinetics. Adherence to the Law of Mass Action is a necessary, but not a sufficient, condition for a reaction to be an elementary step.

Dynamically interesting reactions of the sort that we shall consider in this book always involve more than one elementary step and more than one independently variable concentration. The reactions we have discussed above, where all the concentrations can be expressed in terms of a single independently varying concentration, are very nice for pedagogical treatments, but they occur infrequently in the real world and cannot generate such behavior as oscillation or chaos. If we are given a multistep mechanism, we can construct a set of differential rate equations that describe the rates of change of each concentration. We simply add the contributions from each of the elementary steps. For example, in the Brusselator model given by eqs. (1.7)–(1.10), the rate equation for X, which appears in all four steps, is

---

[1]The special case in which the initial stoichiometric excess u_0 is zero is much easier and is left as an exercise for the reader.

$$dX/dt = k_1 A - k_2 BX + k_3 X^2 Y - k_4 X \tag{2.15}$$

where we have left off the square brackets and represented concentrations just by the symbols for the various species.

In general, a mechanism for a reaction leads to as many differential equations as there are independently variable species in the elementary steps that constitute the mechanism. The number of variable species may be reduced, for example, by the fact that one species is in such high excess that its concentration is not significantly affected by the progress of the reactions in which it is involved. Species A, the grass in the Lotka-Volterra model of eqs. (1.1)–(1.3), is usually treated as such a species, as are the precursors A and B in the Brusselator model. Often, we need not consider the time evolution of a species that appears only as a product, like dead lynxes, P, in the Lotka–Volterra model (though sometimes the whole point is to figure out how rapidly a product is being generated). Finally, there may be stoichiometric constraints that reduce the number of independent species. For example, if I^- and I_2 are the only iodine-containing species that appear in the mechanism for a reaction that occurs in a closed system, so that the number of iodine atoms is constant, then it must always be true that

$$2[I_2]_0 + [I^-]_0 = 2[I_2] + [I^-] \tag{2.16}$$

where the subscript 0 signifies initial concentrations.

Even with the assistance of these approaches to limiting the number of variables, it is a rare system indeed that can be described by a single rate equation without further "tricks." One of the most useful approaches utilizes the insight that any process that involves a sequence of subprocesses can proceed only as fast as the slowest subprocess in the sequence. Imagine water flowing through a set of pipes of varying diameter or traffic moving along a highway that is sometimes three lanes wide, sometimes only one or two. The rate at which water travels from the beginning to the end of the pipes is determined by the rate at which it flows through the narrowest of the pipes. Similarly, it is the length of the one-lane stretch that will play the major role in determining how long is spent on the highway. A similar principle applies to a sequence of chemical reactions. If one step is significantly slower than the rest, then the rate of the entire reaction will be determined by that of the slowest step. The reaction has a single rate law: the rate law of this *rate-determining step*.

For example, autocatalysis rarely, if ever, arises from a single elementary step. Far more likely than the simple reaction $A + X \rightarrow 2X$ is a sequence like

$$A + X \rightarrow Y \quad \text{(slow)} \tag{2.17}$$
$$Y \rightarrow 2X \quad \text{(fast)} \tag{2.18}$$

If reaction (2.17) is rate-determining, then the rate of reaction is proportional to $[A]$ and $[X]$, so the net reaction is autocatalytic. The intermediate Y does not appear in the rate law. If, on the other hand, reaction (2.18) were the slow one, then the rate of the overall reaction would be proportional to $[Y]$, and in order to write a single rate law we would need to find a way to express $[Y]$ in terms of $[A]$ and $[X]$. Of course, we could also deal with the full set of rate equations for the three species and attempt to solve these numerically using some of the techniques we shall describe in Chapter 7.

Another simplification that often arises in complex mechanisms occurs when an elementary step and its reverse both occur at relatively rapid rates. Suppose, for example, that reaction (2.17) is reversible and that both the forward and reverse reactions are fast compared with reaction (2.18). In this case, the rate law is

$$d[X]/dt = k_2[Y] \tag{2.19}$$

where k_2 is the rate constant of reaction (2.18). However, since reaction (2.17) proceeds rapidly in both directions, we have a *rapid pre-equilibrium* established in that reaction, so that we can set the forward and reverse rates equal to obtain an expression for $[Y]$ in terms of $[A]$, $[X]$, and the equilibrium constant $K_1 = k_1/k_{-1}$ of reaction (2.17):

$$[Y] = K_1[A][X] \tag{2.20}$$

and, thus,

$$d[X]/dt = k_2 K_1[A][X] \tag{2.21}$$

The reaction is still autocatalytic in X.

One more scenario is worth mentioning. Often, an intermediate in a mechanism is present at a very low concentration that changes little during most of the reaction. Typically, the intermediate is absent initially, then builds up rapidly to a nearly constant level at which it remains before dropping back to zero at the completion of the reaction. For such intermediates, I (there may be more than one in a mechanism), it is often useful to invoke the *steady-state approximation* and set $d[I]/dt = 0$. This condition provides an algebraic relation among the concentrations, which enables us to solve for one concentration, say, $[I]$, in terms of the others. In our example above, Y is the only intermediate. If we assume that reaction (2.17) is reversible, but make no other assumptions about the relative rates of reactions (2.17) and (2.18), we can set $d[Y]/dt = 0$ and obtain

$$d[Y]/dt = 0 = k_1[A][X] - k_{-1}[Y] - k_2[Y] \tag{2.22}$$
$$[Y] = k_1[A][X]/(k_{-1} + k_2) \tag{2.23}$$

Since the rate of the overall reaction must equal the rate of step (2.18), we have

$$d[X]/dt = k_2[Y] = k_2 k_1[A][X]/(k_{-1} + k_2) \tag{2.24}$$

Note that this more general rate law reduces to eq. (2.19) if reaction (2.17) is irreversible or if $k_{-1} \ll k_2$ and to eq. (2.21) if $k_{-1} \gg k_2$.

One alternative to working with a full elementary step mechanism that has proved useful in the study of a number of complex reactions involves the use of an *empirical rate law model*. In this approach, we describe the overall reaction not in terms of a complete set of elementary steps, but in terms of a set of processes that add up to the full reaction and for each of which a rate law is known experimentally. The set of rate equations that describes the system is then the set of rate laws for all of these processes. Typically, there will be far fewer rate equations than for a full mechanism, but each equation will have a more complex form, since we are no longer limited to uni- and bimolecular steps. The behavior of the system will be correctly described by such a model only if the intermediates in each subprocess do not interact with the species in any other subprocess; that is, if there is no "cross talk" among subprocesses.

The iodate oxidation of arsenous acid is conveniently described in terms of two component processes (De Kepper et al., 1981a; Hanna et al., 1982): process A, the Dushman reaction (Dushman, 1904), and process B, the Roebuck reaction (Roebuck, 1902).

$$IO_3^- + 5I^- + 6H^+ = 3I_2 + 3H_2O \tag{A}$$

$$H_3AsO_3 + I_2 + H_2O = H_3AsO_4 + 2I^- + 2H^+ \tag{B}$$

When arsenous acid is in stoichiometric excess $((As(III))_0 > 3(IO_3^-)_0)$, the net reaction is given by (A) + 3(B) or

$$IO_3^- + 3H_3AsO_3 + 5I^- \to 6I^- + 3H_3AsO_4 \tag{2.25}$$

Under the usual conditions, Process A is rate-determining, so that the overall rate is determined by the rate law for Process A:

$$\frac{d[I^-]}{dt} = (k_a + k_b[I^-])[I^-][IO_3^-][H^+]^2 \tag{2.26}$$

where k_a and k_b are phenomenological rate constants. Because the total number of iodide atoms is fixed by the initial iodate concentration,

$$[I^-] + [IO_3^-] = [IO_3^-]_0 \tag{2.27}$$

we can obtain a single equation for the rate of iodide production:

$$\frac{d[I^-]}{dt} = (k_a + k_b[I^-])[I^-]([IO_3^-]_0 - [I^-])[H^+]^2 \tag{2.28}$$

This equation can be solved either analytically or numerically to yield the curve in Figure 2.1, which shows the typical "explosion" in the concentration of the autocatalytic species I^-.

2.1.3 Feedback

Systems that exhibit complex temporal and spatial behavior almost always have some form of *feedback* in their kinetics; that is, the concentration of some species affects the rate of its own production. Autocatalysis is a frequently observed form of *positive* feedback in the systems we will be interested in. Although we shall focus primarily on isothermal systems, it is worth pointing out that one of the most common examples of autocatalysis is thermal feedback, in which the heat evolved in an exothermic chemical reaction increases the rate of that reaction, thereby increasing the rate of heat production, and so on. This type of feedback plays a key role in combustion reactions.

In general, the rate constant k for a chemical reaction depends on temperature, according to the *Arrhenius equation*:

$$k = A \exp{(-E_a/RT)} \tag{2.29}$$

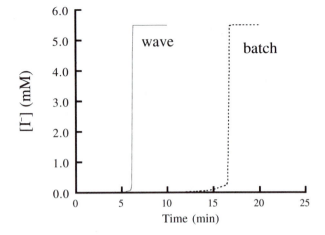

Figure 2.1 Iodide concentration as a function of time in a reaction mixture containing excess arsenous acid. Solid line shows the concentration at a single point in an unstirred system as a chemical wave passes by; dashed line shows the concentration in a well-stirred system. Solution composition: $[NaIO_3]_0 = 5.00 \times 10^{-3}$ M, $[H_3AsO_3]_0 = 5.43 \times 10^{-2}$ M, $[H^+]_0 = 7.1 \times 10^{-3}$ M. (Adapted from Hanna et al., 1982.)

where A is the *pre-exponential factor* determined by the structural parameters of the reactant molecules, R is the gas constant, and E_a is the *activation energy*, which is a measure of the energy required to initiate the reaction. Since the activation energy is, in general, a positive quantity, the rate will increase with temperature.

Many biochemical reactions, as well as chemical oscillators, exhibit *negative* feedback, in which the rate of production of a species decreases as the concentration of that species rises. This form of regulation has obvious advantages in living systems, which need to be able to cut off production of proteins, hormones, and other key molecules when their concentrations reach appropriate levels.

2.1.4 A Final Note on Mechanisms

A chemical mechanism is a theoretical construct. One can perform experiments to test it. The results of the experiments may or may not confirm the predictions of the mechanism. It is not possible to *prove* a chemical mechanism, or, for that matter, any scientific theory. It is only possible to *disprove* one. If the experiments disagree with the mechanism's predictions, the mechanism needs to be discarded, or at least modified. If the experiments and the mechanism agree, we can maintain the mechanism and try some more experiments. The fact that we have found a mechanism that agrees with all of our observations does not mean that we have found *the* mechanism. Although we may have struggled mightily and been extremely clever in coming up with our mechanism, there are always other mechanisms consistent with the same set of data, and only further experiments will enable us to choose among them. In Chapter 5, we shall study several examples of mechanisms for complex reactions, examine how to construct such mechanisms, and consider how well they do in explaining the results of experiments.

2.2 Stability Analysis

In analyzing the dynamical behavior of a chemical system, it is useful to begin at the end. This seemingly paradoxical observation is meant to point out that it is essential to understand the asymptotic, or long-term, behavior of a system if we are to characterize its evolution in time. Most chemical systems ultimately arrive at a state in which concentrations are no longer changing; a handful are able to maintain indefinitely a state in which the concentrations change periodically; still fewer end up in a state in which the concentrations vary aperiodically. We shall focus here on the time-independent or steady states, though, as we will discuss later on, the same concepts can be applied to time-varying periodic or aperiodic states.

In addition to finding the concentrations that make all the time derivatives in the rate equations vanish, it is useful to have another piece of information about such a time-independent or steady state. If the system starts at the steady state and is then subjected to a small perturbation, for example, injection or removal of a pinch of one of the reactants, we may ask whether the system will return to the original state or will evolve toward some other asymptotic behavior. The question we are asking here is whether or not the state of interest is *stable*. One of the basic tools of nonlinear chemical dynamics is stability analysis, which is the determination of how a given asymptotic solution to the rate equations describing a system will respond to an infinitesimal perturbation.

If our system is a closed one, such as a reaction in a beaker, it can be shown that the only asymptotic state is the equilibrium state, and that the system must eventually reach equilibrium. If the equilibrium concentrations are slightly perturbed, either by an inherent fluctuation or by the addition of small amounts of reactants and/or products, the system will return to equilibrium. We then say that chemical equilibrium is *globally stable*. To obtain more interesting behavior, it is necessary to work with an *open system*—that is, one which can exchange matter and/or energy with its surroundings. For example, one can imagine a reactor that allows us to pump fresh reactants continuously into the system and to pump reacted solution out so as to maintain a constant volume. Such an apparatus is known as a flow reactor, or, more formally, as a continuous-flow stirred tank reactor (CSTR). The CSTR has played a major role in nonlinear chemical dynamics; its characteristics are discussed in more detail in Chapter 3. If we run a reaction in an open system like a CSTR, it becomes possible for the system to have one or more steady states, as well as other, time-dependent asymptotic states. Unlike the case of equilibrium in a closed system, the stabilities of these states cannot be predicted a priori. This is where stability analysis comes in. Although unstable states are easy enough to calculate on a piece of paper or with a computer, they are never observed in experiments, because the system is always subject to small fluctuations that will drive it away from an unstable state, ultimately to an asymptotically stable state. Thus, knowing whether or not an asymptotic state is stable tells us whether or not we can expect to see it in an experiment. This is not to say that unstable states are chimeras, irrelevant to the real world. The existence of an unstable state can be crucial in determining the dynamics of a system. Also, cleverly designed experiments can stabilize certain unstable states, making them directly observable.

2.2.1 One-Variable Systems

As we did with rate equations, we start with the simplest possible systems—that is, those described by a single concentration variable. Consider such a system with a first-order irreversible reaction:

$$A \rightarrow B \qquad \text{rate constant } k_1, \qquad a = [A], b = [B] \qquad (2.30)$$

$$da/dt = -k_1 a \qquad (2.31)$$

In a closed system, the (globally stable) asymptotic solution is obviously $a = 0$, $b = a_0$, where a_0 is the initial concentration of A. Not very interesting! Now we want to study this reaction in a CSTR. Assume that we let flow in a solution with concentration a_0 at a flow rate of F mL s^{-1} and that the reactor has a volume V mL. In order to include the effects of the flow in a fashion that does not explicitly depend on the reactor geometry, we define the *reciprocal residence time k_0* by

$$k_0 = \frac{F}{V} \qquad (2.32)$$

The quantity k_0 has units of reciprocal time, in this case s^{-1}. A little thought will convince you that the reciprocal of k_0 is the average time that a molecule spends in the reactor. Note that a fast flow rate through a large reactor is equivalent, in dynamical terms, to a slow flow rate through a small reactor. Keeping the contents of the reactor at a constant volume requires that solution be removed continuously. Therefore, the flow will remove not only the product B, but also some unreacted A. The change in reactant concentration will be determined by the rate of the chemical reaction, eq. (2.31), and by the difference between the reactant concentration and the feedstream concentration. The process is described by the following rate law:

$$\frac{da}{dt} = -k_1 a + k_0 a_0 - k_0 a = k_0(a_0 - a) - k_1 a \qquad (2.33)$$

We can find the steady state a_{ss} of eq. (2.33) by setting da/dt equal to zero. We then obtain

$$a_{ss} = \frac{k_0 a_0}{k_0 + k_1} \qquad (2.34)$$

The problem can also be studied graphically (Figure 2.2) by plotting the rate of consumption, $k_1 a$, and the rate of production from the flow, $k_0(a_0 - a)$, as func-

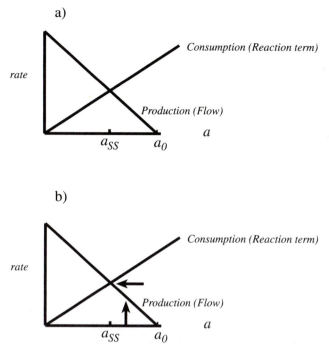

Figure 2.2 (a) The production and consumption curves for reaction (2.30) in a CSTR. (b) Graph that shows what happens when the steady state is perturbed by an increase in the concentration of a.

tions of a. Gray and Scott (1990) have developed this technique to great advantage. The steady state is located at the intersection of the production and consumption curves.

We now wish to determine whether or not the steady state is stable. What happens if we are at the steady state and then perturb the system a little? Imagine that a is increased from a_{ss} to $a_{ss} + \delta$. Graphically, we move over to the vertical arrow on the x axis in Figure 2.2b. Now, the rate of consumption is greater than the rate of production (from the inflow), and the concentration will decrease, as indicated by the horizontal arrow. The opposite will happen if a_{ss} is reduced. The rate of production will exceed the rate of consumption, and a will increase back to the steady state value. Either way, the system returns to the steady state; it is stable.

What led to stability in the above example was the fact that if we start at the steady state and increase a, the net rate of production of a decreases, while if we decrease a, we generate a net increase in its rate of production. To develop a more quantitative approach, which generalizes to systems with more than one variable, we will examine the derivative of the rate of production of a with respect to its concentration. We shall refer to this quantity as the *Jacobian, J*:

$$J = \frac{\partial(da/dt)}{\partial a} \tag{2.35}$$

where the derivative is evaluated at the steady state; that is, eq. (2.34). In this case, $J = -k_0 - k_1$, independent of a. For more complex kinetics, we would have to use the steady-state concentration a_{ss} to determine the value of J. The fact that J is negative means that an infinitesimal perturbation of the steady state will decay. In fact, the Jacobian even tells us how fast it will decay: the perturbation will decrease as $\exp(Jt)$. To see this, let

$$a(t) = a_{ss} + \delta(t) \tag{2.36}$$

Then,

$$d\delta/dt = da/dt = k_0 a_0 - (k_0 + k_1)a = k_0 a_0 - (k_0 + k_1)(a_{ss} + \delta) = -(k_0 + k_1)\delta \tag{2.37}$$

Solving the differential equation (2.37) for $\delta(t)$ yields

$$\delta(t) = \delta(0)\exp\left(-(k_0 + k_1)t\right) = \delta(0)\exp(Jt) \tag{2.38}$$

The system we have just analyzed had linear kinetics, a situation that made for easy analysis, but one that prevented the occurrence of any interesting dynamics. We will now increase the complexity by considering a quadratic, and therefore nonlinear, rate law. To make life interesting, we will look at an autocatalytic system in a CSTR:

$$A + B \rightarrow 2B \qquad k_2 \tag{2.39}$$

$$da/dt = k_0(a_0 - a) - k_2 ab = k_0(a_0 - a) - k_2 a(a_0 - a) = (k_0 - k_2 a)(a_0 - a)$$
$$= k_0 a_0 - (k_0 + k_2 a_0)a + k_2 a^2 \tag{2.40}$$

where we assume that only a is flowed in, and we have utilized the fact that $a + b = a_0$.[2] If we set $da/dt = 0$ in eq. (2.40), we obtain two steady states, $a_{ss} = a_0$ and $a_{ss} = k_0/k_2$. The first solution corresponds to the reaction never getting going—it obviously requires some of the autocatalyst B to start things off. The second one may or may not be physically reasonable—we cannot end up with a steady-state concentration of A that exceeds a_0. The following condition must hold in order for there to be two plausible steady-state solutions:

$$k_2 a_0 > k_0 \tag{2.41}$$

Physically, eq. (2.41) tells us that the steady state in which there is some A and some B can exist only if reaction (2.39) produces B at a rate that is sufficiently rapid compared with the rate at which B flows out of the CSTR.

Figure 2.3 shows that we can perform our analysis graphically as well as algebraically.

We can calculate the Jacobian by differentiating eq. (2.40) with respect to a:

$$J = 2k_2 a - k_2 a_0 - k_0 = \pm(k_0 - k_2 a_0) \tag{2.42}$$

[2]If b is flowed in as well, we would need to replace a_0 by $a_0 + b_0$ here.

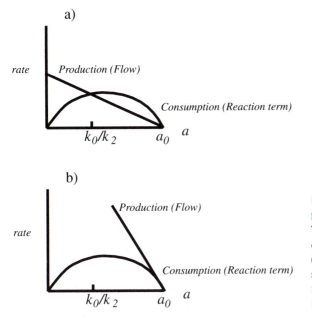

a)

Production (Flow)

rate

Consumption (Reaction term)

k_0/k_2 a_0 a

b)

Production (Flow)

rate

Consumption (Reaction term)

k_0/k_2 a_0 a

Figure 2.3 The flow diagram for the quadratic reaction. (a) The case of two steady states of which only one is stable. (b) The case of only one steady state when the flow rate is higher than the reaction rate.

where the plus and minus signs refer to the values of J obtained by substituting the two steady-state values of a_{ss}, k_0/k_2, and a_0, respectively. Equation (2.42) shows that only one of the two possible steady states can be stable for any given flow rate k_0. If $k_2 a_0 > k_0$, that is, if the reaction is fast enough for a reacted steady state to exist, then the unreacted steady state becomes unstable (the consumption curve lies above the production curve at a_0), and the reacted state is stable. In the opposite case, in which eq. (2.41) does not hold, only the unreacted steady state exists, and it is stable.

Thus, for quadratic autocatalysis, the system has a single stable steady state. If we imagine starting from a situation in which only the unreacted state exists and then either increasing a_0 or decreasing k_0 until eq. (2.41) holds, we observe a continuous change in the steady-state concentration, because at the transition point, where the unreacted state loses stability and the reacted state becomes stable, the two states have identical concentrations: $a_{ss} = a_0 = k_0/k_2$.

We reach the next level of complexity by analyzing a system with cubic autocatalysis in a CSTR:

$$A + 2B \rightarrow 3B \qquad k_3 \qquad (2.43)$$

$$da/dt = k_0(a_0 - a) - k_3 ab^2 = (a_0 - a)[k_0 - k_3 a(a_0 - a)] \qquad (2.44)$$

Again, the system possesses an unreacted state at $a_{ss} = a_0$. Dividing the right-hand side of eq. (2.44) by $a_0 - a$ yields an equation which we solve for the remaining steady states:

$$k_3 a^2 - k_3 a a_0 + k_0 = 0 \qquad (2.45)$$

$$a_{ss} = [k_3 a_0 \pm (k_3^2 a_0^2 - 4k_0 k_3)^{1/2}]/2k_3 \qquad (2.46)$$

Since a_{ss} must be real and positive, we have a total of three steady states if

$$k_3 a_0^2 > 4k_0 \qquad (2.47)$$

If the inequality goes the other way, that is, if the rate of reaction is too small compared with the flow, then only the unreacted steady state exists.

Differentiating eq. (2.44) shows that the Jacobian is given by

$$J = -k_0 - k_3(a_0 - a)(a_0 - 3a) \qquad (2.48)$$

For the unreacted state, we have $J = -k_0 < 0$, so this state is always stable. Algebraic analysis of the sign of J at the other two steady states is more tedious. As an alternative, we can plot the production and consumption curves as shown in Figure 2.4, where it is clear that when three steady states exist, the middle one is always unstable, while the other two are stable.

The model that we have been analyzing has one unrealistic feature. Condition (2.47) for the existence of three steady states provides only a single limit on the experimentally accessible parameters, k_0 and a_0; there should be three steady states for all inflow concentrations exceeding $(4k_0/k_3)^{1/2}$ and at flow rates approaching zero (where only a single steady state, equilibrium, can exist). In actual systems, one almost always observes a finite range within which behavior like multiple steady state occurs. What have we neglected?

Consider eq. (2.43). The net reaction is $A \rightarrow B$, but it is catalyzed by two molecules of B. A catalyst does not change the thermodynamic driving force of a reaction; it only changes the rate of the reaction. Therefore, if reaction (2.43) proceeds, then so should the uncatalyzed reaction

$$A \rightarrow B \qquad (2.49)$$

We leave it as an exercise for the reader to show that adding eq. (2.49) to our scheme for cubic autocatalysis creates a situation in which only a single steady state arises if the reaction rate is *either* too high or too low. In the former case, the steady state is a reacted one, with a high concentration of B. The reader may find a graphical approach more congenial than an algebraic one.

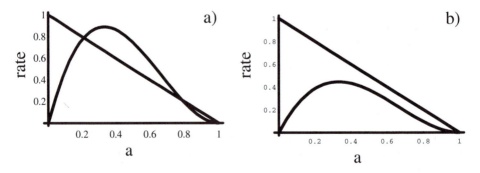

Figure 2.4 Cubic autocatalysis. (a) Equation (2.47) is satisfied and there are three steady states. (b) Equation (2.47) is violated and only the unreacted steady state exists.

2.2.2 Hysteresis and Bistability

Suppose that for a system like the one we have been considering, there are three steady states, two of which are stable, for $\alpha < k_0 < \beta$. Now we do an experiment in which we vary the flow rate k_0 slowly from below α to above β and back (we shall provide more details on how this is done in Chapter 3), allowing the concentrations to reach their steady-state values at each step before the flow rate is changed further. What do we see?

Each line in Figure 2.5 indicates a different flow rate. Figure 2.6 shows what we actually measure experimentally.

Notice that in certain ranges of the flow rate, there are two stable steady states. Each state is stable to small perturbations, so when it is in one steady state the system does not "see" the other stable steady state unless it experiences a large perturbation. The situation is analogous to the double-well potential illustrated in Figure 2.7.

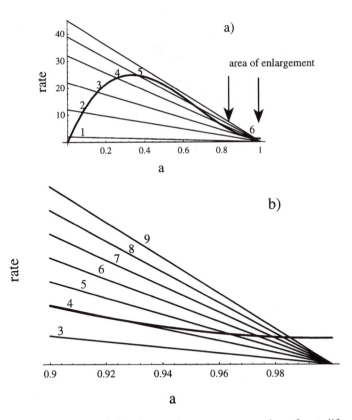

Figure 2.5 Determining the steady-state concentration of a at different flow rates: (a) starting from low flow rates, (b) starting from high flow rates. [Note that only the range of a between the arrows in part a is shown, so that the slopes, i.e., the flow rates in part b are significantly higher than those in part a.]

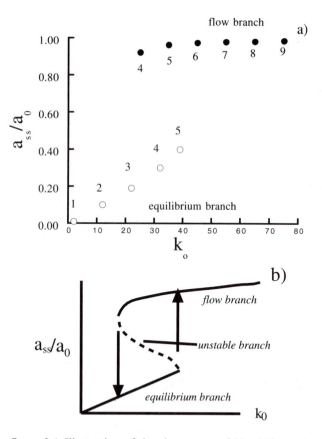

Figure 2.6 Illustration of the phenomena of bistability and hysteresis. (a) The numbered points correspond to the steady states for the flow rates indicated in Figure 2.5. (b) The three branches of steady states. Solid lines are stable branches; dashed line shows unstable branch.

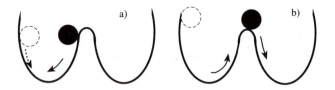

Figure 2.7 A double-well potential constitutes a mechanical analog of a bistable chemical system. (a) If the ball is perturbed by a small amount from its stable steady state at the bottom of one of the wells, it returns to its original position. (b) A large perturbation can cause the ball to make the transition from one well or state to the other. The maximum between the two wells corresponds to the unstable steady state of the chemical system.

The situation that we have just described is known as *bistability*. The solid lines in Figure 2.6 indicate stable *branches*, while the dashed line is an unstable branch of steady states. In a flow system, the lower branch is called the *equilibrium or thermodynamic branch*, because it extends from the thermodynamic equilibrium point at zero flow rate. The upper or *flow branch* is so named because it can be accessed from high flow rate. Typically, the composition of steady states on the equilibrium branch resembles that of the products, while states on the flow branch look more like the reactants in the input flow. Obviously, in the limits of zero or infinite flow, we obtain the compositions of the equilibrium state and the pure reactants, respectively. The arrows in Figure 2.6b indicate the direction that the system will take when the flow rate reaches a particular value. Notice that the system "remembers" its history. Which state the system resides in depends on where it came from—that is, low or high flow rate. This memory effect, associated with the switching between branches of states in a bistable system, is known as *hysteresis*.

Hysteresis is an extremely important phenomenon, closely connected with oscillatory behavior, as we shall see in Chapter 4. Hysteresis is familiar to physicists from the study of low-temperature magnetic phenomena, but is less commonly encountered in chemistry. It occurs in cooperative reactions like certain types of adsorption, where occupancy of one site makes a neighboring site easier to occupy. Cooperativity may be viewed as a form of autocatalysis in that the more occupied sites we have, the faster the remaining sites become occupied.

2.2.3 Some Other Possibilities

We have established that if the Jacobian is negative, small perturbations to the steady state decay away as $\exp(-|J|t)$, while if $J > 0$, perturbations grow as $\exp(Jt)$ until the system leaves the neighborhood of the initial state and approaches some other, stable state. What happens if $J = 0$? The system is then said to be *marginally stable*. How does it behave? The answer to this question depends on the details of the individual system, particularly on the nonlinear terms, since the utility of the Jacobian derives from an analysis in which only linear terms are kept in analyzing the effect of the perturbation on the steady state. Marginal stability is an unusual situation that typically occurs as a parameter changes and a state goes from being stable to unstable or vice versa. In the Lotka–Volterra model of Chapter 1, it turns out that the steady state is marginally stable for any choice of the parameters, and that this state is surrounded by an infinite number of periodic solutions. In that case, perturbations neither grow nor decay, but simply carry the system on to another periodic orbit, where it remains until it is perturbed again.

Another possibility is that a steady state may be stable, so that small perturbations decay away smoothly, but larger perturbations may grow for a while before the system returns to the original state. Such states, which are referred to as *excitable*, are stable, but they give rise to more interesting dynamics than "ordinary" stable states. Excitability is a property of heart muscle and nerve cells, as well as certain chemical systems like the BZ reaction. We shall consider excitability in more detail in Chapter 6.

2.2.4 Systems with More than One Variable

We have been able to illustrate a number of points by examining models with only a single concentration variable. Such systems are obviously easier to deal with, both algebraically and intuitively. Unfortunately, however, the real world contains very few one-variable systems, and those that do exist display a very limited range of dynamics. Periodic oscillation requires at least two independently varying concentrations, and chaos takes three. The level of effort required to analyze a model goes up very rapidly as the number of concentration variables increases. We shall focus on two-variable systems, where a number of results are available. A great deal is known analytically about the behavior of the three-variable Oregonator model (Noyes and Field, 1974) for the BZ reaction (see, e.g., Field and Noyes, 1977) and about a handful of nonchemical models that give rise to chaos (Lorenz, 1973; Sparrow, 1982). Generally, though, models with three or more variables are best dealt with using the sorts of numerical methods that we will discuss in Chapter 7.

One very important mathematical result facilitates the analysis of two-dimensional (i.e., two concentration variables) systems. The *Poincaré–Bendixson theorem* (Andronov et al., 1966; Strogatz, 1994) states that if a two-dimensional system is confined to a finite region of concentration space (e.g., because of stoichiometry and mass conservation), then it must ultimately reach a steady state or oscillate periodically. The system cannot wander through the concentration space indefinitely; the only possible asymptotic solution, other than a steady state, is oscillations. This result is extremely powerful, but it holds only for two-dimensional systems. Thus, if we can show that a two-dimensional system has no stable steady states and that all concentrations are bounded—that is, the system cannot explode—then we have proved that the system has a stable periodic solution, whether or not we can find that solution explicitly.

To examine the issue of stability in two or more dimensions, we need to generalize the notion of the Jacobian introduced in eq. (2.35). We will consider a model with two independent concentrations, α and β, whose time derivatives are represented by two functions, f and g, respectively:

$$d\alpha/dt = f(\alpha, \beta) \tag{2.50a}$$

$$d\beta/dt = g(\alpha, \beta) \tag{2.50b}$$

To test the stability of a steady state $(\alpha_{ss}, \beta_{ss})$, we add a perturbation to each variable:

$$\alpha = \alpha_{ss} + \delta\alpha \tag{2.51a}$$

$$\beta = \beta_{ss} + \delta\beta \tag{2.51b}$$

We substitute the expressions (2.51) into equations (2.50) and expand the functions f and g in the Taylor series about the steady-state point $(\alpha_{ss}, \beta_{ss})$, where $f = g = 0$. If the perturbations are small enough that we may neglect second and higher order terms, our equations become

$$\frac{d\,\delta\alpha}{dt} = (\partial f/\partial\alpha)_{ss}\,\delta\alpha + (\partial f/\partial\beta)_{ss}\,\delta\beta \qquad (2.52a)$$

$$\frac{d\,\delta b}{dt} = (\partial g/\partial\alpha)_{ss}\,\delta\alpha + (\partial g/\partial\beta)_{ss}\,\delta\beta \qquad (2.52b)$$

Equations (2.52) are just linear differential equations (because we have dropped the nonlinear terms in the Taylor series) in $\delta\alpha$ and $\delta\beta$. Equations of this form have solutions that are sums of exponentials, where the exponents are found by assuming that each variable is of the form $c_i \exp(\lambda t)$. (For more information on methods of solving differential equations, see Boyce and DiPrima, 1977.) Let

$$\delta\alpha(t) = c_1 e^{\lambda t}, \qquad \delta\beta(t) = c_2 e^{\lambda t} \qquad (2.53)$$

We now define the *Jacobian matrix* **J** as

$$\mathbf{J} = \begin{bmatrix} \dfrac{\partial f}{\partial\alpha} & \dfrac{\partial f}{\partial\beta} \\[2mm] \dfrac{\partial g}{\partial\alpha} & \dfrac{\partial g}{\partial\beta} \end{bmatrix}_{ss} \qquad (2.54)$$

The result of substituting eqs. (2.53) into eqs. (2.52) and dividing by $\exp(\lambda t)$ can be written in compact form as

$$(\mathbf{J} - \lambda\mathbf{I})\mathbf{C} = \mathbf{0} \qquad (2.55)$$

where **J** is the Jacobian matrix defined in eq. (2.54), **C** is the vector of coefficients (c_1, c_2) in eq. (2.53), **I** is the 2×2 identity matrix, and **0** is a 2×1 vector of zeros. A standard result of linear algebra (Lang, 1986) is that equations like eq. (2.55) have nontrivial solutions—that is, solutions other than all the coefficients c being zero—only when λ is an *eigenvalue* of the matrix **J**—that is, only when λ is a solution of the determinantal equation

$$\det \begin{bmatrix} \dfrac{\partial f}{\partial\alpha} - \lambda & \dfrac{\partial f}{\partial\beta} \\[2mm] \dfrac{\partial g}{\partial\alpha} & \dfrac{\partial g}{\partial b} - \lambda \end{bmatrix}_{ss} = \left(\dfrac{\partial f}{\partial\alpha} - \lambda\right)_{ss}\left(\dfrac{\partial g}{\partial\beta} - \lambda\right)_{ss} - \left(\dfrac{\partial g}{\partial\alpha}\right)_{ss}\left(\dfrac{\partial f}{\partial\beta}\right)_{ss} = 0 \qquad (2.56)$$

Equation (2.56) can be expanded to give

$$\lambda^2 - \lambda\,\mathrm{tr}\,(\mathbf{J}) + \det(\mathbf{J}) = 0 \qquad (2.57)$$

where $\mathrm{tr}\,(\mathbf{J})$ is the trace of the Jacobian matrix, that is, the sum of the diagonal elements. Equation (2.57) is a quadratic in the exponent λ, which has two solutions and whose coefficients depend on the elements of the Jacobian, that is, the steady-state concentrations and the rate constants. The general solution to eqs. (2.52) for the time evolution of the perturbation will be a linear combination of the two exponentials. The stability of the system will be determined by whether or not the perturbation grows or decays. If either eigenvalue λ has a positive real part, the solution will grow; the steady state is unstable. If both λ values have negative real part, the steady state is stable.

The above analysis can be generalized to systems with any number of variables. If we write the rate equations as

$$dx_i/dt = f_i(x_1, x_2, \ldots, x_n) \qquad i = 1, 2, \ldots, n \tag{2.58}$$

we can define the elements of the $n \times n$ Jacobian matrix associated with the steady state $\mathbf{x}_{ss}$ as

$$J_{ij} = (\partial f_i/\partial x_j)_{ss} \tag{2.59}$$

The steady state will be stable if all the eigenvalues of $\mathbf{J}$ have negative real parts. If any of them has a positive real part, the state is unstable.

To see where all this leads, we will follow the analysis presented by Gray and Scott (1990) for classifying the behavior of two-dimensional systems based on the nature of the solutions to eq. (2.57); these are easily obtained by applying the quadratic formula. We will consider several possibilities for the signs of the trace, determinant, and discriminant (tr $(\mathbf{J})^a - 4$ det $(\mathbf{J})$) of the Jacobian matrix.

(a)   tr $(\mathbf{J}) < 0$,   det $(\mathbf{J}) > 0$,   tr $(\mathbf{J})^2 - 4$ det $(\mathbf{J}) > 0$

If these inequalities hold, then both eigenvalues are negative real numbers. Any perturbation to this steady state will monotonically decrease and disappear. The steady state is a *stable node*, and nearby points in the concentration space are drawn to it.

(b)   tr $(\mathbf{J}) < 0$,   det $(\mathbf{J}) > 0$,   tr $(\mathbf{J})^2 - 4$ det $(\mathbf{J}) < 0$

The last inequality implies that the eigenvalues will be complex conjugates, that is, of the form $\lambda_{1,2} = a \pm ib$ where $i = (-1)^{1/2}$. The real parts of both eigenvalues are negative, meaning that the perturbation will decay back to the steady state. The imaginary exponential is equivalent to a sine or a cosine function, which implies that the perturbation will oscillate as it decays. This steady state is called a *stable focus*.

(c)   tr $(\mathbf{J}) > 0$,   det $(\mathbf{J}) > 0$,   tr $(\mathbf{J})^2 - 4$ det $(\mathbf{J}) < 0$

The eigenvalues in this case are complex conjugates with a positive real part. Perturbations will grow and spiral away from the steady state, which is an *unstable focus*.

(d)   tr $(\mathbf{J}) > 0$,   det $(\mathbf{J}) > 0$,   tr $(\mathbf{J})^2 - 4$ det $(\mathbf{J}) > 0$

The eigenvalues are both real and positive, so the steady state is unstable. Any perturbation will grow exponentially away from the steady state, which is an *unstable node*.

(e)   det $(\mathbf{J}) < 0$

When the determinant is negative, the eigenvalues will be real, but with one positive and one negative. Trajectories approach the steady state along the eigenvector corresponding to the negative eigenvalue, but then move away from the steady state along the transverse direction. The steady state is called a *saddle point* because of the similarity of the trajectories in phase space to a riding saddle (Figure 2.8).

(f) $\text{tr}\,(\mathbf{J}) = 0,$ $\det\,(\mathbf{J}) > 0$

When the determinant is positive and the trace is zero, the eigenvalues become purely imaginary. Such conditions indicate the onset of sustained oscillations through a *Hopf bifurcation*. As a system parameter (e.g., the flow rate in a CSTR) is varied so that the system passes through the Hopf bifurcation, a *limit cycle* or periodic orbit develops surrounding the steady state. If the limit cycle is stable, in which case the bifurcation is said to be *supercritical*, the steady state loses its stability, and any small perturbation will cause the system to evolve into a state of sustained oscillations. In a *subcritical* Hopf bifurcation, the steady state maintains its stability, but becomes surrounded by a pair of limit cycles, the inner one being unstable and the outer one being stable. Such a situation gives rise to bistability between the steady state and the stable oscillatory state.

Schematic representations of the trajectories in the neighborhood of the steady states that we have been describing are shown in Figure 2.9. The same types of steady states can occur in more than two dimensions. In general, the character of a steady state is determined by the eigenvalue(s) of the Jacobian with the most positive real part. If all the eigenvalues have negative real parts, the point will be stable; a single eigenvalue with a positive real part is sufficient for instability.

If a system has more than one stable asymptotic solution or *attractor*—steady state, limit cycle, or something more complex—then where the system will end up depends on its initial conditions. The concentration, or phase, space is divided into a set of *basins of attraction*, which are those regions of the phase space for which a system starting in that region will ultimately arrive at the particular attractor. Figure 2.10 shows a case with two attractors. The dashed line, which is called a *separatrix*, indicates that a system that starts in the left side will end up on the limit cycle; a system starting in the right-hand region will be attracted to the steady state.

In Figure 2.11, a trajectory appears to cross itself. This is an illusion caused by our efforts to portray a higher dimensional system in two dimensions. A trajectory cannot cross itself because solutions to the sorts of ordinary differential equations that occur in chemical kinetics are *unique*. If a trajectory seems to cross itself in two dimensions, then the system actually has higher dimension (it is not completely determined by two variables). A two-variable plot projects the higher dimensional trajectory onto a plane, giving the *appearance* of a crossing of the trajectory. Figure 2.11b shows a two-dimensional projection of a three-

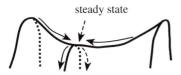

steady state

Figure 2.8 A saddle in phase space.

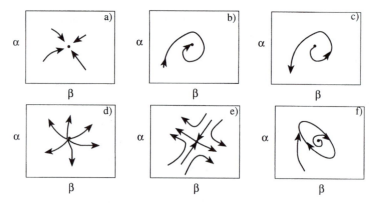

Figure 2.9 Phase plane representations of the local instabilities discussed in terms of the conditions on the trace and determinant of the Jacobian matrix.

dimensional system. Notice how the trajectories cross in the two-dimensional view (part b) but not in the three-dimensional version (part a). The only points at which trajectories can intersect are steady states, where trajectories end. A special case in which a single trajectory begins *and* ends at a single (saddle) steady state is known as a *homoclinic* point and the trajectory is known as as a homoclinic trajectory. Homoclinic points have important consequences for the dynamics of a system (Guckenheimer and Holmes, 1983; Strogatz, 1994).

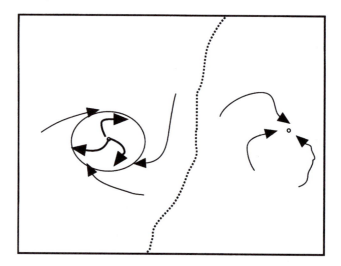

Figure 2.10 Two basins of attraction exist in this diagram. If a system is initialized to the right of the dashed line, it will asymptotically proceed to a stable steady state. If started to the left of the line, it will oscillate.

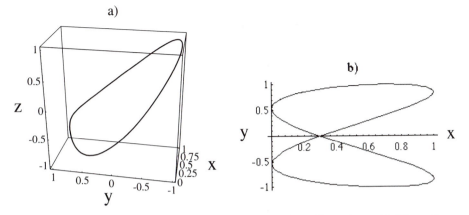

a)

b)

Figure 2.11 Projections of a trajectory that exists in three dimensions. (a) Three-dimensional view projected onto a plane. (b) Two-dimensional view in which the trajectory appears to cross itself

2.2.5 Application of Stability Analysis: The Brusselator

In Chapter 1, we introduced the Brusselator model as the first "chemical" model to demonstrate oscillations and traveling waves. We will now analyze the Brusselator to illustrate how one might use the techniques that we have discussed to establish the behavior of a two-dimensional system. We recall that the equations are:

$$A \rightarrow X \tag{2.60a}$$
$$B + X \rightarrow Y + D \tag{2.60b}$$
$$2X + Y \rightarrow 3X \tag{2.60c}$$
$$X \rightarrow E \tag{2.60d}$$

Let the concentrations, like the species, be represented by capital letters, and call the time T. Then, the rate equations for X and Y corresponding to eqs. (2.60), with the concentrations of A and B held constant, are

$$\frac{dX}{dT} = k_1 A - k_2 BX + k_3 X^2 Y - k_4 X \tag{2.61a}$$

$$\frac{dY}{dT} = k_2 BX - k_3 X^2 Y \tag{2.61b}$$

There are a lot of rate constants floating around in eqs. (2.61). Wouldn't it be nice if they were all equal to unity? This seems like too much to hope for, but it turns out that if we define a unitless time variable and unitless concentrations, which we shall represent with lower-case letters, then eqs. (2.61) take a much simpler form. We set

$$X = \alpha x, \qquad Y = \beta y, \qquad T = \gamma t, \qquad A = \delta a, \qquad B = \varepsilon b \tag{2.62}$$

and substitute eqs. (2.62) into eqs. (2.61) to obtain, after multiplying through by α/γ and α/β in eqs. (2.63a) and (2.63b), respectively,

$$\frac{dx}{dt} = (k_1\delta\gamma/\alpha)a - (k_2\varepsilon\gamma)bx + (k_3\alpha\beta\gamma)x^2y - (k_4\gamma)x \tag{2.63a}$$

$$\frac{dy}{dt} = (k_2\varepsilon\alpha\gamma/\beta)bx - (k_3\gamma\alpha^2)x^2y \tag{2.63b}$$

However, it looks like we have only made things worse, much worse, because eqs. (2.63) contain not only the rate constants but also the *scaling factors*, α, β, γ, δ, and ε that we introduced in eq. (2.62). Now, here comes the trick! We are free to choose the scaling factors any way we like. Let us pick them so as to make all the expressions in parentheses in eqs. (2.63) equal to unity. That is, we shall require that

$$k_1\delta\gamma/\alpha = k_2\varepsilon\gamma = k_3\alpha\beta\gamma = k_4\gamma = k_2\varepsilon\alpha\gamma/\beta = k_3\gamma\alpha^2 = 1 \tag{2.64}$$

Equation (2.64) seems to represent six conditions on our five unknown scaling factors, but one condition turns out to be redundant, and a little algebra gives us our solutions:

$$\alpha = \beta = (k_4/k_3)^{1/2}, \qquad \gamma = 1/k_4, \qquad \delta = (k_4/k_1)(k_4/k_3)^{1/2} \qquad \varepsilon = k_4k_2 \tag{2.65}$$

If we now substitute eqs. (2.65) into eqs. (2.63), we obtain a much prettier version of the Brusselator in terms of non-dimensional or unitless variables. This *rescaling* procedure is often exceedingly useful for reducing the number of parameters (in this case to zero!) before analyzing the properties of a model. We recommend it highly. Our equations are now

$$\frac{dx}{dt} = a - bx + x^2y - x \tag{2.66a}$$

$$\frac{dy}{dt} = bx - x^2y \tag{2.66b}$$

To obtain the steady state(s) of the Brusselator, we set eqs. (2.66) equal to zero and solve for x and y. We find a single solution:

$$x_{ss} = a, \qquad y_{ss} = b/a \tag{2.67}$$

To analyze the stability of this state, we must calculate the elements of the Jacobian matrix:

$$\mathbf{J} = \begin{bmatrix} \dfrac{\partial(dx/dt)}{\partial x} & \dfrac{\partial(dx/dt)}{\partial y} \\ \dfrac{\partial(dy/dt)}{\partial x} & \dfrac{\partial(dy/dt)}{\partial y} \end{bmatrix} \tag{2.68}$$

The elements of the Jacobian matrix are

$$\frac{\partial(dx/dt)}{\partial x}\bigg|_{ss} = -b + 2a(b/a) - 1 = b - 1 \tag{2.69a}$$

$$\frac{\partial(dx/dt)}{\partial y}\bigg|_{ss} = a^2 \tag{2.69b}$$

$$\frac{\partial(dy/dt)}{\partial x}\bigg|_{ss} = b - 2a(b/a) = -b \tag{2.69c}$$

$$\frac{\partial(dy/dt)}{\partial y}\bigg|_{ss} = -a^2 \tag{2.69d}$$

We need to obtain the eigenvalues of the matrix whose elements are given by eqs. (2.69) by solving the characteristic equation:

$$\det\begin{bmatrix} b - 1 - \lambda & a^2 \\ -b & -a^2 - \lambda \end{bmatrix} = 0 \tag{2.70}$$

or equivalently

$$\lambda^2 + (a^2 + 1 - b)\lambda + a^2 = 0 \tag{2.71}$$

At this point, although it would not be terribly difficult, we do not need to solve explicitly for λ because we are interested in the qualitative behavior of the system as a function of the parameters a and b. When a system undergoes a qualitative change in behavior (such as going from a stable steady state to an unstable one) as a parameter is varied, it is said to have undergone a *bifurcation*. Bifurcations, which we shall discuss further in the next section, occur when the roots of the characteristic equation change sign. A knowledge of the bifurcations of a system is often sufficient to afford a qualitative understanding of its dynamics.

Comparing eqs. (2.57) and (2.71), we observe that tr $(\mathbf{J}) = -a^2 + b - 1$, which can be either positive or negative, while det $(\mathbf{J}) = a^2$, which is always positive. The stability of the steady state will depend on the sign of the trace. As we vary a and/ or b, when the trace passes through zero, the character of the steady state will change; a bifurcation will occur. If

$$b > a^2 + 1 \tag{2.72}$$

the sole steady state will be unstable and, thus, by the Poincaré–Bendixson theorem the system must either oscillate or explode. It is possible to prove that the system cannot explode, but is confined to a finite region of the x–y phase space. Equation (2.72) therefore determines the boundary in the a–b constraint space between the region where the system will asymptotically approach a stable steady state and the region where it will oscillate periodically.

2.3 Bifurcation Theory

There are a number of ways to analyze a differential equation or a set of differential equations. The most familiar and direct approach is to attempt to obtain an explicit solution, either analytically, which is rarely possible, or numerically. An alternative technique, which we have touched on earlier in this chapter, is to examine how the nature of the solutions changes as a parameter in the equations is varied. In chemical systems of the sort that we are interested in, this *control parameter* might be the temperature or the flow rate or an input concentration in a CSTR.

A qualitative change in the nature of the solution to a set of equations is known as a *bifurcation*—literally, a splitting. Examples include changes in the number of steady states, changes in stability, or the appearance of a periodic or chaotic solution as a control parameter is varied. Bifurcation theory attempts to enumerate and catalog the various bifurcations that are possible in a set of differential equations with a given number of variables. In section 2.2.3, our analysis of the types of steady states that can occur in a two-dimensional system constitutes a primitive example of bifurcation theory. Mathematicians have developed an extremely sophisticated and powerful theory (Guckenheimer and Holmes, 1983; Golubitsky, 1985), especially for two-dimensional systems.

One useful technique is an extension of the method of analyzing the behavior of a system in the neighborhood of a steady state by linearizing the equations in that neighborhood. If one keeps the lowest order essential nonlinear terms as well, then one obtains the so-called *normal form*, which displays not only the stability of the steady state, but also how the properties of that state change as the control parameter passes through the bifurcation point.

We shall not attempt to present a more detailed discussion of bifurcation theory here, though we shall look briefly at numerical bifurcation techniques in Chapter 7. The interested reader should consult one of the excellent references on the subject, and all readers should be aware that one can often learn more about a system by focusing on the qualitative aspects of its solutions over a range of parameters than by looking only at its quantitative behavior for a single parameter set.

2.4 Modes of Representing Data

It is possible to accumulate a great deal of data when studying the dynamics of a chemical system. Modern instrumentation, particularly with the use of microcomputers, enables an experimenter to obtain enormous numbers of data points. A calculation on a computer can generate as much data as we are willing to allow it to. How much actual *information* we get from these data depends, of course, on how, and how cleverly, they are analyzed. Over the years, workers in the field have found a number of particularly helpful methods for displaying the data obtained in a typical nonlinear chemical dynamics experiment or calculation. We will discuss these techniques in this section.

2.4.1 Time Series

The most natural, and most common, method to look at and present one's data is the way in which those data are taken. In a typical experiment, we measure some function of concentration (e.g., electrode potential or absorbance) as a function of time at one set of constraints. A plot of signal vs. time is known as a *time series*. Time series can be exceedingly dull, for example, in the case of a steady state, or they can be quite difficult to interpret, as in the case of a system that may or may not be chaotic. Nevertheless, they can yield valuable information, and they are certainly the first thing one should look at before proceeding further. Figure 2.12 shows a time series that establishes the occurrence of bistability in the arsenite–iodate reaction.

2.4.2 Constraint–Response Plot

In order to really understand a system, we must study it under a variety of conditions, that is, for many different sets of control parameters. In this way, we will be able to observe whether bifurcations occur and to see how the responses of the system, such as steady-state concentrations or the period and amplitude of oscillations, vary with the parameters. Information of this type, which summarizes the results of a number of time series, is conveniently displayed in a *constraint–response plot*, in which a particular response, like a concentration, is plotted against a constraint parameter, like the flow rate. If the information is available, for example, from a calculation, unstable states can be plotted as well. Bifurcations appear at points where the solution changes character, and constraint–response plots are sometimes called *bifurcation diagrams*. An experimental example for a system that shows bistability between a steady and an oscillatory state is shown in Figure 2.13.

2.4.3 Phase Diagrams

Those who have studied thermodynamics will recall that the properties of a single substance or of a mixture of substances can be concisely summarized in a *phase*

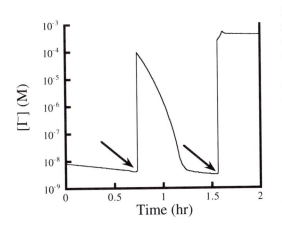

Figure 2.12 Time series showing the iodide concentration in the arsenite–iodate reaction in a CSTR. The system is initially in an oxidized steady state (the potential shows a slight downward drift because of an experimental artifact). At the times indicated by the arrows, a measured amount of acid is injected. With a small injection, the system returns to the steady state, demonstrating the stability of that state. With a larger injection, there is a transition to a second, reduced, steady state. (Adapted from De Kepper et al., 1981a.)

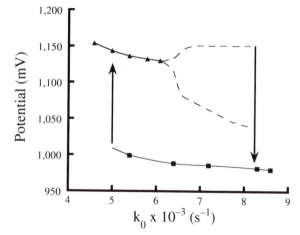

Figure 2.13 This constraint–response plot indicates how the bromate–iodide reaction exists in steady states for high and low values of the residence time but oscillates for intermediate values. The distance between the dashed lines indicates the amplitude of the oscillations.

diagram, which shows the state of the system as a function of the temperature, pressure, and/or composition. A similar construct, the dynamical phase diagram, is useful in describing the dynamical behavior of chemical systems. The results of many time series can be shown on a plot whose axes are the two (or sometimes three) constraint variables of interest. Each point on the plot indicates what kind of state (or states) (steady, oscillatory, or chaotic) was observed at that particular set of parameters. Lines separating different regions indicate bifurcations where the nature of the solutions changes. Notice that quantitative information, such as the concentrations at the steady state or the period and amplitude of the oscillations, is not represented. An example is shown in Figure 2.14.

2.4.4 Phase Portraits

One advantage of the time series over the constraint–response plot or the phase diagram is that the time series tells us how our dynamical system evolves in time. Another way of looking at the evolution of a system is to view it in phase space. The *phase portrait* is similar to a time series in that we look at a fixed set of constraints (i.e., initial concentrations, flow rates, and temperature), but now we plot two or three dependent (concentration) variables to obtain a picture of the system's trajectory, or path through the phase space. A series of phase portraits at several sets of parameter values can provide a comprehensive picture of the system's dynamics. A calculated phase portrait for the Lotka–Volterra model and an experimental one for the BZ system are shown in Figure 2.15.

2.5 Conclusions

In this chapter, we reviewed some basic concepts of chemical kinetics. We saw how to construct and analyze mechanisms for complex chemical reactions, and we

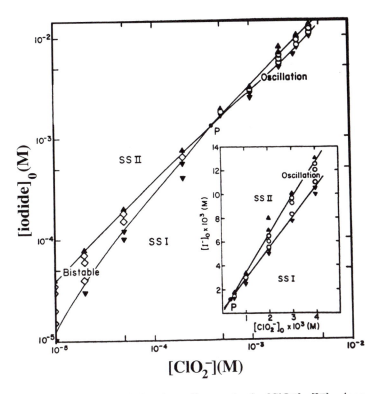

Figure 2.14 Section of the phase diagram in the $[ClO_2^-]_0$–$[I^-]_0$ plane with pH = 2 and reciprocal residence time $k_0 = 1.1 \times 10^{-3}$ s^{-1}. ▲, High iodide state; ▼, low iodide state; ◇, bistability; ○, oscillation. (Reprinted with permission from De Kepper, P.; Boissonade, J.; Epstein, I. R. 1990. "Chlorite–Iodide Reaction: A Versatile System for the Study of Nonlinear Dynamical Behavior," *J. Phys. Chem. 94*, 6525–6536.)

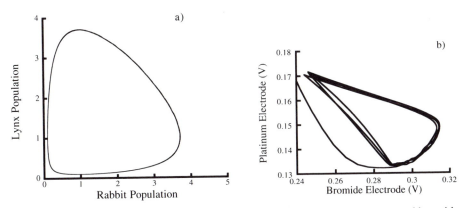

Figure 2.15 Phase portraits for (a) the Lotka–Volterra model and (b) platinum and bromide electrode data for the BZ reaction

introduced the notions of feedback and open systems, which are key to the development of complex dynamics in chemical systems. We learned how to analyze the stability of steady states. In a flow reactor with a nonlinear chemical reaction, we saw the possibility of multiple steady states, hysteresis, and bistability. We established the notion of a bifurcation and looked briefly at some bifurcations in two-dimensional systems. Finally, we considered some of the ways that have proved most useful in plotting and analyzing the data typically obtained in experiments and calculations. This is only the beginning of our exploration of the strange and exciting behavior that can occur in nonlinear chemical systems. In the chapters that follow, we will go into considerably more detail about the ideas and phenomena introduced in this chapter.

3

Apparatus

In the previous chapter, we developed a set of conceptual and mathematical tools for analyzing the models and experimental data that form the subject matter of nonlinear chemical dynamics. Here, we describe some of the key items of experimental apparatus used to obtain these data so that the reader can better appreciate the results discussed in the following chapters and can learn how to begin his or her own investigations. The first several sections are devoted to measurements of temporal behavior, with emphasis on the techniques used to monitor reactions in time and on the reactors in which these reactions are studied. The final section focuses on the study of spatial patterns and waves in chemical systems.

3.1 Analytical Techniques

It is possible, by methods that we shall discuss later, to reconstruct the qualitative dynamics of a system from the measurement of only a single variable. However, the more species whose concentrations can be measured, the easier it is to elucidate a mechanism and the more rigorously that mechanism can be tested. The most impressive study of multiple species in a chemical oscillator was carried out by Vidal et al. (1980), who were able, by a combination of techniques, to monitor the concentrations of Ce^{4+}, Ce^{3+}, Br_2, Br^-, bromomalonic acid, O_2, and CO_2 in the BZ reaction. In the following sections, we will look at the most widely employed techniques: spectroscopic and potentiometric methods. In principle, and occasionally in practice, essentially any technique that can be used to detect changes in concentration can be utilized to monitor the systems that we are interested in. Approaches that have been employed to date include polarography, high-pressure liquid chromatography, and calorimetry.

3.1.1 Spectroscopic Methods

If there are absorbing species, ultraviolet and/or visible (UV/vis) spectroscopy offers rapid response time and high sensitivity for monitoring concentrations, particularly if the species of interest have spectra with relatively little overlap. Measurements can be made in a cuvette placed in a standard UV/vis spectrophotometer, but this configuration has several limitations. Good mixing and temperature control are difficult to achieve, only a single species can be followed in each experiment, and we are limited to working with a closed system. It is possible, depending on the geometry of the spectrophotometer's sample compartment and the competence of the local machine shop, to construct a cuvette that allows for a temperature jacket, a magnetic stirrer, and even input and output flows. Some instrument makers now offer a flow cell or a stirred cell as standard options for their spectrophotometers.

Diode array spectrophotometers can be especially useful because a complete spectrum can be collected in a short interval of time, making it possible to follow several species in a single experiment by taking repeated spectra. Also, several diode array instruments come with an open sample compartment, which makes it relatively easy to construct a flow reactor that can be placed in the light path of the instrument. The metal ion catalysts used in the BZ system absorb in the visible spectrum, so that UV/vis studies of the cerium-catalyzed or ferroin-catalyzed reaction are easy to perform. Iodine is another strongly absorbing species that is found in a number of chemical oscillators. Frequently, iodide ion is present as well, and the system is monitored at the iodine–triiodide isosbestic point at 471 nm.

Nuclear magnetic resonance (NMR) spectroscopy of protons has been used to study the enolization reaction in the BZ system (Hansen and Ruoff, 1989). Nuclear magnetic resonance is a powerful technique because it has the potential to provide information on many compounds simultaneously, particularly in organic systems. Unfortunately, NMR has limited sensitivity if one needs to gather data rapidly enough for kinetic studies. Under these conditions, the detection limit is no better than about 10^{-4} M. Tzalmona and coworkers (Tzalmona et al., 1992; Su et al., 1994) recently studied spatial wave propagation in the manganese-catalyzed BZ reaction by following the effects of the paramagnetic Mn nucleus on the protons in the surrounding water molecules, and Balcom et al. (1992) used magnetic resonance imaging (MRI) to observe propagating fronts of polymerization. Relatively little use has been made of nuclei other than protons in NMR studies of nonlinear chemical dynamics. One challenge is the construction of an open reactor that will fit into the cavity of an NMR spectrometer.

Electron spin resonance (ESR) spectrometry is another technique that has the potential to provide monitoring of multiple species simultaneously. One is limited to species that have unpaired electrons, but many chemical oscillators have radical intermediates. Venkataraman and Sørensen (1991) used ESR to study the behavior of malonyl radicals in the BZ reaction. An ESR study of the cobalt-catalyzed air oxidation of benzaldehyde (Roelofs and Jensen, 1987) suggests that it may be worth devoting further attention to this method for studying oscillating reactions.

3.1.2 Potentiometric Methods

The most commonly employed and convenient methods for studying nonlinear chemical dynamics employ potentiometric techniques. Electrochemical methods offer speed, low cost and, in many cases, excellent selectivity and sensitivity. A platinum electrode and a reference electrode are all that is required for any system with a species that changes its oxidation state during the reaction. The potential E is given by the Nernst equation:

$$E = E^0 - \frac{RT}{nF} \ln Q \qquad (3.1)$$

where Q is the reaction quotient = [Reduced]/[Oxidized], E^0 is the standard reduction potential for the species of interest, R is the gas constant, T is the absolute temperature, F is Faraday's constant, and n is the number of electrons transferred in the reaction. At 25 °C,

$$E = E^0 - \frac{0.059 \text{ V}}{n} \log Q \qquad (3.2)$$

which means that in a one-electron reaction, the cell potential will change by 59 mV for a factor of ten change in Q. For example, if we consider the cerium-catalyzed BZ reaction, the change in the [Ce(IV)]/[Ce(III)] ratio during each cycle can be calculated from the recorded amplitude of oscillation in the potential of a platinum redox electrode. Rearranging the Nernst equation (3.2), we obtain:

$$\Delta[\text{Ce(IV)}/\text{Ce(III)}] = 10^{\Delta E/59 \text{ mV}} \qquad \text{at 25 °C} \qquad (3.3)$$

Notice that the reduction potential E^0 cancels, and the type of reference electrode used does not matter. Early calomel electrodes, which employed Hg/Hg_2Cl_2 in contact with saturated KCl solution, were found to stop oscillations in the BZ reaction (Zhabotinsky, 1964a; Field et al., 1972). Significant amounts of chloride ions could leak into the reacting solution and interfere with the chemistry (Jacobs and Epstein, 1976). Therefore, most investigators have employed a Hg/Hg_2SO_4 reference electrode. Modern calomel electrodes do not have this problem (Rosenthal, 1991). There are even pH electrodes containing a built-in reference that can be used with a platinum electrode (Pojman et al., 1994). The reference lead from the pH electrode is connected to the recorder along with the platinum electrode. (For more information on reference electrodes, see Ives and Janz, 1961.)

Körös, an analytical chemist, brought his knowledge of ion-selective electrodes to Oregon in 1972, which proved extremely useful to Field and Noyes in studying the BZ reaction. Bromide- and iodide-selective electrodes, which consist of a silver wire and a crystal of the appropriate silver halide, have played key roles in the study of nonlinear chemical dynamics. For reactions with significant changes in pH, a combination glass electrode is routinely used. Oxygen electrodes have occasionally been employed (Orbán and Epstein, 1989a).

The usual bromide electrode utilizes the equilibrium:

$$Ag^+ + Br^- \leftrightarrow AgBr_{(s)} \tag{3.4}$$

At 25 °C, a silver bromide-coated silver wire (or AgBr/AgS-impregnated plastic electrode, which is the standard commercial preparation from Orion) responds to [Br$^-$] according to

$$E = E^0 + 59 \ \log \frac{K_{sp}}{[Br^-]} \ mV \tag{3.5}$$

where K_{sp} is the solubility product of silver bromide.

The electrode potential changes by 59 mV for every decade change in bromide ion concentration. Noszticzius et al. (1984b) described a simple procedure for making a very inexpensive bromide-sensitive electrode. Both the commercial and the homemade varieties reliably measure bromide above a concentration of 10^{-6} M in the absence of competing ions, such as sulfide, that form silver compounds with solubilities comparable to or less than that of AgBr. The actual concentration of bromide in the BZ reaction can go as low as 10^{-9} M, which is below the solubility of AgBr unless a large amount of Ag^+ is present. Under these conditions, a bromide-selective electrode does not respond accurately to [Br$^-$]. Nonetheless, electrodes in the BZ reaction still provide a signal. Noszticzius et al. (1982, 1983) showed that the electrode is actually being corroded by HOBr. The bromide is removed from the membrane, leaving an excess of Ag^+. A similar phenomenon occurs with iodide-selective electrodes in the presence of HOI. Bromide-selective electrodes are not effective in BZ systems with very high sulfuric acid concentrations (Försterling et al., 1990).

The electrodes in an experiment can be configured in either of two ways. They can be connected to a strip chart recorder, or the voltages can be digitized by connecting the electrodes to a computer via an analog-to-digital (A/D) converter. The strip chart recorder is the cheaper of the two alternatives and is easy to use for demonstration or preliminary work. More sophisticated dynamic analysis is best performed using digitized data.

Platinum, ion-selective, and reference electrodes can be connected directly to the recorder's inputs. However, this straightforward procedure does not work well with an A/D board. Figure 3.1 shows the signal obtained from connecting the electrodes in an unstirred cerium(IV)/sulfuric acid solution to an A/D board. The signal should be constant, but in part a the signal drifts downward because ceric ions are reduced by the current that is flowing. Connecting the board through a high-impedance device prevents this drift, as shown in Figure 3.1b.

To understand this phenomenon, we observe that a redox reaction away from equilibrium plus a pair of electrodes constitutes a galvanic cell. The potential difference between the platinum and the reference electrode can cause a current to flow. The current will cause an electrochemical reaction at the platinum electrode, locally changing the ratio of oxidized to reduced ions. The concentration at the electrode now differs from that of the bulk solution, though homogeneity will be restored by stirring and/or diffusion. An erratic signal results because the mixing is neither perfect nor instantaneous.

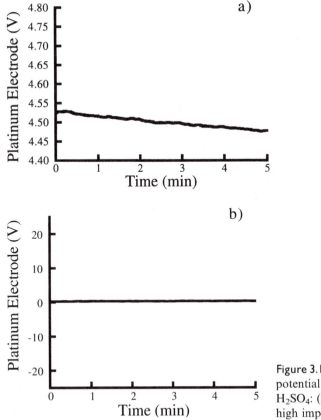

Figure 3.1 Platinum electrode potential in 0.27-M Ce(IV) in 1-M H_2SO_4: (a) no impedance, (b) with high impedance.

The resolution of the problem is to use a high-impedance input to the A/D board (Figure 3.2), so that negligible current can flow between the electrodes. The chart recorder has this type of input built in. The high impedance ensures that the potential of the solution is measured at zero current flow. A straightforward method to bring about this situation when using a platinum and a bromide electrode, for example, is to connect the bromide electrode to a pH meter (commercial bromide electrodes have BNC connections to fit the meter) in series with the platinum and reference electrodes. The negative terminals on the A/D board must be connected. The high impedance of the pH meter functions for both electrodes.

3.2 Batch Reactors

A batch reactor is the simplest configuration for studying a chemical reaction. A solution containing the reactants is placed in a beaker and the reaction proceeds. The reaction is often thermostated with a circulating jacket to maintain constant temperature. No chemicals enter or leave the system once the reaction begins; it is a closed system. Relatively few reactions undergo oscillations in a batch reactor, the BZ reaction being one of these. Oscillations in a batch reactor are necessarily

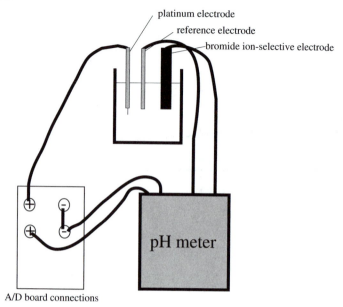

Figure 3.2 Experimental configuration to include a high-impedance input.

transient. Although they may continue for hours and tens of cycles, the Second Law of Thermodynamics requires that a closed system must finally reach equilibrium as its free energy decreases toward its eventual minimum value.

Batch reactors are convenient for lecture demonstrations. Most people who have witnessed an oscillating reaction have seen either the BZ or the Briggs–Rauscher reaction (see Appendix 1 for recipes) in a beaker. For research purposes, batch reactors can be useful for studying the basic kinetics of a system. For example, a reaction that exhibits oscillations in a flow reactor may only exhibit clock reaction behavior in a batch reactor; that is, a sharp change in concentration occurs very rapidly after an initial induction period during which the concentrations remain nearly constant. Studying the factors that affect the induction period can be the first step in developing a comprehensive model of the dynamics (Figure 3.3). Occasionally, one may be able to study one of the elementary steps in a proposed mechanism in a batch reactor.

In developing a mechanism, it is essential to obtain rate constants for as many elementary steps as possible. Often, some of the elementary reactions have rates that are so high that a normal batch reactor cannot be used to study them because the reaction proceeds quite far before mixing of the reactants is complete. Special experimental techniques for studying fast reactions with rapid mixing become essential (Bernasconi, 1986). One technique that has played a major role in establishing the rates of many rapid reactions of importance in nonlinear dynamics is stopped flow, which makes possible the mixing of reactants on a scale of milliseconds. Two solutions are rapidly forced into a spectrophotometer cell by spring-

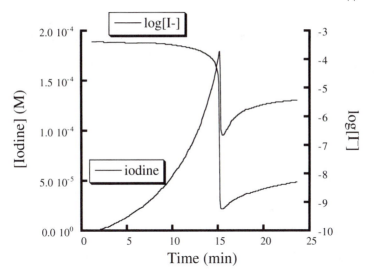

Figure 3.3 In developing a model of the chlorite–iodide reaction, Epstein and Kustin studied the reaction in a batch reactor. (Adapted from Epstein and Kustin, 1985.)

driven pistons, and the cell is configured so that the collision of the streams of solution with each other and with the cell wall causes them to mix.

3.3 Semibatch Reactors

One alternative to a simple closed system is to have a continuous inflow of reactant solutions into a beaker. Typically, no outflow is provided, and the beaker gradually fills as the reaction proceeds. This configuration is called a *semibatch reactor* (Rábai and Epstein, 1992). It has characteristics intermediate between those of a batch and a flow reactor. The input flow can be accomplished either with a pump or, much less expensively, with gravity feed, as we describe below for flow systems. Semibatch reactors can provide useful data for testing mechanisms, as one can vary which species are placed in the beaker and which are flowed in; the behavior of the system can change dramatically with different configurations. Some variants of the BZ reaction oscillate in batch only when bromine is removed by bubbling a stream of gas through the solution (Noszticzius and Bódiss, 1979; Sevcik and Adamčíková, 1982; Noszticzius et al. 1985). Pojman et al. (1992b) found that the manganese-catalyzed BZ reaction with acetone would oscillate only if a gas–liquid interface was present to allow bromine to leave. These systems are, in effect, semibatch reactors with no inflow, but with an outflow of Br_2.

Semibatch reactors are especially practical for studying enzyme reactions because one wishes to avoid flowing expensive enzymes out in the waste stream. Hocker et al. (1994) described another arrangement, specifically designed for studying enzymatic reactions, in which an ultrafiltration membrane is used in conjunction with a flow system. The membrane allows small molecules to flow through the system, but traps the enzyme within the reactor, so there is no need to add enzyme continuously.

3.4 Flow Reactors

The best way to study the phenomena of nonlinear chemical dynamics—oscillations, chaos, waves, and spatial patterns—is to work in an open system. These phenomena can occur in closed systems like batch reactors, but only as transients on the way to the final state of equilibrium. In a closed system, we have to study them on the run, as their properties are changing. For example, to capture periodic oscillation under conditions where the amplitude and period are truly constant, we must have a flow of reactants into the system and a flow of products out of it.

The major experimental breakthrough that made it possible to develop new chemical oscillators in the 1980s was the introduction of the CSTR into nonlinear chemical dynamics. This device, pioneered by chemical engineers, allows us to maintain a system far from equilibrium by pumping fresh reactants into the reactor and reacted materials out in such a way as to maintain constant volume. In a sense, a CSTR is a caricature of a living organism, arguably the most important example of a far-from-equilibrium system that exhibits the phenomena in which we are interested. Living systems maintain oscillations and patterns by taking in high-free-energy reactants (eating) and expelling low-free-energy products (excreting). If either of these flows is cut off for any significant length of time, the system will come to equilibrium (die). If the flows are maintained, the system can continue to oscillate (live) for decades, until disease or time destroys the effectiveness of the complex machinery responsible for the flows.

The characteristics of a CSTR are specified by several parameters. One obvious set consists of the concentrations of the input species that are fed into the reactor. These quantities, usually denoted $[X]_0$, are typically specified as the concentration in the reactor after mixing but before any reaction takes place. That is, the concentration given is not the concentration in the reservoir but that concentration divided by the number of feedstreams, assuming that all streams enter the reactor at the same rate. Another important parameter, whose effect will be discussed at the end of this section and again in Chapter 15, is the rate at which the various input streams mix, which is determined by the stirring rate and the geometry of the reactor.

The final determinant of how a system in a CSTR behaves is the flow rate k_0 or, equivalently, the residence time of the reactor, τ. The average time a molecule spends in the reactor is

$$\langle \tau \rangle = \frac{1}{k_0} = \frac{\text{Volume (mL)}}{\text{Flow rate (mL s}^{-1})} \tag{3.6}$$

The distribution of times is given by an exponential function:

$$f(\tau) = k_0 \, e^{-k_0\tau} d\tau \tag{3.7}$$

which specifies the fraction of molecules that spend time between τ and $\tau + d\tau$ in the reactor (Figure 3.4).

We can test this assertion by checking that the distribution function of eq. (3.7) is normalized:

$$\int_0^\infty k_0 \, e^{-k_0\tau} \, d\tau = 1 \tag{3.8}$$

and that it gives the correct average residence time:

$$\langle \tau \rangle = \int_0^\infty k_0\tau \, e^{-k_0\tau} \, d\tau = \frac{1}{k_0} \tag{3.9}$$

3.4.1 The Pump

A CSTR has four components: a means to pump solutions into the reactor, the reactor itself, a means to remove unreacted and reacted components, and a means to stir the solutions. Two types of pumps are commonly used, although a handful of experiments have been carried out using a gravity feed, in which the input solutions are held in reservoirs located above the reactor and then allowed to flow into the system under the influence of gravity (Rastogi et al., 1984). Such an arrangement has the advantages of being extremely inexpensive and of having no moving parts to break down. The disadvantages are that the flow rate tends to change as the reservoirs drop and that the range of flow rates available is extremely limited.

The *peristaltic pump* consists of a series of rollers around a central axis. Plastic tubes are held against the rollers by clamps so that as the central axis rotates, the tubes are alternately compressed and released. Solutions are thus forced through the tubes in the same way that the contents of the intestine are forced along by peristaltic motion. Figure 3.5 shows a typical pump.

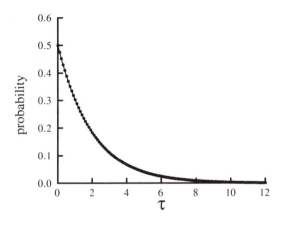

Figure 3.4 The distribution of residence times in a reactor with $\langle \tau \rangle = 2$.

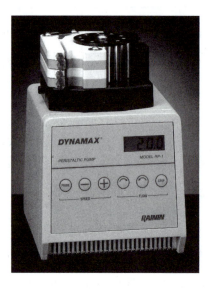

Figure 3.5 A peristaltic pump. (Courtesy of
Rainin Instrument Co., Inc.)

These pumps are relatively inexpensive (around US$2000 for a four-channel device), can pump from several reservoirs at once, and can do so almost indefinitely and over a wide range of flow rates (0–15 mL min^{-1}). The flow rate is a function of the tubing diameter and the rate of rotation of the pump. A typical pump can handle tubes whose diameters range over a factor of 8–10 and has a variable rotation rate that gives another factor of 10–20. The pump must be calibrated; that is, the flow in milliliters per second at each rotation rate must be determined for each tube size. This is easily done by pumping water through the reactor at a fixed rotation rate for a few minutes and collecting the liquid that flows through in a graduated cylinder. If data are collected at about half a dozen rotation rates covering the entire range, interpolating a smooth curve should yield a sufficiently accurate calibration. The tubing degrades with use and must be calibrated and/or replaced regularly. The maximum operating backpressure is 5 atm.

The main disadvantage of peristaltic pumps is that the flow is not constant, but occurs in small pulses. Occasionally, a system that is not truly oscillatory, but is quite sensitive to small perturbations, will exhibit spurious oscillations in response to the periodic bursts of fresh reactants from the pump. These "pump oscillations" can be detected by varying the flow rate since the frequency of oscillations is directly proportional to the rate of rotation of the pump and hence to the flow rate. In simulations of the BZ reaction, Bar-Eli (1985) found that pump pulsations could also affect the period of genuine reaction oscillations. The flow can be smoothed considerably by pumping the fluid first into a vertical tube that has an outflow (Figure 3.6). It is better to have the input tubes below the level of the solution because dripping introduces another periodic mechanical perturbation.

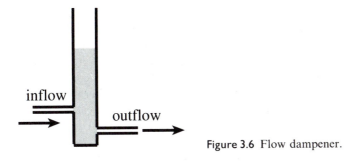

Figure 3.6 Flow dampener.

Syringe pumps have a large syringe whose plunger is driven by a stepper motor. A stepper motor has a shaft that rotates a small fraction of a degree when the motor is pulsed by a fixed voltage. Such a motor can be controlled by a computer that delivers a programmed number of voltage pulses. The main advantage of a syringe pump is that it produces a very smooth flow that can be controlled and does not vary over long times. The flow can also be produced against a significant backpressure. Disadvantages include limited volume (unless a reciprocal pump is used in which one syringe pumps while the other refills) and relatively high cost (several thousand dollars *per channel*). However, if a precise, controllable flow rate is needed, syringe pumps are the pumps of choice. It is nearly impossible to obtain exactly equal flow rates through all the tubes of a multichannel peristaltic pump because of small differences in tube and clamp characteristics. If a precise ratio of feedstreams must be maintained, then syringe pumps are essential.

3.4.2 The Reactor

If the reactions to be studied are relatively slow and insensitive to oxygen (and do not involve volatile components that can be lost), the reactor need be no more complicated than a beaker with a cap to hold the electrodes and flow tubes. The outflow can be accomplished with a glass tube fixed at a set height in the beaker and attached to a water aspirator. The suction removes fluid, maintaining a reasonably constant volume. This arrangement has been used successfully by Rábai et al. (1989a, b) to study pH oscillators. These reactions have very long period oscillations, on the order of hours, so the relatively inefficient mixing, pump perturbations, and small volume variations are not important.

For quantitative dynamics, controlling the volume and achieving efficient mixing are essential. Leaving the solution open to the atmosphere can affect the reaction in two ways: volatile components can leave, and oxygen can enter. Both processes affect the BZ reaction. The rate of bromine loss across the solution–gas interface can have a significant effect on the reaction (Sevcik and Adamčiková, 1988, 1989; Noszticzius et al., 1990; López-Tomás and Sagués, 1991). In batch BZ systems, high stirring rates in batch reactors entrain oxygen and can completely suppress oscillations (Pojman et al., 1994). Therefore, if a beaker is used, a Teflon or polyethylene cap should be machined to fit precisely so that there is no air space between the solution surface and the cap. In this way, we can maintain a fixed volume and prevent liquid–gas exchange.

Precision work is carried out with reactors like that shown in Figure 3.7. These reactors can be machined from Lucite or Plexiglas. The reactors are relatively small (about 30 mL) in order to reduce the quantity of reagents consumed. If spectrophotometric detection is to be used, then it is necessary either to build the reactor from an optically transparent material like quartz, or, more economically, to build in a pair of transparent (and flat) windows for the light path of the spectrophotometer.

One of the hardest aspects of the reactor to quantify and control is the mixing. (We will deal with mixing effects in Chapter 15.) A simple magnetic stirrer at the bottom of the reactor is rarely sufficient. Stirring motors attached to propeller-shaped mixers that rotate at 1000 + rpm are considerably better. Baffles in the reactor are also useful. Whatever configuration is used, keeping the stirring rate constant from run to run is important, since the stirring rate is one of the parameters that governs the behavior of the system.

3.5 Reactors for Chemical Waves

The easiest way to study chemical waves is to pour the reactants into a test tube and watch the patterns develop. Extremely thin tubes (capillaries) can be used to approximate a one-dimensional geometry. This technique is adequate for studying fronts in which a chemical reaction propagates, converting reactants into products, much like a burning cigarette. The velocity can be determined by noting the position of the front as a function of time. Many systems can be studied visually, or a video camera may be used, especially if very narrow tubes are used. Infrared (IR) cameras can provide useful information on temperature distributions (Figure 3.8), but the walls prevent direct measurement of the solution temperature.

If an excitable medium is studied, which we will consider in Chapter 6, patterns can develop in two and three dimensions. Most commonly, a Petri dish is used (see Figure 1.7).

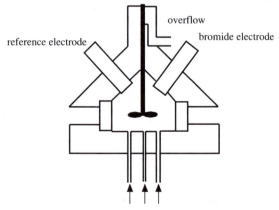

Figure 3.7 Type of reactor used by Györgyi et al. to confirm chaos in the BZ reaction. (Adapted from Györgyi et al., 1992.)

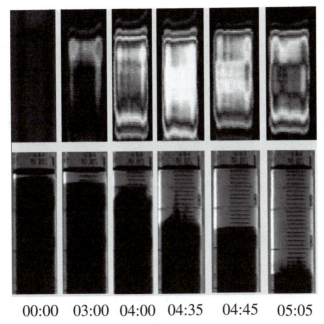

00:00   03:00   04:00 04:35   04:45   05:05

minutes

Figure 3.8 Simultaneous visible and IR images of fronts in
the bromate–sulfite system (Nagy et al., 1995) propagating
in a 100-mL graduated cylinder.

Waves can sometimes be initiated chemically, as with pH fronts in which
hydrogen ion is the autocatalyst; a drop of sulfuric acid is enough to initiate a
front. In the BZ system, waves will spontaneously appear, usually on some local
inhomogeneity like a dust particle or the dish wall. Waves can be initiated electrochemically (Showalter et al., 1979). A hot wire or a drop of acid can also be
used.

Any propagating front or chemical wave generates concentration and temperature gradients that create differences in the density of the solution. In Chapter 9,
we will consider this issue in detail, but the important point is that natural convection can arise from buoyancy. Such convective flows will distort the front and
change the wave velocity. Several methods can be used to eliminate convection.
Extremely narrow capillary tubes will often suppress convection, but it becomes
difficult to observe the front and two-dimensional phenomena cannot be studied.

The most common method is to use a gel. Yamaguchi et al. (1991) compared a
wide variety of gels for studying BZ waves. One of the simplest methods is to add
ultrafine silica gel to the solution. This does not create a true crosslinked gel, but
for wave studies in a Petri dish, especially for demonstrations (see Appendix 1 for
recipes and Appendix 2 for lab experiments), the technique is ideal. For quantitative work, polyacrylamide gels are popular because they can be used to prepare
the spatial analog of a CSTR.

3.5.1 Continuously Fed Unstirred Reactor (CFUR)

Swinney and coworkers developed the continuously fed unstirred reactor (CFUR) in 1987 (Tam et al., 1988a). It consists of a gel that is in diffusive contact with an input from the CSTR, as shown in Figure 3.9. The gel prevents convection, and the CSTR keeps the system away from equilibrium. With this device, transitions between well-defined patterns can be studied. The CFUR does for the study of pattern formation and chemical waves what the CSTR did for oscillating reactions. If the gel is in the shape of a ring (an annular gel reactor), then, with special initiation conditions, "pinwheels" of wavefronts chasing each other can be created (Nos+ticzius et al., 1987). As we will see in Chapter 14, gel reactors proved essential for the discovery of Turing patterns.

3.5.2 Couette Flow Reactor

An effectively one-dimensional system can be created in which the diffusion coefficients of all the species are the same and can be varied from 10^{-2} to 1 $cm^2 s^{-1}$. The reactor consists of two coaxial cylinders with a narrow gap between them filled with the reactive solution (Ouyang et al., 1991, 1992). The inner cylinder is rotated at speeds up to 150 rad s^{-1}. Figure 3.10 shows a schematic of the reactors developed in Texas and Bordeaux. Reagents are introduced and products removed from each end so that there is no net axial flow. Below a critical rotation rate, the flow organizes into pairs of toroidal vortices. At higher rates, turbulent flow occurs inside each vortex cell. Although the transport is convective, over distances greater than the gap between the cylinders, it follows Fick's law for transport between the cells. The effective diffusion coefficient is linearly proportional to the rotation rate and can be made considerably higher than the coefficient for ordinary molecular diffusion.

3.5.3 Immobilized Catalyst on Beads and Membranes

Maselko and Showalter used cation-exchange beads that immobilized the ferroin catalyst for the BZ reaction. They were able to study chemical waves in an inhomogeneous medium with beads of 38–1180 mm (Maselko et al., 1989;

Figure 3.9 Schematic diagram of a CFUR. Polyacrylamide gel is 0.076 cm thick. The glass capillary array, consisting of 10-μm diameter tubes, is 0.1 cm thick and has diameter 2.54 cm. The CSTR has 5.3 cm^3 volume and is stirred at 1900 rpm. (Adapted from Tam et al., 1988a.)

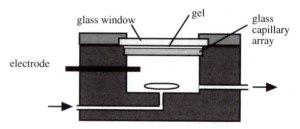

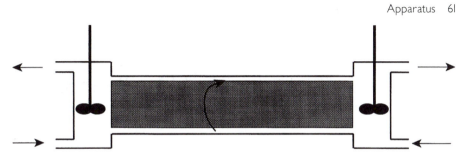

Figure 3.10 Couette flow reactor. The reaction occurs in the narrow region between the inner (shaded) and outer (solid lines) cylinders that connect the two CSTRs.

Maselko and Showalter, 1991). They were even able to study BZ waves on the surface of an individual bead (Maselko and Showalter, 1989)!

Showalter's group developed a new type of reactor (Figure 3.11) in which two CSTRs interacted via a Nafion membrane (a sulfinated Teflon material from Du Pont) (Winston et al., 1991). Such cross-membrane coupling led to unusual and sustained patterns.

3.6 Summary

In this chapter, we have considered the analytical techniques that can be used in monitoring chemical oscillations and waves, from standard spectrophotometry to spatial NMR methods. The most commonly used methods have been potentiometric because of their low cost and ease of use. Of course, to measure something, you need a reaction in a vessel. We evaluated the types of reactors used for homogeneous reactions, of which the CSTR is the most common. Spatial phenomena can be studied in simple test tubes or Petri dishes, but careful work requires open systems using gel or membrane reactors.

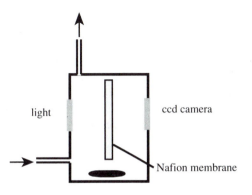

light   ccd camera

Nafion membrane

Figure 3.11 Cross-membrane coupling reactor of the type used by Winston et al. (1991).

4

Synthesis of Chemical Oscillations

There is no doubt that the phenomenon of chemical oscillation—the periodic or nearly periodic temporal variation of concentrations in a reacting system—provided the initial impetus for the development of nonlinear chemical dynamics, and has continued to be the most thoroughly studied of the phenomena that constitute this rich field. In our opening chapter, we alluded to the early skepticism that experimental observations of chemical oscillation engendered. We also noted that the first chemical oscillators were discovered accidentally, by researchers looking for other phenomena. It is relatively easy to understand intuitively why a typical physical oscillator, like a spring, should behave in a periodic fashion. It is considerably more difficult for most of us to see how a chemical reaction might undergo oscillation. As a result, the thought of building a physical oscillator seems far more reasonable than the notion of designing an oscillatory chemical reaction.

In this chapter, we will examine how chemical oscillation can arise, in general, and how it is possible to create chemical reactions that are likely to show oscillatory behavior. In the next chapter, we will discuss how to take a chemical oscillator apart and analyze why it oscillates—the question of mechanism. We also look in detail there at the mechanisms of several oscillating reactions.

4.1 Phase-Plane Analysis

In order to gain some insight into how oscillation might arise in a chemical system, we shall consider a very simple and general model for a reaction involving two concentrations, u and v. Two independent concentration variables is the smallest number that can generate oscillatory behavior in a chemical system. The basic idea, however, is applicable to many-variable systems, because the essential features of the dynamics are often controlled by a small number of variables, and the other variables simply follow the behavior of the key species.

The equations of our model are

$$du/dt = (1/\varepsilon) f(u, v) \tag{4.1a}$$

$$dv/dt = g(u, v) \tag{4.1b}$$

where the functions f and g, which contain all the chemistry, are defined for nonnegative values of u and v in such a way that they never become infinite. Note also, that f is multiplied by $1/\varepsilon$, where ε is a small number. The introduction of a scaling factor, such as ε, is a common technique in analyzing dynamical systems. Here, it is a way of specifying that u changes much more rapidly than v. In chemical terms, u rapidly reaches a steady-state concentration, while v changes much more slowly.

Since there are only two variables, we can picture the evolution of the system in a phase space whose coordinates are u and v and then utilize a technique called *phase-plane analysis*. The idea is to plot out, at least qualitatively, the *vector field*—that is, the direction in phase space that the system is moving for any values of u and v. To do this, we need only calculate, from eqs. (4.1), du/dt and dv/dt and attach to each point (u, v) a vector with components $(du/dt, dv/dt)$. A less laborious and more intuitively revealing approach is to focus on the curves on which the rate of change of each variable is zero, that is, the *null clines*. For example, we refer to the curve on which $f(u, v) = 0$ as the u-null cline or the $f = 0$ null cline. The intersection(s) of the two null clines constitute(s) the steady state(s) of the system because, at any such points, both rates, du/dt and dv/dt, are zero. The time derivative of the variable corresponding to the null cline changes sign as we cross the null cline. Thus, if we know where the null clines are and we know the signs of the derivatives on either side of each null cline, we have both the steady states and a qualitative picture of the flow in phase space. We will be interested in the case of an s-shaped u-null cline [i.e., $f(u, v) = 0$ is something like a cubic equation] and a v-null cline that intersects it once, as shown in Figure 4.1.

We can get some idea of the dynamics of the system just by using our intuition. Because u changes much more rapidly than v, if we are off the $f(u, v) = 0$ curve, the system moves faster horizontally than vertically. Consider two cases. In the first, illustrated in Figure 4.1, the steady state that falls on the left branch (the right branch gives the same picture) is stable, because all trajectories in its neighborhood point in toward the intersection of the two null clines. This state has an interesting and important property, particularly if it lies near the bottom of the s-curve. This property, which we alluded to briefly in Chapter 2, is called *excitability*. If an excitable system is perturbed a small amount, it quickly returns to the steady state, but if it is given a large enough perturbation, it undergoes a large amplitude excursion before returning. Let us see how that works.

Figure 4.2a shows a small perturbation to the steady state. The value of u is increased to point A. The system jumps back to the u-null cline at D because the $1/\varepsilon$ term in eq. (4.1a) makes u change much more rapidly than v. Now, the system

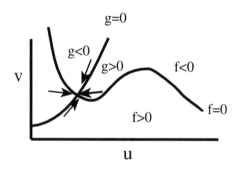

Figure 4.1 The null clines for the two-variable model of eqs. (4.1). Arrows indicate the direction of the flow in phase space. The steady state occurring at the intersection of $g = 0$ and $f = 0$ is stable.

moves along the u-null cline from D back toward the original steady state S as v slowly decreases. Thus, S is stable. In Figure 4.2b, a larger perturbation is applied to the steady-state concentration of u, so that the initial concentrations after the perturbation now place point A in the positive region for $f(u, v)$. Because f is positive, the perturbation continues to grow, pushing the system toward the right-hand branch of the $f = 0$ null cline. The concentration of v changes slowly compared with that of u. When the system reaches the $f = 0$ null cline at B, it begins to move up the null cline as v increases ($g(u, v) > 0$), until it comes to the maximum at C, where it jumps rapidly to the left-hand branch of the u-null cline. The system then slides along this null cline back to the steady state. Figure 4.3 shows the temporal evolution of u during these sequences.

Excitability is an important property in many biological systems, such as neurons, and it is also closely connected with traveling waves in chemical systems. Excitability has two important characteristics: a *threshold of excitation* and a *refractory period*. The threshold is the minimum perturbation that will cause

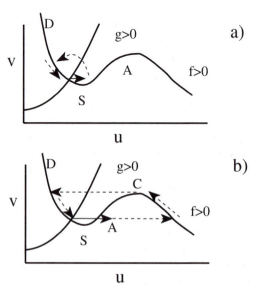

Figure 4.2 Phase-plane analysis of perturbations to the steady state in the excitable two-variable model of eqs. (4.1). Solid arrows indicate perturbation in the steady-state concentration of u; dashed arrows indicate the path of the system following perturbation. (a) Small perturbation; (b) large perturbation.

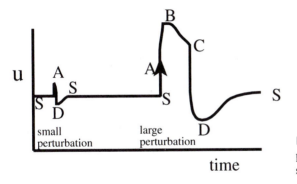

Figure 4.3 Temporal evolution of perturbations to an excitable system.

the system to jump to the far right branch of the null cline.[1] The refractory period is the minimum time that the system takes to return to the steady state and be ready for another stimulation.

If the intersection of the null clines lies on the middle branch, then the trajectories point away from the steady state; it is unstable. In this case, illustrated in Figure 4.4, we get oscillations, called *relaxation oscillations*. Since the dynamics are constructed so that u changes very rapidly until the system reaches the $f = 0$ null cline, we shall begin our analysis on that null cline, at point A in Figure 4.4a. The system will move slowly along the null cline as v decreases from A to B. Upon reaching B, u will rapidly increase as the system jumps to the other branch of the $f = 0$ null cline (C). This branch is in the positive region of $g(u, v)$, so the system will move along the u-null cline toward D as v slowly increases. Upon reaching D, the system again "falls off" this branch of the u-null cline and makes a rapid transition to E. It then proceeds back toward A, and the cycle repeats. We have periodic temporal oscillation, as illustrated in Figure 4.4b. The behavior of the system, after an initial transient period during which the system approaches the u-null cline, will be the same no matter what initial conditions we choose.

The above idealized analysis can be applied to an actual experimental system, the Briggs–Rauscher reaction, as shown in Figure 4.5. Here, the iodide concentration corresponds to the variable u and the iodine concentration to the variable v. We have labeled the points A through E in Figure 4.5 to be in rough correspondence with the labeled points in Figure 4.4. The points in Figure 4.5a were taken at fixed time intervals; hence, when the points are close together in Figure 4.5a, the concentrations are changing slowly, but in Figure 4.5b, periods of slow change are marked by well-separated points.

4.2 The Cross-Shaped Diagram Model

We have just seen that oscillation can arise in a two-variable system in which one variable changes significantly faster than the other and has relatively complex

---

[1]Note that a perturbation (a decrease) that brings the value of v below the minimum in the u-null cline curve also causes the system to undergo a large excursion before returning to the steady state.

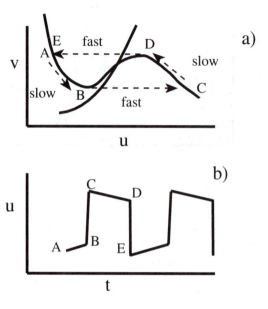

Figure 4.4 (a) Phase-plane representation of relaxation oscillations. (b) Temporal evolution of u corresponding to phase-space behavior in part a.

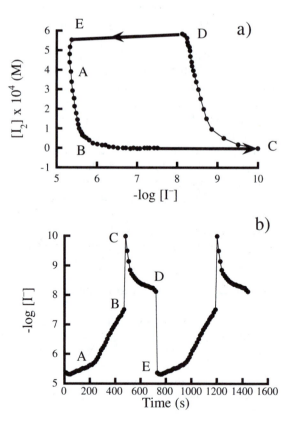

Figure 4.5 (a) Phase portrait of an oscillation in the Briggs–Rauscher reaction. The time interval between points is 12 s. (b) Temporal evolution of u ($[I^-]$) corresponding to phase-space behavior in part a. (Adapted from De Kepper and Epstein, 1982.)

(e.g., cubic) kinetics. This is not, of course, the only way in which chemical oscillation can come about, but it turns out to occur more frequently than one might think.

So, one way to design a chemical oscillator might be to look for systems that obey rate equations like eqs. (4.1). Let us ignore, for the moment, the question of where one might hope to find real systems that have the right kinetics and look first at a beautifully simple model that captures the essence of eqs. (4.1) and points us toward a way of designing real chemical oscillators. The model, proposed by Boissonade and De Kepper (1980), consists of two rate equations for the species x and y:[2]

$$dx/dt = -(x^3 - \mu x + \lambda) - ky \qquad (4.2a)$$

$$dy/dt = (x - y)/\tau \qquad (4.2b)$$

Clearly, eq. (4.2a) gives a cubic for the x-null cline, and eq. (4.2b) yields a straight line for the y-null cline. Chemically speaking, x is produced autocatalytically (the μx term) and, in the absence of the feedback provided by the ky term, it can have one unstable and two stable steady states for appropriate values of the parameters μ and λ. Thus, the x variable describes a bistable system. As we shall see, the feedback term ky has the effect of moving the system along the hysteresis loop generated by the x dynamics. Recall that for oscillations to occur, eqs. (4.1) require, via the scaling factor ε, that x change much more rapidly than y. In eqs. (4.2), this separation of time scales is accomplished by choosing a sufficiently large value for τ, which is the relaxation time for y to approach x, that is, for the system to reach the y-null cline. It turns out that "sufficiently large" means that

$$\tau > 1/\mu \qquad (4.3)$$

which makes sense, since $1/\mu$ is the characteristic time for the autocatalytic production of x, which is the key process in determining the dynamics of the x subsystem.

A linear stability analysis of eqs. (4.2) requires some algebraic dexterity, but it pays handsome rewards in terms of both results on this system and tricks that are generally useful, so let us try it. Finding the steady states is easy. Setting eq. (4.2b) to zero immediately tells us that $x = y$. Equation (4.2a) then becomes a cubic equation in x:

$$q(x) = x^3 - (\mu - k)x + \lambda = 0 \qquad (4.4)$$

The steady states are the solutions of eq. (4.4) with $y = x$. Since a cubic equation can have either one or three real solutions, we first need to settle how many solutions there are. In Figure 4.6, we plot eq. (4.4) for fixed μ and k at several values of λ. Note that increasing λ simply shifts the curve upwards.

We see that there are three roots if, and only if, $\lambda_\alpha < \lambda < \lambda_\beta$. The trick is to calculate λ_α and λ_β given μ and k. To do this, we observe that λ_α and λ_β are the

---

[2]Note that in this abstract model, the "concentrations" x and y can be negative. It is possible, by some clever trickery (Samardzija et al., 1989), to transform any polynomial set of rate equations, like eqs. (4.2), to a set with variables that remain nonnegative, but we need not bother with such gyrations for our present purpose.

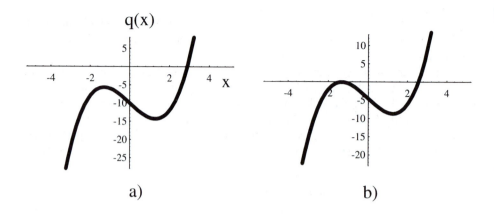

a)

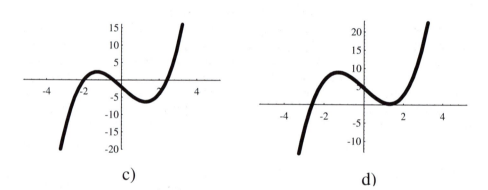

c)

d)

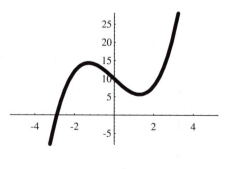

e)

Figure 4.6 Plots of $q(x)$ in eq. (4.4) for several values of λ: (a) $\lambda < \lambda_\alpha$, (b) $\lambda = \lambda_\alpha$, (c) $\lambda_\alpha < \lambda < \lambda_\beta$, (d) $\lambda = \lambda_\beta$, (e) $\lambda > \lambda_\beta$.

special values of λ at which the cubic equation is tangent to the x axis, so that eq. (4.4) has exactly two roots, or, more precisely, it has one double root and one single root. If we call these roots a and b, respectively, then, at these particular values of λ, eq. (4.4) can be written as

$$(x - a)^2(x - b) = 0 \tag{4.5}$$

The trick now is to expand out eq. (4.5) and equate the coefficients of like powers of x in eqs. (4.4) and (4.5). Equation (4.5) becomes

$$x^3 - (2a + b)x^2 + (2ab + a^2)x - a^2b = 0 \tag{4.6}$$

Equating like powers gives

$$x^3: \quad 1 = 1 \tag{4.7a}$$
$$x^2: \quad 0 = -(2a + b) \tag{4.7b}$$
$$x: \quad k - \mu = 2ab + a^2 \tag{4.7c}$$
$$1: \quad \lambda = a^2b \tag{4.7d}$$

The last three of eqs. (4.7) constitute three conditions on the two unknowns, a and b. We can solve any two of them for a and b in terms of k, λ, and μ, but in order for all three equations to be consistent, there must be an additional relationship among the parameters. It is this relationship that gives us our desired expression for λ_α and λ_β. Equation (4.7b) tells us that $b = -2a$, which we substitute into eq. (4.7c) to obtain

$$k - \mu = -3a^2 \tag{4.8a}$$
$$a = \pm[\mu - k)/3]^{1/2} \tag{4.8b}$$

We first observe that having three solutions means that k must be less than μ, since otherwise eq. (4.8a) could not be satisfied. Substituting our results into eq. (4.7d), we find that

$$\lambda = -a^2b = 2a^3 = \pm 2[\mu - k)/3]^{3/2} \tag{4.9}$$

The values of λ_α and λ_β are given by the negative and positive solutions, respectively, in eq. (4.9).

We now know how many steady states there will be. What about their stability? We have to calculate the Jacobian matrix as we learned to do in Chapter 2. Since we are already using λ as a parameter in our equations, we will use ω for the eigenvalues of the Jacobian. The Jacobian matrix for eqs. (4.2), where x refers to the steady-state value whose stability we are interested in, is

$$\mathbf{J} = \begin{bmatrix} -3x^2 + \mu & k \\ 1/\tau & -1/\tau \end{bmatrix} \tag{4.10}$$

The equation for the eigenvalues, $\det(\mathbf{J} - \omega\mathbf{I}) = 0$, is

$$\omega^2 + (3x^2 - \mu + 1/\tau)\omega + (1/\tau)(3x^2 - \mu + k) = 0 \tag{4.11}$$

For the steady state to be stable, both roots of the quadratic in eq. (4.11) must have negative real parts. This can occur only if the coefficients of both the ω term and the constant term are positive, that is, if

$$3x^2 - \mu + 1/\tau > 0 \tag{4.12}$$

and

$$3x^2 - \mu + k > 0 \tag{4.13}$$

Looking back at eq. (4.4) and Figure 4.6, we observe that (1) the left-hand side of eq. (4.13) is just $dq(x)/dx$, and (2) the slope of $q(x)$ is always positive at its first and third (if there is a third) crossings of the x axis. Therefore, eq. (4.13) always holds for the largest and smallest roots when there are three roots of eq. (4.4) and for the single root when only one exists. Equation (4.13) is always violated for the middle root when eq. (4.4) yields three steady states. The middle state is thus always unstable. We are interested in finding conditions under which *all* steady states are unstable, because then, according to the Poincaré–Bendixson theorem, the system must oscillate (or diverge). Thus, we need to know under what conditions eq. (4.12) can be violated.

Let us look for a value of λ that, for fixed μ and k, makes eq. (4.12) an equality. If such a λ value exists, then eq. (4.12) should be violated either just above or just below it. Of course, we will have to figure out which way the inequality goes on each side. We solve eq. (4.12) for x:

$$x = \pm[\mu - 1/\tau)/3]^{1/2} \tag{4.14}$$

We then solve eq. (4.4) for λ:

$$\lambda = (\mu - k)x - x^3 \tag{4.15}$$

and substitute eq. (4.14) into eq. (4.15) to obtain

$$\begin{aligned} \lambda &= \pm\{(\mu - k)[(\mu - 1/\tau)/3]^{1/2} - [(\mu - 1/\tau)/3]^{3/2}\} \\ &= \pm[1/3^{3/2}(\mu - 1/\tau)^{1/2}(3k - 2\mu - 1/\tau)] \end{aligned} \tag{4.16}$$

Let us label the negative and positive values of λ in eq. (4.16) as λ_γ and λ_δ, respectively. At these values, eq. (4.12) becomes an equality. How do we decide which way the inequality goes as we move away from these special parameters? We go back to Figure 4.6, which shows how the curve $q(x)$ shifts as we change the value of λ. The case we are most interested in occurs when there is only a single root, that is, Figures 4.6a and 4.6e. We see that in these regions, making λ very negative makes the single steady state x very positive, while making λ very positive makes x very negative; that is, large $|\lambda|$ means large $|x|$ when there is just one steady state. Large $|x|$ means large x^2, which means that eq. (4.12) will be satisfied. Thus, we conclude that eq. (4.12) will be violated and we will have a single unstable steady state, which implies that the system must oscillate when $\lambda < \lambda_\alpha$ or $\lambda > \lambda_\beta$ and $\lambda_\gamma < \lambda < \lambda_\delta$.

In Figure 4.7 we summarize our results by plotting the critical values of λ, which we have just calculated on a k–λ phase diagram at fixed μ and τ. Some additional results obtained by Boissonade and De Kepper for the case when three

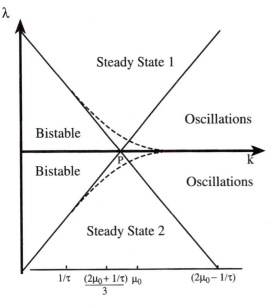

Figure 4.7 The cross-shaped phase diagram for $\tau \gg 1/\mu$. Shown here are the four major regions and the critical values of λ (solid lines) for the system of eqs. (4.2). Dashed curves are obtained from more detailed nonlinear stability analysis. (Adapted from Boissonade and De Kepper, 1980.)

steady states exist are included. Although more details can be obtained from further analysis of the system of eqs. (4.2), its dynamical behavior is basically described in terms of the four regions shown in the figure: two in which a single stable steady state exists; one in which there is bistability between the two stable steady states; and one in which neither state is stable, so that the system oscillates. The four regions come together at the cross point P, where $k = (2\mu + 1/\tau)3$ and $\lambda = 0$. Because of its shape and its generic occurrence in a wide variety of systems characterized by dynamics similar to the system in eqs. (4.1), a plot like the one in Figure 4.7 is called a *cross-shaped phase diagram*. It plays a major role in theories of nonlinear chemical dynamics and was instrumental in the development of methods to design chemical oscillators.

4.3 A Design Algorithm

Real systems are not quite as simple as the one that we have just finished analyzing, but fortunately they do not have to be. What is crucial is that the essential topological character of the cross-shaped phase diagram, the meeting of the four regions at a single point, is found in a wide variety of systems, even if the actual diagram is rotated or distorted in other ways from the simple picture in Figure 4.7. If we work from this phase diagram and the other insights that we have gained from the analysis we have carried out, we can develop a systematic approach to designing new chemical oscillators, *even though we may have no clue what the actual rate equations of the system look like.*

To initiate our phenomenological approach to oscillator construction, we want to start with a bistable system. We can think of this step as providing either one piece of the cross-shaped phase diagram or as constituting one of our underlying rate equations, like eq. (4.1a) or eq. (4.1b). As we have already pointed out, bistability can occur only in an open system, so the first thing we need to do is to work in such conditions—for example, by using a flow reactor of the sort described in Chapter 3. As luck would have it, there are lots of reactions that show bistability when run in a flow reactor; most of them are autocatalytic (Epstein et al., 1981).

We can think of the concentration associated with the bistable behavior as the x variable of a system like eqs. (4.2). Now we need to introduce a feedback species in order to make the system oscillate. Suppose that we can find a substance that consumes x at a relatively slow rate. The requirement that the feedback reaction be slow is equivalent to the specification we made about time scales in our models, that is, ε must be much less than 1 or μ must be much greater than $1/\tau$. But what do we mean by slow in the present case, where we do not have rate equations, only experiments? The relevant times are the times required for the x system to relax back to the bistable steady states when it is perturbed away from them and the time required for the feedback reaction to change x substantially. Both of these quantities are experimentally measurable, and we want the time of the feedback reaction to be significantly longer than either of the relaxation times for the unperturbed, feedback-free system. We are not implying, by the way, that the feedback species—that is, the substance that consumes x—is the same as v in eq. (4.1b) or y in eq. (4.2b). In fact, the amount of this species that is fed into the reactor is more closely analogous to the parameter k, the strength of the feedback, in eq. (4.2a). The nice thing about this approach is that we never need to know what y is; we can work completely from our experimental observations.

Let us assume that we have managed to find a bistable system, as confirmed by our experimental observation of a hysteresis loop like the one shown in Figure 2.6, and that we have picked a candidate, let us call it z, for our feedback species. How do we know if this choice is likely to lead to oscillations? We look for two things to happen. First, adding a small amount of z should cause the range of bistability to narrow. Experimentally, we add a flow of z to our reactor input stream and measure the hysteresis loop as a function of some conveniently tunable parameter, often the flow rate. This parameter plays a role analogous to that of λ in eq. (4.2a). Second, although the *range* of the parameter over which bistability occurs should become smaller when the feedback species is introduced, the *character* of the states (i.e., the concentrations of major species) should change little, if at all. In particular, the states should retain their individual identities and should not begin to approach one another in character, as this would indicate that the feedback reaction was not sufficiently slow compared with the reactions that establish the fundamental bistability.[3]

Another way of seeing this important point is shown in Figure 4.8, where we depict (a) the underlying hysteresis loop associated with the species x, (b) the effect of adding the feedback species z, (c) the oscillatory behavior for a particular

---

[3]If the feedback is too rapid, it will cause the two bistable states to merge into a single, stable steady state, and no oscillation will occur.

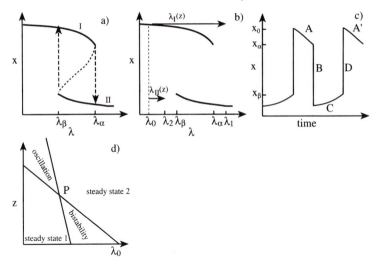

Figure 4.8 Systematic design of a chemical oscillator. (a) The fundamental bistable system, with the steady-state concentration of x shown as a function of the parameter λ. Steady states SSI and SSII are two distinct, stable steady states. Dashed line shows third, unstable steady state. (b) The system in part a perturbed by a feedback species z. The actual value of λ is λ_0. The arrows indicate effective increase in λ caused by the perturbation. (c) Time course followed by x corresponding to the values of λ_0 and z illustrated in part b. (d) Phase diagram obtained when experiments like that shown in part b are performed at different levels of z. Panel (a) corresponds to $z = 0$.

value λ_0, and (d) the resulting cross-shaped phase diagram. Two points deserve special mention. First, the orientation of the diagram in Figure 4.8d is something more than 90° rotated from the diagram shown in Figure 4.7. Second, the effect of the feedback species, illustrated by the arrows in Figure 4.8b, must be considerably greater in one steady state than in the other. Is this latter condition plausible? Yes, because, as we see in the hysteresis loop in Figure 4.9 or the potential traces in Figure 2.12, pairs of bistable states can have concentrations that differ by as much as a factor of 100,000! It is therefore quite likely that adding the same amount of z to the system in two different steady states will yield very different magnitudes of change in the concentration of x or, as we have shown in Figure 4.8b, in the effective value of the parameter λ. If we think of λ as inversely related to the input concentration of x (or of some species that generates x), what we are saying in Figure 4.8b is that adding a certain amount of z when the actual value of λ is λ_0 decreases x in steady state SSI by as much as if we had increased λ to λ_1, while adding the same amount of z when the system is in steady state SSII decreases x by a much smaller amount, equivalent to raising the input control parameter to λ_2.

What we hope will happen is shown in Figure 4.8c and d. As we add more z, the range of bistability narrows as shown in Figure 4.8d, with SSI losing stability more

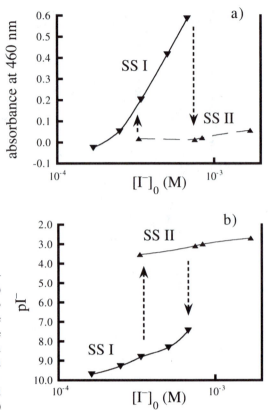

Figure 4.9 Bistability in the chlorite–iodide reaction as measured by (a) 460-nm absorbance (proportional to $[I_2]$) and (b) an iodide-selective electrode. Dashed arrows indicate spontaneous transitions between states. Reciprocal residence time in CSTR $= 5.4 \times 10^{-3}$ s^{-1}, $[ClO_2^-]_0 = 2.4 \times 10^{-4}$ M, pH $= 3.35$. (Adapted from Citri and Epstein, 1987.)

rapidly with increasing z than SSII because the feedback has a larger influence (longer arrow in Figure 4.8b) on the former state. Eventually, at the cross point P, the bistability range shrinks to zero, and if we increase z still further, the system begins to oscillate. Our drawing suggests that the range of oscillation continues to grow as z increases, but in real systems additional effects cause the region to close, sometimes only a short distance (in terms of z) beyond the cross point.

In Figure 4.8c, we show how the system is expected to behave when z and λ lie in the oscillatory region. In Figure 4.8b, λ_0 is the actual value of λ, and the system begins in SSI with $x = x_0$. Because of the added z, the system begins to "think" that the actual value of λ is λ_1. The level of x begins to drop, but because relaxation to the steady states is much faster than the feedback reaction, x decreases relatively slowly along the branch of steady states SSI. This slow decrease constitutes segment A of Figure 4.8c. The drop in x continues until x reaches x_α. The concentration of x still needs to decrease because the system has not yet reached λ_1, the value of λ appropriate for SSII to the amount of added z. When x falls below x_α, the system undergoes a rapid transition to SSI (segment B of Figure 4.8c), the only state stable when the effective value of λ is below λ_α. When the system finds itself in SSII, it now "wants" to reach a level of x corresponding to λ_2, and x therefore begins to rise slowly along the SSII curve of steady states (segment C). Again, however, the system reaches the end of the branch, this time at λ_β, before achieving its objective, and it is forced to jump again, now back up to SSI (segment D). The story now repeats itself, since x will decrease along segment A' until the system again reaches the next jump at x_α. Thus, we have periodic relaxation oscillations in which the feedback-perturbed system repeatedly traces out the hysteresis loop corresponding to the fundamental bistable system.

We have emphasized that the approach described above, although it was constructed using models and theoretical considerations, is totally based on experimental observation. No knowledge of the underlying kinetics is required, though obviously the more one knows, the more likely one is to succeed. We can summarize our oscillator construction algorithm as follows:

1. Choose a reaction that produces a species x autocatalytically.
2. Run the reaction in a flow reactor and find conditions under which the system is bistable as a function of some parameter λ, such as the flow rate.
3. Add to the flow a species z that perturbs the system on a time scale that is slow with respect to the relaxation of the bistable system to its steady states and by different amounts in the two bistable states.
4. Increase the amount of z added and see if the range of bistability narrows and the steady states maintain their character.
5. If step 4 succeeds, continue increasing z until the bistability vanishes and oscillation appears. If it fails, go back to step 3 and try a different feedback species.

In the next section, we look at the first actual application of this procedure.

4.4 A Systematically Designed Chemical Oscillator

The design algorithm outlined above was first applied to a real chemical system in 1980. De Kepper et al. (1981b) chose two autocatalytic reactions, the chlorite–iodide and arsenite–iodate reactions, to work with. Chlorite reacts with iodide in two stages. The first, and dynamically more important, step produces iodine autocatalytically (Kern and Kim, 1965) according to

$$ClO_2^- + 4I^- + 4H^+ = Cl^- + 2I_2 + 2H_2O \qquad (4.17)$$

Any remaining chlorite reacts with the iodine generated in reaction (4.17) to yield iodate:

$$5ClO_2^- + 2I_2 + 2H_2O = 5Cl^- + 4IO_3^- + 4H^+ \qquad (4.18)$$

Iodate reacts with arsenite in a classic autocatalytic reaction known as a Landolt-type reaction (Eggert and Scharnow, 1921). The rate-determining process is the Dushman reaction (Dushman, 1904):

$$IO_3^- + 5I^- + 6H^+ = 3I_2 + 3H_2O \qquad (4.19)$$

which is followed by the Roebuck reaction (Roebuck, 1902):

$$I_2 + H_3AsO_3 + H_2O = 2I^- + H_3AsO_4 + 2H^+ \qquad (4.20)$$

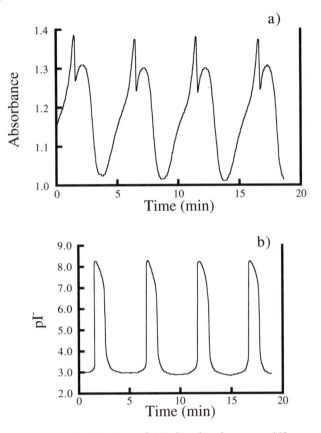

Figure 4.10 Oscillations of (a) the absorbance at 460 nm (proportional to [I_2]) and (b) the iodide concentration in the chlorite–iodate–arsenite system in a CSTR. [KIO_3]$_0$ = 24 × 10^{-3} M, [As_2O_3]$_0$ = 2 × 10^{-3} M, [$NaClO_2$]$_0$ = 2 × 10^{-3} M, [Na_2SO_4]$_0$ = 0.1 M, [H_2SO_4]$_0$ = 0.01 M, residence time = 400 s. (Adapted from Citri and Epstein, 1987.)

If we add eq. (4.19) and three times eq. (4.20) to eliminate the intermediate I_2, we obtain the following net reaction, which, because eq. (4.19) is rate-determining, is autocatalytic in the product iodide:

$$IO_3^- + 3H_3AsO_3 = I^- + 3H_3AsO_4 \tag{4.21}$$

Thus, we have two autocatalytic reactions that have the species I^- and I_2 in common. In late 1980, Patrick De Kepper, who had developed the cross-shaped phase diagram model while working with Jacques Boissonade at the Paul Pascal Research Center in Bordeaux, arrived at Brandeis University to join forces with Irving Epstein and Kenneth Kustin, who had independently come to the conclusion that chemical oscillators could be built from autocatalytic reactions and had targeted the chlorite–iodide and arsenite–iodate systems as promising candidates. The collaboration quickly bore fruit.

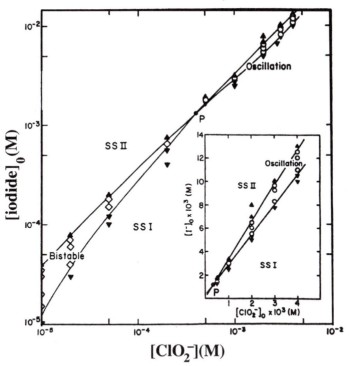

Figure 4.11 Phase diagram of the chlorite–iodide reaction in a CSTR. Inset shows the same data on a linear scale, focusing on the oscillatory region. P indicates the cross point of this cross-shaped phase diagram. pH = 2.04, k_0 = 1.1 $\times$ 10^{-3} s^{-1}. (Adapted from De Kepper et al., 1990.)

The team first established that the arsenite–iodate reaction showed bistability in a CSTR (De Kepper et al., 1981a). They then introduced chlorite as a feedback species and obtained oscillations as shown in Figure 4.10. Note, particularly in the trace of the iodide-selective electrode, the alternation between periods of slow concentration change and rapid jumps between pseudo-steady-state levels, similar to the behavior shown schematically in Figure 4.8c.

The first systematic design of a chemical oscillator had been achieved! There remained some ambiguity, however. Since two autocatalytic reactions had been employed, it was not immediately clear which constituted the fundamental autocatalytic reaction and which provided the feedback in the model scheme. Historically, the arsenite–iodate system had been chosen for the former role, since its bistable behavior had been established first. More careful investigation revealed that, in fact, it was the chlorite–iodide reaction that provides the essential dynamical features of this system. The evidence comes in two forms. First, the chlorite–iodide reaction is also bistable in a CSTR (Dateo et al., 1982) and the relaxation to its steady states is more rapid than the relaxation behavior of the arsenite–iodate system. According to our theory, the slower reaction constitutes the feedback to the underlying, faster, fundamental subsystem.

Even more convincing, and of greater ultimate significance, is the fact that the chlorite–iodide reaction, as shown in Figure 4.11, turns out to oscillate *all by itself* in a flow reactor, though in a much narrower range of conditions than the full chlorite–iodate–arsenite system (Dateo et al., 1982). Thus, in some way, the chlorite–iodide reaction generates sufficient feedback internally to allow it to oscillate. The addition of other species like arsenite merely enhances the feedback so as to make the system oscillate more easily. In fact, a wide range of species can be appended to the chlorite–iodide (Orbán et al., 1982a) or chlorite–iodate (Orbán et al., 1981) backbones to produce a family of nearly two dozen chlorite-based oscillators (Epstein and Kustin, 1984). In the next section, we discuss how this notion of families of chemical oscillators has helped to guide the design of new systems and the mechanistic analysis of known oscillating reactions.

4.5 Families of Oscillators—Chemical Taxonomy

Before the systematic design procedure described in section 4.3 became available, the easiest way (in fact just about the only way) to look for a new chemical oscillator was to start with one of the few known, accidentally discovered, oscillating reactions and to make some small change in one of the reactants. If you were lucky, replacing the ceric–cerous catalyst couple in the BZ reaction with ferroin–ferriin, which has a very similar redox potential, or substituting the chemically similar methylmalonic acid for malonic acid, would give you a reaction that still oscillated and maybe even had some new and interesting behavior. If you were still more daring, you might figure out how to replace both the metal catalyst and the organic acid with a single organic species that played both roles (Orbán et al., 1978) or you might succeed in mixing some of the ingredients from the BZ reaction with the reactants of the Bray–Liebhafsky system to obtain a more colorful and more robust oscillator (Briggs and Rauscher, 1973).

4.5.1 A Family of Chlorite-Based Oscillators

The advent of an algorithm for building new chemical oscillators did not eliminate the approach of starting from an existing reaction and varying one or more of its reactants. In fact, the use of the CSTR made it more likely that such an approach would succeed. The family of chlorite-based oscillators mentioned in the previous section, and summarized in Table 4.1, is just one example of a group of oscillating reactions that have key reactants in common. Just before the first chlorite-based oscillators were discovered, Noyes (1980) attempted to provide a general mechanistic framework for the family of bromate-based oscillators. The notion of a family of chemical oscillators is a useful one, and here we explore one way to classify oscillating reactions based on chemical and mechanistic relationships. Ross and coworkers (Eiswirth et al., 1991) have proposed an alternative categor-

Table 4.1 Chlorite-Based Chemical Oscillators in a CSTR

Oxidant	Reductant	Special Features	Reference
	I^-	Bistablity between stationary and oscillatory states	Dateo et al., 1982
IO_3^-	I^-		Orbán et al., 1982a
	I^-, malonic acid	Batch oscillation, spatial wave patterns	De Kepper et al., 1982
$Cr_2O_7^{2-}$	I^-		Orbán et al., 1982a
MnO_4^-	I^-		Orbán et al., 1982a
BrO_3^-	I^-	Bistability between stationary and oscillatory states	Orbán et al., 1982a
IO_3^-	H_3AsO_3	First chlorite oscillator discovered	De Kepper et al., 1981b
IO_3^-	$Fe(CN)_6^{4-}$		Orbán et al., 1981
IO_3^-	SO_3^{2-}		Orbán et al., 1981
IO_3^-	$S_2O_3^{2-}$	Batch oscillation	De Kepper et al., 1982
IO_3^-	$CH_2O \cdot HSO_2$		Orbán et al., 1981
IO_3^-	Ascorbic acid		Orbán et al., 1981
IO_3^-	$I^- +$ malonic acid	Batch oscillation	De Kepper et al., 1982
IO_3^-	$I^- + H_3AsO_3$	Tristability	Orbán et al., 1982a
I_2	$Fe(CN)_6^{4-}$		Orbán et al., 1982a
I_2	SO_3^{2-}		Orbán et al., 1982a
I_2	$S_2O_3^{2-}$		Orbán et al., 1982a
BrO_3^-	Br^-		Orbán and Epstein 1983
BrO_3^-	I^-	Birhythmicity, compound oscillation	Alamgir and Epstein, 1983
BrO_3^-	SO_3^{2-}		Orbán and Epstein, 1983
BrO_3^-	$Fe(CN)_6^{4-}$		Orbán and Epstein, 1983
BrO_3^-	AsO_3^{3-}		Orbán and Epstein, 1983
BrO_3^-	Sn^{2+}		Orbán and Epstein, 1983
I_2	$I^- + S_2O_3^{2-}$	Chaos, tristability	Maselko and Epstein, 1984
	Thiourea	Chaos, spatial patterns	Alamgir and Epstein, 1985a
	SCN^-		Alamgir and Epstein, 1985b
	Br^-		Alamgir and Epstein, 1985b
	S^{2-}	Cu(II)-catalyzed	Orbán, 1990

ization of chemical oscillators based on dynamical considerations, which nicely complements the approach outlined here.

The chlorite family contains several subfamilies, most notably the iodate, iodide, and bromate branches. Several oscillators—for example, the chlorite–bromate–iodide reaction—constitute links between subfamilies, with a foot in each camp.

4.5.2 The Notion of a Minimal Oscillator

The chlorite family of oscillators, with its more than two dozen brothers, sisters, and cousins, is a remarkably fecund one, being the largest known to date. One can see quite easily that some groups of reactions are more closely related than others. For example, the chlorite–iodide–oxidant and the chlorite–bromate–reductant systems appear to constitute well-defined branches of the larger family. Is it possible to identify more precise relationships among the family members, and to find, for example, a patriarch or matriarch whose chemistry gives birth, in some sense, to that of other oscillators in the family? This question leads to the concept of a *minimal oscillator* in a family or subfamily of oscillating reactions (Orbán et al., 1982b). The idea is that in at least some, and possibly every, family or sub-family, there exists a set of components that (1) are sufficient to give rise to oscillations when introduced into a flow system without additional reactants; and (2) appear, either as reactants or as intermediates, in every member of the family.

The chlorite–iodide reaction is clearly the minimal oscillator of the subfamily of chlorite–iodide–oxidant oscillating reactions, and, since iodate and iodine appear to be produced in all of them, it can be considered to be the progenitor of a still larger family that includes the chlorite–iodate–reductant and chlorite–iodine–reductant groups listed in Table 4.1. Clearly, the chlorite–iodide reaction cannot be minimal for the entire family of chlorite-based oscillators since many of them contain no iodine species. It is not at all obvious that a minimal oscillator for the entire chlorite family exists.

A minimal oscillator typically, but not invariably, has a narrower range of oscillation than its progeny. Note the relatively narrow region of oscillation of the chlorite–iodide oscillator in Figure 4.11. Figure 4.12 shows a phase diagram for the first minimal oscillator to be discovered—the reaction of bromate, bromide, and a catalyst in a CSTR—the minimal bromate oscillator (Orbán et al. 1982b). This system, which underlies the BZ, as well as all other bromate-driven oscillators, has such a small range of parameters over which it oscillates that the aging of the tubes in the peristaltic pump, over several days of experiments, can alter the flow rate enough to make oscillations disappear. One must then use the cross-shaped phase diagram algorithm to relocate the tiny oscillatory region in the phase space!

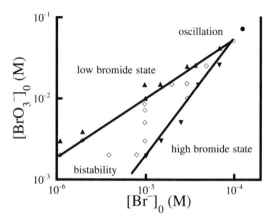

Figure 4.12 Phase diagram of the minimal bromate oscillator with flow rate $k_0 = 0.0128$ s^{-1}. $[Mn^{2+}]_0 = 1.02 \times 10^{-4}$ M, $[H_2SO_4] = 1.5$ M, $T = 25\,^\circ$C. (Adapted from Orbán et al., 1982b.)

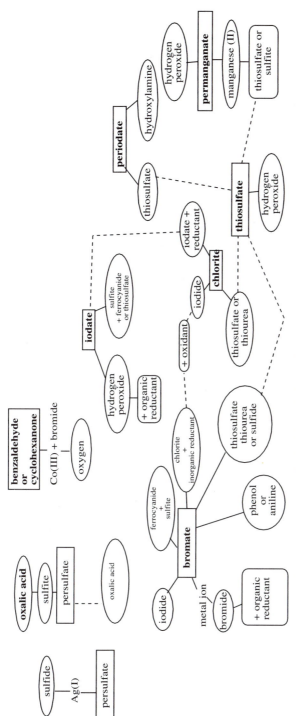

Figure 4.13 A taxonomy of chemical oscillators. Dashed lines indicate family relationships, that is, oscillating reactions that fall into both families.

In addition to the chlorite–iodide and minimal bromate systems, one other minimal oscillator has been identified and characterized. The reaction of manganous ion with permanganate in the presence of a stabilizing species such as phosphate (Orbán and Epstein, 1989b) gives rise to a sizable family of permanganate-based oscillators (Orbán and Epstein, 1990; Doona et al., 1991; Melichercik et al., 1992), as well as at least one system involving the oxidation of Mn(II) to generate permanganate (Orbán et al., 1991).

4.5.3 Oscillator Taxonomy

The number of chemical oscillators known is growing rapidly, both through the expansion of existing families of oscillators and by the discovery of new families. For a long time, it appeared that only the oxyhalogens provided a sufficient variety of oxidation states coupled with autocatalytic kinetics to generate chemical oscillation; the iodate-based Bray reaction and the bromate-driven BZ reaction and their derivatives were the only systems to have been characterized until well into the 1980s. With the advent of systematic search procedures and the growing interest in chemical oscillation, families based on sulfur, carbon (of course), and transition-metal chemistry soon emerged. Oscillation was spreading through the Periodic Table! In Figure 4.13, we summarize the current state of knowledge about chemical oscillators, including relationships among families and subfamilies. We have no doubt that by the time you read these words, many more oscillators will have been discovered, and the picture we have drawn will be obsolete. This is the mark of a vital field of research.

The classification of chemical oscillators shown in Figure 4.13 rests on an implicit hypothesis that the systems within a family are related by more than just sharing an element in common. After all, just about every oscillating reaction contains oxygen. Why not call them all members of the oxygen family? The assumption we are making is that the members of a family overlap not only in composition, but also in *mechanism*. In the following chapter, we will explore what is known about the mechanisms of chemical oscillators and how one goes about establishing an understanding of the underlying chemistry of such a complex chemical reaction.

5

Analysis of Chemical Oscillations

Many of the most remarkable achievements of chemical science involve either synthesis (the design and construction of molecules) or analysis (the identification and structural characterization of molecules). We have organized our discussion of oscillating reactions along similar lines. In the previous chapter, we described how chemists have learned to build chemical oscillators. Now, we will consider how to dissect an oscillatory reaction into its component parts—the question of *mechanism*.

A persuasive argument can be made that it was progress in unraveling the mechanism of the prototype BZ reaction in the 1970s that gave the study of chemical oscillators the scientific respectability that had been denied it since the discovery of the earliest oscillating reactions. The formulation by Field, Körös, and Noyes (Field et al., 1972) of a set of chemically and thermodynamically plausible elementary steps consistent with the observed "exotic" behavior of an acidic solution of bromate and cerium ions and malonic acid was a major breakthrough. Numerical integration (Edelson et al., 1975) of the differential equations corresponding to the FKN mechanism demonstrated beyond a doubt that chemical oscillations in a real system were consistent with, and could be explained by, the same physicochemical principles that govern "normal" chemical reactions. No special rules, no dust particles, and no vitalism need be invoked to generate oscillations in chemical reactions. All we need is an appropriate set of uni- and bimolecular steps with mass action kinetics to produce a sufficiently nonlinear set of rate equations.

Just as the study of molecular structure has benefited from new experimental and theoretical developments, mechanistic studies of complex chemical reactions, including oscillating reactions, have advanced because of new techniques. Just as any structural method has its limitations (e.g., x-ray diffraction cannot achieve a resolution that is better than the wavelength of the x-rays employed), mechanistic studies, too, have their limitations. The development of a mechanism, however, has an even more fundamental and more frustrating limitation, sometimes referred to as the fundamental dogma of chemical kinetics. It is not possible to *prove* that a reaction mechanism is correct. We can only *disprove* mechanisms. There are always additional candidates for the title of the "true mechanism."

As chemists have become increasingly sophisticated in their efforts to construct mechanisms, this fundamental limitation has become increasingly apparent. A quarter century ago, the prospect of creating even a single chemically reasonable mechanism for the BZ reaction seemed so daunting that FKN may be pardoned for the hubris implicit in their claim to have employed the method of (Sherlock) Holmes: when the impossible has been eliminated, whatever remains, however improbable, must be the truth. Since the 1970s, clever scientists, aided by powerful computers, have become increasingly adept at assembling mechanisms, often several for a single reaction, that give rise to oscillation. Controversies, occasionally heated ones, have arisen between proponents of rival mechanisms. Even the venerable FKN mechanism has been challenged (Nosizticzius et al., 1984a), which led to careful additional study and reassessment of several of the rate constants for the component steps of the BZ reaction (Field and Försterling, 1986).

To date, mechanisms have been proposed for about half of the known oscillators. These mechanisms vary in the level of detail that they attempt and in the quality of the agreement between their predictions and the experimental data. In this chapter, we describe how one goes about constructing a mechanism for a chemical oscillator. We then examine a few examples of actual mechanisms. We conclude with a brief look at the possibility of using mechanistic ideas to categorize chemical oscillators in a fashion complementary to the approach described in Chapter 4.

5.1 Models and Mechanisms

To be precise, the term *mechanism* or *reaction mechanism* should only be used to designate a molecular level description: that is, a complete set of elementary steps that specifies how a chemical reaction takes place. However, the word is often employed in reference to less detailed characterizations of reactions. These descriptions, which can involve several levels of simplification or assumption, are more properly referred to as *models*, rather than mechanisms. Models are not necessarily inferior to, or less useful than, mechanisms. They simply offer a different type of description. For certain purposes, one may gain more insight into how a system works from a three-step model than from a thirty-step mechanism.

One class of models that has played a key role in the development of nonlinear chemical dynamics consists of *abstract models*. Typically, these have a small number of variables, perhaps two or three, and are meant to elucidate a particular phenomenon, reaction, or class of reactions. Models of this type can be derived from a full mechanism by focusing on the most important overall processes and using intuition about rapid equilibria, quasi-steady states, or relative rates of competing steps to reduce the number of species and steps needed to yield the behavior of greatest interest. The Oregonator (Field and Noyes, 1974b) is the earliest and most important example of this approach. It succeeds in describing the oscillatory behavior and pattern formation in the BZ reaction with only three variable species concentrations and five "elementary steps." Another example is the Lengyel–Epstein model (Lengyel and Epstein, 1991) of the chlorine dioxide–iodine–malonic acid reaction in which, by abandoning mass action kinetics and adopting a nonpolynomial rate law to describe iodide inhibition, the authors are able to model not only the temporal kinetics of the system, but also the development of spatial Turing patterns (see Chapter 14).

Another category of generic abstract models is not derived from the mechanism of any particular reaction, but rather is constructed to illustrate some specific form of dynamical behavior. Mechanisms of this type include the venerable Lotka–Volterra model (Lotka, 1920b), the first "chemical" model to give rise to sustained oscillations; the Brusselator (Prigogine and Lefever, 1968), which was used to illustrate a variety of spatial and temporal pattern-formation phenomena (Nicolis and Prigogine, 1977); and a series of models due to Rössler (1976) in which different forms of chaos appear. The virtue of abstract models is that, because they contain only a handful of variables, they lend themselves to detailed, often exact, mathematical analysis without the total reliance on numerical methods that is necessitated by more realistic mechanisms containing perhaps dozens of coupled differential equations. A particularly nice treatment of some simple abstract models can be found in the book by Gray and Scott (1990).

Intermediate in complexity and accuracy between true mechanisms and abstract models are *empirical rate law models*. Here, the modeler eschews the identification of elementary steps and, instead, works with experimentally established rate laws for the component overall stoichiometric processes that make up a particular reaction. Each process may consist of several elementary steps and involve many reaction intermediates, but it enters the model only as a single empirical rate equation, and only those species that appear in the rate equation need be included in the model. Assuming that the empirical rate laws have been accurately determined, this approach will give results for the species contained in the rate laws that are identical to the results from the full mechanism, so long as no intermediate builds up to a significant concentration and so long as the component processes are independent of one another. This last requirement implies that no intermediate that is omitted from the model is involved in more than one process, and that there are no cross-reactions between intermediates involved in different processes.

An elegant example of the empirical rate law approach is the model developed by Gáspár and Showalter (1990) for the mixed Landolt or Edblom–Orbán–Epstein (EOE) reaction (Edblom et al., 1986). The reaction of iodate, sulfite, and ferrocyanide ions in a CSTR gives rise to sustained oscillation. In a flow reactor, the concentrations of two of these input reactants—iodate and ferrocyanide—remain nearly constant, while the pH and the concentrations of sulfite, bisulfite, and iodine oscillate over a wide range. Thus, the variables used in the empirical rate law model are: $A = SO_3^{2-}$, $X = HSO_3^-$, $Y = H^+$, and $Z = I_2$.

Gáspár and Showalter identify five overall stoichiometric processes that make up the total reaction. These are summarized in Table 5.1. The rate laws used for processes A, B, and C have been simplified from the full empirical multiterm rate laws (Reynolds, 1958; von Bünau and Eigen, 1962; Liebhafsky and Roe, 1979) by taking into account the conditions of the oscillatory EOE reaction. When the stoichiometries of the model equations for processes A and D are simplified by dividing by 3, we obtain the following set of pseudo-elementary reactions as our final empirical rate law model. By chance, the model obeys mass action kinetics, though the "rate constants" are products of rate constants for the nonelementary component processes and the constant concentrations.

$$A + Y \leftrightarrow X \tag{5.1}$$

$$X \to Y \tag{5.2}$$

$$2Y \to Z \tag{5.3}$$

$$Z + X \to 3Y \tag{5.4}$$

$$Z \to \tag{5.5}$$

Integrating the four differential equations obtained from eqs. (5.1)–(5.5) gives pH oscillations in good agreement with those obtained experimentally. The model can be simplified still further by noting that A and Z change on time scales much faster than those on which X and Y vary. Thus, one can use the steady-state approximation to express A and Z as functions of X and Y, thereby obtaining a two-variable model with nonpolynomial kinetics. This model, which gives

Table 5.1 Empirical Rate Law Model of the EOE Reaction[a]

Empirical Stoichiometry and Rate Law	Model Stoichiometry and Rate Law
Process A	
$\{IO_3^-\} + \{5I^-\} + 6H^+ \to 3I_2 + \{3H_2O\}$	$6Y \to 3Z$
$v_a = \{k_a[IO_3^-][I^-]^2\}[H^+]^2$	$v_a = k_a Y^2$
Process B	
$I_2 + HSO_3^- + \{H_2O\} \to \{2I^-\} + \{SO_4^{2-}\} + 3H^+$	$Z + Y \to 3Y$
$v_b = k_b[I_2][HSO_3^-]$	$v_b = k_b ZX$
Process C	
$I_2 + \{2Fe(CN)_6^{4-}\} \to \{2I^-\} + \{2Fe(CN)_6^{3-}\}$	$Z \to$
$v_c = \{k_c[Fe(CN)_6^{4-}]\}[I_2]$	$v_c = k_c Z$
Process D	
$\{IO_3^-\} + 3HSO_3^- \to \{I^-\} + \{3SO_4^{2-}\} + 3H^+$	$3X \to 3Y$
$v_d = \{k_d[IO_3^-]\}[HSO_3^-]$	$v_d = k_d X$
Process E	
$SO_3^{2-} + H^+ \leftrightarrow HSO_3^-$	$A + Y \leftrightarrow X$
$v_e = k_e[SO_3^{2-}][H^+]$	$v_e = k_e AY$
$v_{-e} = k_{-e}[HSO_3^-]$	$v_{-e} = k_{-e} X$

[a]Species and concentrations in {} are taken to be constant, either because they change very little during the reaction or because they are inert products.

results nearly identical to those of the four-variable model, in effect condenses steps (5.3) and (5.4) into a single elementary step, thereby pointing up its resemblance to the abstract autocatalator model of Gray and Scott (1986), which contains a similar termolecular autocatalytic step.

5.2 Building a Mechanism

The art, and it is an art, of constructing a mechanism for a reaction complicated enough to exhibit chemical oscillation requires skill, care, and intuition. Nonetheless, there are a number of principles that provide valuable guidance in assembling a set of elementary steps to describe a chemical oscillator. Above all, every aspect of a mechanism must be consistent with the rules of chemistry and physics. For example, the mechanism must not violate macroscopic reversibility; rate constants cannot exceed the diffusion-controlled limit; an activated complex that is a bottleneck for the forward step of a reaction must also be a bottleneck for the reverse step. It does no good to formulate a system of differential equations whose solutions behave just like our experiments if those equations require steps that lead to implausible stoichiometries or have huge positive free-energy changes or contain an intermediate with oxygen bonded to four other atoms.

Once a tentative sequence of steps has been devised, the investigator needs to address mechanistic questions such as: Is this reaction fast enough to contribute to the rate? Is some form of catalysis, for example, general acid catalysis, required? Does a particular step occur via atom transfer or electron transfer? The builder of a mechanism can draw on many theories and empirical rules to help answer questions like these. For example, if electron transfer reactions are thought to be important, the Marcus theory of outer-sphere electron transfer (Marcus, 1993) may be relevant. If metal ions are involved, the Eigen mechanism of substitution reactions (Eigen, 1963a, 1963b) may be invoked. If the plausible reactions involve free radicals, the National Institute of Science and Technology (NIST) compilations (Neta et al., 1988) can be invaluable. For enzymatic reactions, Hammes' work (1982) on the theory of diffusion-controlled reactions of macromolecules provides a useful guide.

The classic mechanistic study of Field, Körös, and Noyes on the BZ reaction is exemplary in its attention to physicochemical principles (Field et al., 1972). We could do far worse than to model our efforts at unraveling new systems on this quarter-century old work. The authors start out by summarizing the experimental phenomena that they seek to explain: primarily batch oscillatory behavior over a range of initial reactant concentrations. Next, they identify three overall stoichiometric processes that can occur in the system and relate these reactions to the major net stoichiometric change, which generates the free-energy decrease that drives the reaction. They then analyze the thermodynamics of a large set of plausible reactions by compiling a list of free energies of formation of the likely bromine and cerium species in the reaction. These analyses, carried out *before* any kinetic considerations are invoked and in the absence of complete knowledge of the thermodynamics of all the relevant oxybromine species (BrO and BrO_2, in particular), already provide a number of significant ideas on which the mechanism can be built and eliminate several potential blind alleys.

Field et al. next examine the kinetics of the system by breaking down the complex overall stoichiometric processes into simpler subreactions, each of which is studied in detail. They make extensive use of the literature, in several cases finding rate constants for the relevant elementary steps that had been measured earlier by a variety of techniques, including pulse radiolysis, isotopic exchange, temperature jump, and stopped flow, as well as by more classic methods. Where necessary, they make activity coefficient corrections, combine kinetic with thermodynamic data or carry out experiments of their own to supplement incomplete data or to decide among conflicting results from the literature. Once they have assembled a full set of reactions, they use the steady-state approximation and chemical intuition to analyze the kinetic behavior of the major reactants, as well as that of the key intermediates. They are able to develop a convincing description of how the system behaves during the pre-oscillatory induction period and during each major portion of the oscillatory cycle, and they show how this description explains changes in behavior at different initial reactant concentrations. The only missing ingredient that would be found in a more modern study is numerical integration of the proposed mechanism to see whether it leads to concentration profiles in agreement with those found experimentally. This last step, which was not possible with the computing power available to the authors, was carried out successfully a short time later in collaboration with workers at Bell Laboratories (Edelson et al., 1975), who had the use of hardware and software then inaccessible to most laboratory chemists, but today easily available on inexpensive desktop machines.

Guided by this model study, we can summarize the steps that one should take in constructing a mechanism for an oscillatory, or other complex, reaction system. While it may not be feasible to follow this recipe step-by-step in every case, adhering to the spirit of this route will help prospective mechanism builders to avoid pitfalls and to come up with plausible schemes for further testing in a relatively expeditious fashion. Here is our recommended approach:

1. Assemble all relevant experimental data that the mechanism should be able to model. The wider the range of phenomena and conditions, the more believable the mechanism that explains them.
2. Identify the primary stoichiometry of the overall reaction and of any major component processes.
3. Compile a list of chemically plausible species—reactants, products, and intermediates—that are likely to be involved in the reaction.
4. Obtain all available thermodynamic data pertaining to these species.
5. Break down the overall reaction into component processes, consider the likely elementary steps in each of these processes, and examine the literature critically for kinetics data relevant to as many of these processes and steps as can be found.
6. Use known rate constants wherever possible. If none exist, guess a value, or try to isolate from the main reaction a smaller subset of reactions and determine their rate laws.
7. Put all the thermodynamically plausible steps and the corresponding kinetics data together to form a trial mechanism. Use analytic methods, for example, the

steady-state approximation, as well as chemical intuition, to derive a qualitative picture of how the system works.

8. Use numerical methods, such as numerical integration and bifurcation analysis, to simulate the experimental results. If serious discrepancies arise, consider the possibility of missing steps, erroneous data, or false assumptions, and return to the previous steps to reformulate the mechanism.

9. Continue to refine and improve the mechanism, testing it against all new experimental results, particularly those carried out under conditions very different from those of the data used to construct the mechanism. The greatest success of a mechanism is not to give agreement with data already in hand, but to predict successfully the results of experiments that have not yet been carried out.

The complex set of reactions studied in the next section demonstrates how these principles can be applied in practice.

5.3 A Recent Example

We have used the FKN mechanism of the BZ reaction as an example to illustrate a general approach to constructing mechanisms. This mechanism was developed at a time when the range of experimental and computational methods was less extensive than it is today. It may be useful to look at a more recent example of a mechanistic study, in which a wide range of techniques was brought to bear on the chlorite–iodide and related reactions (Lengyel et al., 1996).

Only the BZ reaction has played a more central role in the development of nonlinear chemical dynamics than the chlorite–iodide reaction (De Kepper et al., 1990). This latter system displays oscillations, bistability, stirring and mixing effects, and spatial pattern formation. With the addition of malonic acid, it provides the reaction system used in the first experimental demonstration of Turing patterns (Chapter 14). Efforts were made in the late 1980s to model the reaction (Epstein and Kustin, 1985; Citri and Epstein, 1987; Rábai and Beck, 1987), but each of these attempts focused on a different subset of the experimental data, and none was totally successful. Since each model contains a different set of reactions fitted to a different set of data, individual rate constants vary widely among the different models. For example, the rate constant for the reaction between HOCl and HOI has been given as zero (Citri and Epstein, 1987), 2×10^3 M^{-1} s^{-1} (Rábai and Beck, 1987), 5×10^5 M^{-1} s^{-1} (Citri and Epstein, 1988), and 2×10^8 M^{-1} s^{-1} (Epstein and Kustin, 1985).

What gives the chlorite–iodide reaction its unusual dynamical properties is the fact that it is not only autocatalytic in the product, iodine, but it is also inhibited by the reactant, iodide. While the earlier models do attempt to deal with these features, they ignore other significant aspects of the reaction, including the fact that incomplete mixing has major effects on the dynamics, the formation of I_3^- when both I_2 and I^- are present, and the protonation of HOI at pH values of 2 and below. Lengyel et al. (1996) attempt to incorporate all of these features and to construct an experimentally based mechanism by re-examining not only the chlorite–iodide reaction itself, but also a set of component reactions that play key roles in this system. They employ primarily spectrophotometric techniques, utilizing a diode array spectrophotometer that permits simultaneous measurements at multiple wavelengths. For fast reactions, the instrument is combined with a stopped-flow apparatus to give rapid mixing with a dead time of about 5 ms. A further improvement over earlier studies is afforded by generating the chlorite in situ from the reaction of chlorine dioxide and iodide. Commercial sources of sodium chlorite are only about 80% pure, and even tedious recrystallization procedures (Nagypál and Epstein, 1986) give less than 99% purity. It is possible to prepare much purer chlorine dioxide (Lengyel et al., 1990b) and to use it to generate ClO_2^- free of impurities that might affect the reaction.

Under these conditions, the reaction occurs in four easily separable stages, as seen in Figure 5.1, where the absorbance at 470 nm (which is proportional to the iodine concentration) is plotted. Each segment of the reaction is labeled with a letter.

Stage a, which is quite rapid, takes only a few seconds. During this period, chlorite and iodine are generated from the starting materials via eq. (5.6),

$$ClO_2 + I^- = ClO_2^- + \tfrac{1}{2}I_2 \qquad (5.6)$$

In stage b, the chlorite produced in stage a reacts with the remaining iodide to produce more iodine in a reaction that is accelerated both by the growing concentration of the autocatalyst, I_2, and by the declining concentration of the inhibitor, I^-. The stoichiometry during this part of the reaction is

$$ClO_2^- + 4I^- + 4H^+ = 2I_2 + Cl^- + 2H_2O \qquad (5.7)$$

Once the iodide is consumed, stage b comes to an end, and stage c, during which the iodine generated in the first two stages reacts with excess chlorite, can begin. The fraction of the iodine so consumed and the stoichiometry of stage c depend critically on the ratio of chlorite to iodine. Under the conditions of interest, $[I^-]_0/[ClO_2]_0$ varies between 3 and 5, so that $[ClO_2^-]/[I_2]$ in stage c is always less than 0.33, which means that iodine is present in excess. If chlorite were in excess, it would react with HOCl to regenerate ClO_2, making the reaction considerably more complicated. With excess iodine, the stoichiometry of stage c is a mixture of the following three processes, written in terms of chlorous acid, which

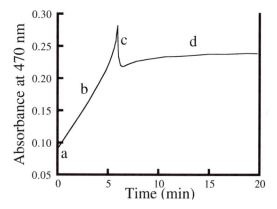

Figure 5.1 Time course of iodine concentration in a closed system starting with chlorine dioxide and iodine. $[ClO_2^-]_0 = 2.47 \times 10^{-4}$ M, $[I^-]_0 = 8.0 \times 10^{-4}$ M, $pH_0 = 2.15$, $[SO_4^{2-}] + [HSO_4^-] = 0.05$ M, ionic strength = 0.3 M. (Adapted from Lengyel et al., 1996.)

is predominant at the pH of interest, with larger excesses of iodine favoring the production of HOI rather than IO_3^-.

$$HClO_2 + 2I_2 + 2H_2O = 4HOI + Cl^- + H^+ \tag{5.8}$$

$$3HClO_2 + 2I_2 + 2H_2O = 4HIO_2 + 3Cl^- + 3H^+ \tag{5.9}$$

$$5HClO_2 + 2I_2 + 2H_2O = 4IO_3^- + 5Cl^- + 9H^+ \tag{5.10}$$

Finally, in stage d a small amount of iodine is regenerated by the slow disproportionation of HOI and HIO_2:

$$5HOI = 2I_2 + IO_3^- + 2H_2O + H^+ \tag{5.11}$$

$$5HIO_2 = I_2 + 3IO_3^- + H_2O + 3H^+ \tag{5.12}$$

By making absorbance measurements at 320 nm (HOI maximum), 350 nm (I_3^- maximum) and 470 nm (I_2–I_3^- isosbestic point) and using the known equilibrium constant for triiodide formation and the conservation of iodine atoms, it is possible to derive information about the concentrations of all major iodine-containing species. The following reactions were studied individually by varying the initial concentrations of the reactants: ClO_2–I^-–H^+, ClO_2–I_2, HOCl–I_2, and the disproportionations of HOI and of HIO_2. Rate constants for each reaction were determined by fitting the kinetics curves measured under a range of conditions using a program that combines numerical integration with nonlinear least-squares fitting and statistical analysis (Peintler, 1995). Data for the simpler systems were analyzed first, and rate constants determined from these systems were fixed and used in the analysis of the more complicated systems, from which further rate constants were established. Thus, the whole mechanism rests on a self-consistent set of rate constants. In all, curves were measured at more than sixty different sets of experimental conditions; three to five replicate determinations were made of each curve; and fifty to 250 points were used from each curve, depending on the complexity of the curve. In many cases, the fitting process led to new experiments under different initial conditions or with simpler subsystems from which a rate constant could be determined with greater sensitivity and/or with lower correlation with other parameters. Figures 5.2 and 5.3 give examples of the kind of data obtained and the quality of the fit.

The final mechanism is summarized in Table 5.2. There are nine independent variables, since several (e.g., I^-, I_2, and I_3^-, or ClO_2^- and $HClO_2$) can be related either by stoichiometry or by the assumed rapid equilibria M14–M16. Two variables, $[Cl_2]$ and $[Cl^-]$, as well as reactions M11–M13, are important only at pH < 2.0 and concentrations of $HClO_2$ above 10^{-3} M, or at high $[Cl^-]_0$ or in the HOCl–I_2 reaction. Otherwise, they can be neglected.

5.4 General Approaches

There have been a number of efforts to identify elements that are common to all, or to groups of, oscillatory mechanisms, thereby providing either a set of necessary and/or sufficient conditions for chemical oscillation or a mechanistically

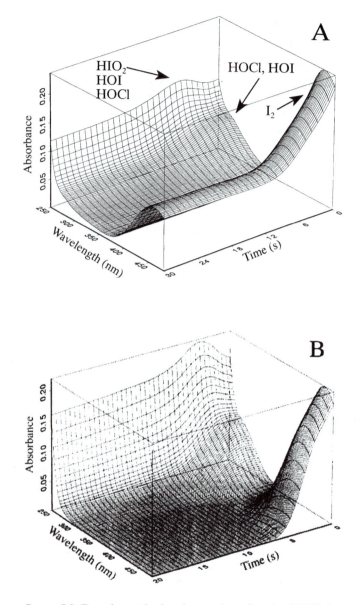

Figure 5.2 Experimental absorbance data for the HOCl–I_2 reaction with (A) iodine and (B) HOCl in stoichiometric excess. $[I_2]_0 = 3.2 \times 10^{-4}$ M, $[HOCl]_0 = 8.03 \times 10^{-4}$ M in (A) and $[I_2]_0 = 2.95 \times 10^{-4}$ M, $[HOCl]_0 = 1.25 \times 10^{-3}$ M in (B). $pH_0 = 1.66$, $[SO_4^{2-}] + [HSO_4^-] = 0.05$ M and ionic strength $= 0.3$ M for both (A) and (B). (Reprinted with permission from Lengyel, I.; Li, J.; Kustin, K.; Epstein, I. R. 1996. "Rate Constants for Reactions between Iodine- and Chlorine-Containing Species: A Detailed Mechanism of the Chlorine Dioxide/Chlorite–Iodide Reaction," *J. Am. Chem. Soc. 118*, 3708–3719. © 1996 American Chemical Society.)

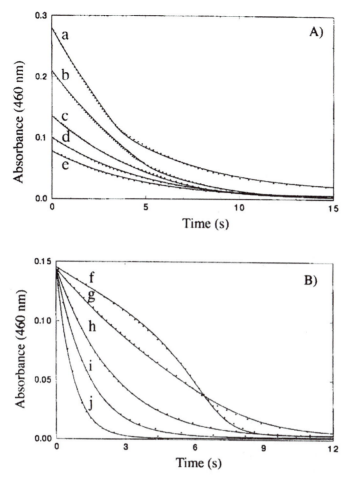

Figure 5.3 Kinetics curves for the $HClO_2$–I_2 reaction, varying (A) initial iodine concentration and (B) initial pH. $[I_2]_0$ = (a) 3.81 × 10^{-4} M, (b) 2.86 × 10^{-4} M, (c) 1.84 × 10^{-4} M, (d) 1.36 × 10^{-4} M, (e) 1.05 × 10^{-4} M, (f–j) 1.96 × 10^{-4} M; $[ClO_2^-]_0$ = (a–e) 9.0 × 10^{-4} M, (f–j) 1.0 × 10^{-3} M; pH_0 = (a–e) 1.89, (f) 1.19, (g) 1.74, (h) 2.23, (i) 2.50, (j) 2.82; $[SO_4^{2-}]$ + $[HSO_4^-]$ = 0.05 M and ionic strength = 0.3 M for (a–j). Dots represent experimental points; solid curves are computer fits. (Reprinted with permission from Lengyel, I.; Li, J.; Kustin, K.; Epstein, I. R. 1996. "Rate Constants for Reactions between Iodine- and Chlorine-Containing Species: A Detailed Mechanism of the Chlorine Dioxide/Chlorite–Iodide Reaction," *J. Am. Chem. Soc. 118*, 3708–3719. © 1996 American Chemical Society.)

Table 5.2 Mechanism of the Chlorite–Iodide and Related Reactions (Lengyel et al., 1996)[a]

Number	Reaction	Rate Law
M1	$ClO_2 + I^- \rightarrow ClO_2^- + \frac{1}{2}I_2$	$v_1 = 6 \times 10^3[ClO_2][I^-]$
M2	$I_2 + H_2O \leftrightarrow HOI + I^- + H^+$	$v_{2a} = 1.98 \times 10^{-3}[I_2]/[H^+] - 3.67 \times 10^9[HOI][I^-]$
		$v_{2b} = 5.52 \times 10^{-2}[I_2] - 3.48 \times 10^9[H_2OI^+][I^-]$
M3	$HClO_2 + I^- + H^+ \rightarrow HOI + HOCl$	$v_3 = 7.8[HClO_2][I^-]$
M4	$HClO_2 + HOI \rightarrow HIO_2 + HOCl$	$v_4 = 6.9 \times 10^7[HClO_2][HOI]$
M5	$HClO_2 + HIO_2 \rightarrow IO_3^- + HOCl + H^+$	$v_5 = 1.5 \times 10^6[HClO_2][HIO_2]$
M6	$HOCl + I^- \rightarrow HOI + Cl^-$	$v_6 = 4.3 \times 10^8[HOCl][I^-]$
M7	$HOCl + HIO_2 \rightarrow IO_3^- + Cl^- + 2H^+$	$v_7 = 1.5 \times 10^3[HOCl][HIO_2]$
M8	$HIO_2 + I^- + H^+ \leftrightarrow 2HOI$	$v_8 = 1.0 \times 10^9[HIO_2][I^-][H^+] - 22[HOI]^2$
M9	$2HIO_2 \rightarrow IO_3^- + HOI + H^+$	$v_9 = 25[HIO_2]^2$
M10	$HIO_2 + H_2OI^+ \rightarrow IO_3^- + I^- + 3H^+$	$v_{10} = 110[HIO_2][H_2OI^+]$
M11	$HOCl + Cl^- + H^+ \leftrightarrow Cl_2 + H_2O$	$v_{11} = 2.2 \times 10^4[HOCl][Cl^-][H^+] - 22[Cl_2]$
M12	$Cl_2 + I_2 + 2H_2PO \rightarrow 2HOI + 2Cl^- + 2H^+$	$v_{12} = 1.5 \times 10^5[Cl_2][I_2]$
M13	$Cl_2 + HOI + H_2O \rightarrow HIO_2 + 2Cl^- + 2H^+$	$v_{13} = 1.0 \times 10^6[Cl_2][HOI]$
	Rapid Equilibria	
M14	$HClO_2 \leftrightarrow ClO_2^- + H^+$	$K_{14} = [ClO_2^-][H^+]/[HClO_2] = 2.0 \times 10^{-2}$
M15	$H_2OI^+ \leftrightarrow HOI + H^+$	$K_{15} = [HOI][H^+]/[H_2OI^+] = 3.4 \times 10^{-2}$
M16	$I_2 + I^- \leftrightarrow I_3^-$	$K_{16} = [I_3^-]/[I_2][I^-] = 7.4 \times 10^2$

[a]All concentrations in M, times in seconds.

grounded categorization of oscillatory reactions. Higgins (1967) derived several significant insights from his analysis of a set of biochemical examples. Tyson (1975) analyzed the "community matrix," that is, the signs of the elements of the Jacobian matrix, and showed that all processes capable of destabilizing the steady state, and thus of leading to oscillatory behavior, fall into three classes. Noyes (1980) developed a comprehensive, mechanistically based treatment of the family of bromate oscillators. Franck (1985) presented a general analysis and categorization of oscillatory mechanisms, which emphasized the role of feedback, but he discussed only abstract models rather than actual mechanisms. In this section, we look at several attempts, starting from mechanisms of real chemical oscillators, to elucidate general features of oscillatory reaction mechanisms.

5.4.1 A Model for pH Oscillators

One approach to understanding the general aspects of oscillatory reactions is to seek key features that are shared by a family of oscillators and to incorporate these features into a general model for the family. The pH oscillators (Rábai et al., 1990) consist of a dozen or more reactions in which there is a large amplitude change in the pH and in which the pH change is the driving force for, rather than a consequence or an indicator of, the oscillation. Luo and Epstein (1991) identified a number of common features in pH oscillators and abstracted from this analysis a scheme that captures the essential elements of these systems. Their

Table 5.3 Mechanism for the Bromate–Sulfite–Ferrocyanide Reaction

Number	Reaction
B1	$BrO_3^- + HSO_3^- \rightarrow HBrO_2 + SO_4^{2-}$
B2	$HBrO_2 + Br^- + H^+ \rightarrow 2HOBr$
B3	$HOBr + Br^- + H^+ \rightarrow Br_2 + H_2O$
B4	$Br_2 + H_2O \rightarrow HOBr + Br^- + H^+$
B5	$2HBrO_2 \rightarrow BrO_3^- + HOBr + H^+$
B6	$Br_2 + HSO_3^- + H_2O \rightarrow 2Br^- + SO_4^{2-} + 3H^+$
B7	$H^+ + SO_3^{2-} \rightarrow HSO_3^-$
B8	$HSO_3^- \rightarrow H^+ + SO_3^{2-}$
B9	$BrO_3^- + 2Fe(CN)_6^{4-} + 3H^+ \rightarrow HBrO_2 + 2Fe(CN)_6^{3-} + H_2O$

approach is best understood by starting from the mechanism of a specific pH oscillator, the bromate–sulfite–ferrocyanide reaction (Edblom et al., 1989), which is shown in Table 5.3

A key element of the mechanism in Table 5.3, which accurately describes both the oscillatory and the bistable behavior found in this system, is the protonation–deprotonation equilibrium of sulfite–bisulfite, reactions B7 and B8, which are represented by eq. (5.13) in the general model:

$$X + H \leftrightarrow HX \tag{5.13}$$

where X corresponds to SO_3^{2-} and H to H^+.

The second essential ingredient is an autocatalytic pathway leading to the formation of hydrogen ion. In the mechanism of Table 5.3, this route is provided by steps B1–B6, which can be summarized by the two abstract reactions (5.14) and (5.15):

$$A + HX + H \rightarrow Y \tag{5.14}$$
$$HX + Y \rightarrow 3H \tag{5.15}$$

where A is BrO_3^- and Y represents one or more of the intermediates, such as $HBrO_2$, $HOBr$, or Br_2. The presence of H in the initiating reaction (5.14) takes into account the participation of hydrogen ion in steps B2 and B3. The three molecules of H formed in reaction (5.15) not only conform to the stoichiometry of step B6, but also constitute the minimum number of hydrogen ions necessary to give autocatalysis, since two are consumed in steps (5.13) and (5.14).

Reactions (5.13)–(5.15) provide the positive feedback for the system, with the equilibrium reaction (5.13) giving rise to a secondary autocatalysis, since as H increases, unreactive X is converted to the reactive form HX, speeding the production of H via reactions (5.14) and (5.15). In order for the system to oscillate, a negative feedback is required as well. This role is served in the real system by steps B4 and B9, which act, respectively, to deplete the system of the intermediate Br_2 and the autocatalyst H^+. In the model, these reactions are represented by

$$Y \rightarrow P \tag{5.16}$$
$$B + H \rightarrow Q \tag{5.17}$$

where we identify B with ferrocyanide, Q with ferricyanide, and P with bromide ion. If reaction (5.16) is fast enough, the system will oscillate without step (5.17). Conversely, if the rate constant for reaction (5.17) is sufficiently high, steps (5.14)–(5.16) can be merged into the single step (5.18), eliminating the intermediate Y as a variable:

$$A + HX + H \rightarrow nH \qquad n \geq 2 \qquad (5.18)$$

In batch conditions, the model behaves (as does the experimental system) as a clock reaction, with an induction period followed by a sudden jump of several orders of magnitude in H and then an exponential decrease in H. Under flow conditions, bistability and oscillations are obtained. The model can easily be adapted to describe a wide variety of pH oscillators, including iodate–sulfite–thiourea (Rábai et al., 1987), iodate–sulfite–thiosulfate (Rábai and Beck, 1988), periodate–thiosulfate (Rábai et al., 1989a), periodate–hydroxylamine (Rábai and Epstein, 1989), and iodate–hydroxylamine (Rábai and Epstein, 1990).

5.4.2 A Feedback-Based Classification

Our analysis of pH oscillators suggests that feedback, both positive and negative, plays a key role in chemical oscillators. Franck (1985, p. 5) describes chemical oscillation as "a result of an antagonistic interaction or interplay between a labilizing positive feedback releasing a conversion of state and a coupled negative feedback, which recovers that conversion and restores the initial state . . ." Luo and Epstein (1990) attempt to classify chemical oscillators according to the kind of feedback that causes them to oscillate. They point out that there is less variation in the nature of the positive feedback than in the negative feedback, and that a necessary feature, in addition to the presence of both positive and negative feedback, is the existence of a time delay between these feedbacks. This last point is elaborated on in Chapter 10.

Some form of positive feedback (typically, though not always, autocatalysis), is necessary in order to destabilize the steady state. Using Tyson's (1975) characterization of the ways in which an instability can arise, Luo and Epstein (1990) identify "direct autocatalysis" as the case in which the Jacobian matrix has a positive diagonal element J_{ii}, for some species i, signifying that increasing the concentration of that species causes it to be generated more rapidly. The most obvious example occurs in the case of simple autocatalysis:

$$A + X \rightarrow 2X \qquad (5.19)$$

but more subtle cases can occur, such as substrate inhibition, where the rate decreases as the concentration of a reactant increases. An example is found in the chlorite–iodide reaction, where one term in the rate law is inversely proportional to the concentration of the reactant I^- (De Meeus and Sigalla, 1966). This dependence is explained (Epstein and Kustin, 1985) mechanistically by a rate-determining step consisting of the hydrolysis of $ICLO_2$, which is formed along with I^- in a rapid equilibrium;

$$I_2 + ClO_2^- \leftrightarrow IClO_2 + I^- \tag{5.20}$$

$$IClO_2 + H_2O \rightarrow HIO_2 + HOCl \tag{5.21}$$

$$\text{Rate} = -d[I^-]/dt = k_{21}[IClO_2] = k_{21}K_{20}[I_2][ClO_2^-]/[I^-] \tag{5.22}$$

A further distinction may be made between "explosive autocatalysis," in which the concentration of A, the precursor of X in reaction (5.19), is held constant, and nonexplosive or "self-limiting autocatalysis," in which (A) is a variable that will naturally decrease as the autocatalytic reaction proceeds. Self-limiting autocatalysis can serve as a positive feedback on one species (X) and a negative feedback on another (A) simultaneously.

An alternative form of positive feedback is "indirect autocatalysis." Here, although J_{ii} and J_{jj} are both negative, the combined effects of species i on species j and of j on i are such that a rise in one of these concentrations ultimately leads to an increase in how rapidly it is produced. Mathematically, we require that $J_{ij}J_{ji} > 0$. There are two possibilities: competition, in which i inhibits j, and j inhibits i, so that both Jacobian elements are negative; and symbiosis, in which the two species activate each other, resulting in positive J_{ij} and J_{ji}. A simple example of symbiosis arises from introducing an intermediate Y and breaking down reaction (5.19) into a pair of steps:

$$A + X \rightarrow Y \tag{5.23}$$

$$Y \rightarrow 2X \tag{5.24}$$

We then have $J_{XX} = -k_{23}A < 0, J_{YY} = -k_{24} < 0, J_{XY} = 2k_{24} > 0$, and $J_{YX} = k_{23}A > 0$. Loops of more than two species $(i, j, k, \ldots, q)$ such that $J_{ij}J_{jk} \ldots J_{qi} > 0$ also result in positive feedback, but these tend to be found in biochemical systems rather than in simple chemical systems.

Luo and Epstein (1990) identified three types of negative feedback. The most common one in chemical oscillators is referred to as "coproduct autocontrol." In this type of negative feedback, the positive feedback step, generally autocatalysis, generates not only more of the autocatalyst, but also a second product, or coproduct. This latter species reacts, either directly or indirectly, in such a way as to remove the autocatalyst and ultimately to shut down the autocatalysis. This pattern can be seen in the Oregonator (Field and Noyes, 1974b):

$$A + Y \rightarrow X + P \tag{5.25}$$

$$X + Y \rightarrow 2P \tag{5.26}$$

$$A + X \rightarrow 2X + Z \tag{5.27}$$

$$2X \rightarrow A + P \tag{5.28}$$

$$Z \rightarrow fY \tag{5.29}$$

Reaction (5.27) constitutes the positive feedback, explosive autocatalysis, since A is assumed to be constant. Although reaction (5.28) provides some negative feedback, analysis shows that it is insufficient to bring the system back to a point where the cycle can start again. The combination of steps (5.29) and (5.26) provides the necessary limitation on X, as Z generates enough Y (bromide ion) to compete successfully with the autocatalysis, at least if the stoichiometric factor f is large enough. The transformation of Z to Y also provides the time delay necessary for the system to oscillate (Epstein and Luo, 1991).

Coproduct autocontrol is the dominant form of negative feedback in mechanisms for the copper-catalyzed hydrogen peroxide–thiocyanate (Luo et al., 1989), benzaldehyde autoxidation (Colussi et al., 1991), and bromate–iodide (Citri and Epstein, 1986) oscillators.

A second type of negative feedback is "double autocatalysis." The essence of this pattern is that a species generated autocatalytically, often explosively, in the primary autocatalysis is consumed by a second autocatalytic reaction in which it serves as the precursor. As the primary autocatalyst builds up, the second autocatalytic reaction accelerates, consuming the primary species at an ever-faster rate and ultimately terminating the "explosion." Franck (1985) constructed a modified version of the original Lotka–Volterra model that nicely illustrates double autocatalysis:

$$A + X \rightarrow 2X \tag{5.30}$$

$$X + Y \rightarrow 2Y \tag{5.31}$$

$$X + Z \rightarrow P \tag{5.32}$$

where A is fixed and X, Y, and Z all flow in and out of the system. The non-explosive autocatalysis reaction (5.31) shuts down the explosive reaction (5.30), and the additional negative feedback reaction (5.32) generates the necessary time lag, since Z [which builds up via the flow during the time that X and Y are depleted by reaction (5.31)] then holds X at a low level even though there is little Y present. Eventually, X and Z are restored by the flow, and the cycle starts again.

Variants of the double autocatalysis mechanism occur in the Limited Explodator model of the BZ reaction (Nosztsiczius et al., 1984a) and the Briggs–Rauscher oscillator (De Kepper and Epstein, 1982).

A final, and in some ways the simplest, mode of negative feedback identified by Luo and Epstein (1990) is termed "flow control." In some systems, the negative feedback provided by the chemical reactions alone is insufficient to generate oscillation. These systems cannot function as batch oscillators, but can give oscillations with appropriate flows in a CSTR. An illustrative example is given by Franck (1985):

$$X + Z \rightarrow 2X \tag{5.33}$$

$$X + Y \rightarrow P \tag{5.34}$$

$$A + X \rightarrow Q \tag{5.35}$$

The concentration of A is taken as constant, while X, Y, and Z flow in and out of the reactor. Both Y and Z are consumed during the autocatalytic growth of X. In the absence of flow, the system would then shut down once X was consumed by reaction (5.35). If the flow rate and input concentrations are chosen so that Y recovers before Z does, then step (5.34) serves to hold X at a low level until Z increases enough for the cycle to begin again. If the parameters do not yield a sufficient delay, reaction (5.33) resumes too soon for a complete cycle to occur, and the system reaches a steady state with perhaps a few damped oscillations.

Flow control occurs in the mechanism of many oscillators that run only in a CSTR. Some of these include the Citri–Epstein mechanism for the chlorite–iodide reaction (Citri and Epstein, 1987), the bromate–sulfite–ferrocyanide system (Edblom et al., 1989), and the Noyes–Field–Thompson (NFT) mechanism for the minimal bromate oscillator (Noyes et al., 1971). The last system is particularly worth noting in that three mechanisms that describe a single system, the BZ reaction, that is, Oregonator, Limited Explodator, and NFT, correspond to the three types of negative feedback—coproduct autocontrol, double autocatalysis and flow control, respectively.

Two other points should be made before moving on. First, the types of feedback identified here may not be exhaustive; there may well be other ways in which feedback can occur and lead to oscillations. It is certainly conceivable that other, possibly more enlightening, classification schemes can be developed. Second, a single system can contain more than one kind of feedback. We have seen several examples, starting with the Lotka–Volterra model, in which explosive and self-limiting feedback exist in the same system. The bromate–chlorite–iodide reaction, which is discussed in detail in Chapter 12 as an example of a chemically coupled oscillator system, exhibits both coproduct autocontrol and flow control in its negative feedback. In analyzing and in constructing mechanisms of chemical oscillators, it is important to identify the sources and types of positive and negative feedback and to understand how the delay between them arises.

5.4.3 Network Approaches

One way to think of a chemical reaction mechanism is as a network that connects the various reactants, intermediates, and products. This point of view has led to important general results about the properties of reaction mechanisms. In nonlinear chemical dynamics, one is interested in the question of *network stability* (Clarke, 1980), that is, whether or not all the steady states of a given network or mechanism are stable. A related question, called by Clarke the *stability diagram problem,* is to find for a given mechanism the boundary (in a space whose coordinates are the rate constants) that separates unstable from stable networks.

Viewing mechanisms as networks lends itself nicely to compact graphical representation of even very complex mechanisms. An example is shown in Figure 5.4, where the stoichiometry and kinetics of an eleven-step mechanism for the BZ reaction, given in Table 5.4 and corresponding to the core of the FKN mechanism, are represented in a single diagram. Only the values of the rate constants are missing. The species A, B, H, Q, and S do not appear in the diagram because their concentrations are considered to be held constant. The information content of the diagram is enhanced by clever use of the feathers and barbs on the arrows, each of which represents a single reaction. Note that these reactions are overall stoichiometric processes and not elementary steps. The number of feathers on the back of an arrow next to a species indicates the number of molecules of that species consumed in the reaction. The number of barbs at the front of the arrow shows the number of molecules of that species produced in the reaction. The kinetics are given by the further convention that the number of feathers on the left of an arrow indicates the order of the reaction with respect to that reactant, with simple first-order mass action kinetics shown by a straight tail with no feathers. For example,

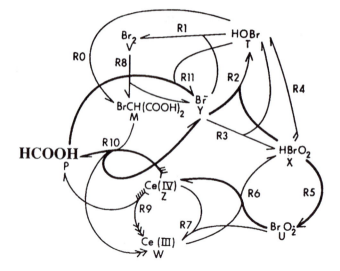

Figure 5.4 Diagram of the mechanism shown in Table 5.4. Darkened lines have no diagrammatic significance, but indicate a destabilizing feedback loop that emerges from the analysis. (Adapted from Clarke, 1976.)

step R7 is first order in both U and Z, while R4 represents the second-order disproportionation of $HBrO_2$ (X).

5.4.3.1 The Zero-Deficiency Theorem

One of the first to recognize the power of network approaches was Feinberg, a chemical engineer. As well as an important result, he and his collaborators derived a useful formalism, known as the *zero-deficiency theorem,* which we summarize

Table 5.4 Schematic Mechanism of the BZ Reaction Shown in Figure 5.4

Symbol	Species	Reaction No.	Reaction
A	$CH_2(COOH)_2$	R0	$A + T \rightarrow M + S$
B	BrO_3^-	R1	$T + Y + H \rightarrow V + S$
H	H^+	R2	$X + Y + H \rightarrow 2T$
M	$CHBr(COOH)_2$	R3	$B + Y + 2H \rightarrow X + T$
P	$HCOOH$	R4	$2X \rightarrow B + T + H$
Q	CO_2	R5	$B + X + H \rightarrow 2U + S$
S	H_2O	R6	$U + W + H \rightarrow X + Z$
T	$HOBr$	R7	$U + Z + S \rightarrow B + W + 2H$
U	BrO_2	R8	$A + V \rightarrow M + Y + H$
V	Br_2	R9	$A + 6Z + 2S \rightarrow P + 2Q + 6W + 6H$
W	$Ce(III)$	R10	$M + 4Z + 2S \rightarrow P + 2Q + 4W + Y + 5H$
X	$HBrO_2$	R11	$P + T \rightarrow Q + Y + H + S$
Y	Br^-		
Z	$Ce(IV)$		

here (Feinberg and Horn, 1974). As we see in Figure 5.5, the diagrams used in Feinberg's analysis are somewhat simpler than the kind shown in Figure 5.4, largely because they contain only stoichiometric, not kinetic, information. An arrow connecting two groups of species, each of which is referred to as a *complex*, implies that the mechanism contains a reaction in which the complex before the arrow is converted to the complex that follows the arrow. The number of complexes in a given mechanism is denoted as n. Thus, mechanism 3 has three complexes; mechanism 10 has nine. The zero complex as in mechanisms 9 and 10 can denote flow into or out of the system, production of a species from a precursor with constant concentration, or generation of an inert product. Only species that enter into the kinetics are shown. A second concept is that of *linkage classes*, designated l. A linkage class is a group of complexes that are connected by reaction arrows. Mechanisms 1–6 each consists of a single linkage class, but mechanisms 7 and 8 have $l = 2$. Finally, we define s, the number of linearly independent reaction vectors in the mechanism. Essentially, s represents the smallest number of reactions required so that all reaction stoichiometries in the mechanism can be constructed as linear combinations of this set. A pair of reversible reactions constitutes a single member of this set; a cycle involving c complexes, like the ones in mechanisms 4–8, contributes $c - 1$ to s. For mechanisms 1 and 2, $s = 1$; for mechanisms 3–6 and 9, $s = 2$; for mechanisms 7 and 8, $s = 3$; and for mechanism 10, $s = 5$. The *deficiency* of a mechanism (which is always ≥ 0) is defined as

$$\delta = n - 1 - s \tag{5.36}$$

In addition to the above definitions, we define a mechanism as *reversible* if for every reaction in the mechanism, its reverse reaction is also included. A mechanism is said to be *weakly reversible* if for any pair of complexes connected by a

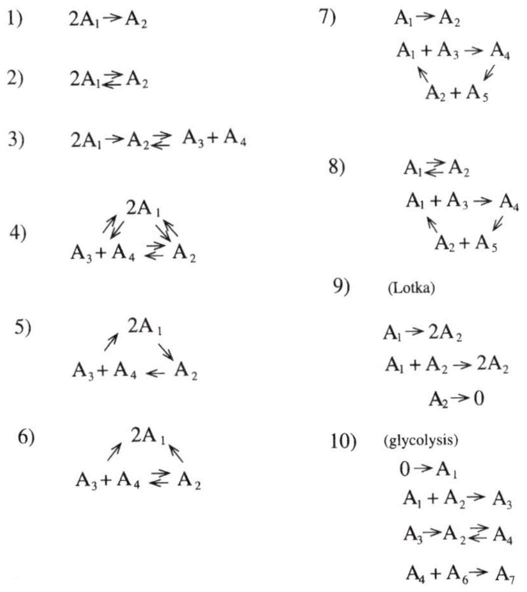

1) $2A_1 \rightarrow A_2$

2) $2A_1 \rightleftarrows A_2$

3) $2A_1 \rightarrow A_2 \rightleftarrows A_3 + A_4$

4) $\begin{array}{c} 2A_1 \\ \nearrow \quad \searrow \\ A_3 + A_4 \rightleftarrows A_2 \end{array}$

5) $\begin{array}{c} 2A_1 \\ \nearrow \quad \searrow \\ A_3 + A_4 \leftarrow A_2 \end{array}$

6) $\begin{array}{c} 2A_1 \\ \nearrow \quad \nwarrow \\ A_3 + A_4 \rightleftarrows A_2 \end{array}$

7) $A_1 \rightarrow A_2$
$\begin{array}{c} A_1 + A_3 \rightarrow A_4 \\ \nwarrow \quad \swarrow \\ A_2 + A_5 \end{array}$

8) $A_1 \rightleftarrows A_2$
$\begin{array}{c} A_1 + A_3 \rightarrow A_4 \\ \nwarrow \quad \swarrow \\ A_2 + A_5 \end{array}$

9)   (Lotka)
$A_1 \rightarrow 2A_2$
$A_1 + A_2 \rightarrow 2A_2$
$A_2 \rightarrow 0$

10)   (glycolysis)
$0 \rightarrow A_1$
$A_1 + A_2 \rightarrow A_3$
$A_3 \rightarrow A_2 \rightleftarrows A_4$
$A_4 + A_6 \rightarrow A_7$

Figure 5.5 Sample mechanisms for Feinberg analysis. (Adapted from Feinberg and Horn, 1974.)

directed arrow path in one direction, there exists a directed arrow path in the other direction that also connects those complexes. For example, mechanism 5 is weakly reversible because the clockwise cyclic path enables us to get "back to" any complex once we have gone from it to any other complex. Mechanism 6 is not weakly reversible because we can get from A_2 to $2A_1$, but we cannot go in the other direction. Of the mechanisms in Figure 5.5, only 2, 4, 5, and 8 are weakly reversible. Clearly, a reversible mechanism is weakly reversible, but the converse is not necessarily true.

The *zero-deficiency theorem,* a proof of which is given by Feinberg and Horn (1977), makes two important statements about reaction mechanisms for which $\delta = 0$ that hold regardless of the values of the rate constants. First, if the network *is not* weakly reversible, then no matter what the kinetics, there is no steady state or equilibrium in which all concentrations are greater than zero, nor is there a periodic trajectory in which all concentrations are positive. If the network *is* weakly reversible and has mass action kinetics, then it has a single equilibrium or single steady state, which is asymptotically stable. In essence, the theorem states that zero-deficiency mechanisms cannot oscillate or show any of the other behaviors of interest in nonlinear dynamics. This result is significant, of course, only if there are many mechanisms with zero deficiency; empirically, this proves to be the case. In Figure 5.5, only mechanisms 9 and 10 are of positive deficiency. While some of the mechanisms might have been easy to analyze by other methods, we contend that it is less than obvious either that mechanism 7 has no steady state with all concentrations positive or that mechanism 8 cannot give rise to oscillation. These results follow immediately from a calculation of the deficiency of the network.

Feinberg (1980) has extended this approach to derive additional results for networks of positive deficiency.

5.4.3.2 Stoichiometric Network Analysis

A still more ambitious approach has been pioneered by Clarke (1976, 1980). His *stoichiometric network analysis* attempts to solve both the network stability and the stability diagram problems by using sophisticated mathematical techniques to identify critical subnetworks within a complex mechanism that can result in instability. The problem can be converted into the geometrical problem of finding the vertices, edges, and higher dimensional faces of a convex polyhedron (polytope) in a high-dimensional space.

Clarke's methods are extremely powerful and yield not only mathematical rigor but also chemical insights. They are also extremely cumbersome, and for this reason they have not been widely used. Unfortunately, no generally available computer program exists to implement them. The one detailed exposition in the literature (Clarke, 1980) runs to over 200 pages and is filled with warnings like "There is much in this chapter that would be of practical value to experimental chemists if they could only understand it" (p. 9) and "The material reads like mathematics" (p. 9). We make no attempt to provide a summary of Clarke's methods here, but encourage the intrepid, mathematically inclined reader to seek the rewards offered by mastering this remarkable theory.

5.4.3.3 A Mechanistically Based Classification Scheme

One of the few successful attempts, by investigators other than Clarke, to utilize stochastic network analysis was carried out by Eiswirth et al. (1991), who divide the known mechanisms of chemical oscillators into four categories: two based on the positive feedback loop, and a threefold division of one of these according to the negative feedback loop. The structure of the classification scheme resembles somewhat the results of the approach described in section 5.4.2, but the analysis is considerably more detailed and rigorous.

Like the Epstein–Luo approach, the treatment focuses on the positive and negative feedback loops associated with the instability—a supercritical Hopf bifurcation—that leads to oscillation. One first identifies species as essential or nonessential with respect to the oscillatory behavior. Although Eiswirth et al. give a more rigorous definition, the basic notion is that a nonessential species is one whose concentration may be maintained constant without eliminating the Hopf bifurcation and hence the oscillation. Such species can be reactants that produce other, essential species; products whose reactions are irrelevant to the oscillatory behavior; or other species, including reactive intermediates, whose concentration can be kept constant without destroying the oscillation, though its quantitative features may be changed significantly. The remaining species, whose concentrations cannot be maintained constant without suppressing oscillation, are termed essential. By eliminating all nonessential species from the full mechanism, one obtains a skeleton mechanism that can be examined using stoichiometric network analysis.

The analysis starts by identifying loops, or, by analogy with electrical networks, *currents* or current cycles in the reaction diagram. A look at Figure 5.4 shows that a complex reaction mechanism can contain many such cycles; the theory allows one to select a set that determines the properties of the network. Current cycles are termed strong, critical, or weak, according to whether the kinetic order of the exit reaction from the cycle is, respectively, lower than, equal to, or higher than the order of the cycle. Figure 5.6 gives a simple example for cycles consisting of a single autocatalytic reaction.

Strong cycles lead to instability and weak ones to stability, while critical cycles can produce instability (and hence the possibility of oscillation), depending on other features of the network. Eiswirth et al. first divide oscillatory reactions into two major classes: those (category 1) whose networks contain a critical current cycle and a suitable, destabilizing, exit reaction; and those (category 2) in which the instability arises from a strong current cycle. Figure 5.7 shows diagrams of the basic elements of category 1 mechanisms. Diagram E_1 constitutes the necessary unstable critical cycle in which X is formed autocatalytically and exits, both with first-order kinetics. The two pairs of E_2 and E_3 diagrams are examples of additional reactions that are necessary to prevent the concentrations of X and Y from increasing indefinitely.

Category 1 can be further subdivided according to the nature of the exit reaction. In type $1B$ oscillators, Y is generated in a chain of reactions (negative feedback loop) via at least one intermediate Z such that $J_{ZX}J_{XY}J_{YZ} < 0$. In type $1C$ systems, no such feedback loop exists, but there is an essential species Z that is

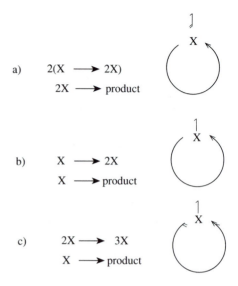

a) $2(X \longrightarrow 2X)$

      $2X \longrightarrow$ product

b) $X \longrightarrow 2X$

      $X \longrightarrow$ product

Figure 5.6 Some simple current cycles: (a) weak (exit order = 2, cycle order = 1), (b) critical (1, 1, respectively), (c) strong (1, 2, respectively). (Adapted from Eiswirth et al., 1991.)

c) $2X \longrightarrow 3X$

      $X \longrightarrow$ product

consumed by the critical current cycle, giving rise to a negative feedback such that $J_{XZ}J_{ZX} < 0$. Examples are shown in Figure 5.8. Since type $1C$ oscillators require an inflow of the essential species Y, while in type $1B$ oscillators Y is generated internally, one may expect $1B$ systems to show oscillatory behavior in batch, while $1C$ oscillators should function only under open conditions. While this distinction holds in most instances, it is possible that, because a poison is produced that must be washed out by an outflow, a type $1B$ system will oscillate only in a CSTR. It is also conceivable that other reactions might generate the necessary flow of Y in a type $1C$ system, thereby making possible batch oscillation.

A final distinction, illustrated by the examples in Figure 5.8, can be made among type $1C$ oscillators. In some oscillators of this type, an additional species,

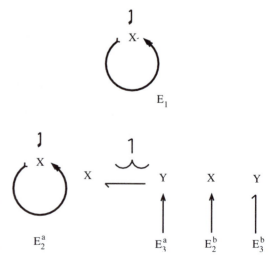

Figure 5.7 Network diagrams for basic elements of category 1 oscillators. (Adapted from Eiswirth et al., 1991.)

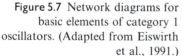

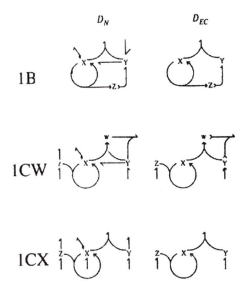

Figure 5.8 Network diagrams for simple examples of the three types of category 1 oscillators. (Adapted from Eiswirth et al., 1991.)

W, is essential. This species is formed in the critical current cycle and reacts with Y (type $1CW$). Alternatively, the system can oscillate without any additional essential species, but a flow of X is required (type $1CX$). For further details regarding how to distinguish the various mechanistic types, we refer the reader to the thorough discussion given by Eiswirth et al. (1991).

Among the category $1B$ oscillators are found the Oregonator and FKN mechanisms, as well as the Briggs–Rauscher reaction. Category $1CX$ includes the EOE reaction of Table 5.1 and the bromate–sulfite–ferrocyanide reaction of Table 5.3, as well as the peroxidase-catalyzed NADH oxidation (Yokota and Yamazaki, 1977). In category $1CW$, we find the bromate–manganous–sulfite, bromate–iodide, and chlorite–iodide reactions. Category 2 contains the Brusselator model as well as the nonisothermal hydrolysis of 2,3-epoxy-1-propanol (Vance and Ross, 1988) and the isothermal oxidation of carbon monoxide on platinum (Eiswirth et al., 1990).

5.4.4 Perturbation Methods

The experimental approaches we have described so far for the construction of mechanisms of oscillating reactions closely parallel the methods used by kineticists to study "simpler" systems—that is, measuring the rate laws and rate constants of the postulated elementary steps and comparing measurements of concentration vs. time data for the full system with the predictions of the mechanism. This is as it should be, since chemical oscillators obey the same laws as all other chemical systems. Another set of methods that has proved useful in kinetics studies involves perturbation techniques. One takes a system, subjects it to a sudden change in composition, temperature, or some other state variable, and then analyzes the subsequent evolution of the system. In particular, relaxation methods, in which a system at equilibrium is perturbed by a small amount and then relaxes back toward equilibrium with a first-order rate constant determined by the forward and reverse rate constants, the rate laws, and the equilibrium concentrations, have proven particularly valuable for studying rapid reactions (Bernasconi, 1976).

It is worth asking whether perturbation methods might yield as much or more information about oscillating reactions, where it might be possible to probe not only constant or monotonically varying concentrations but also amplitude and phase relationships. Schneider (1985) reviewed a variety of model calculations and experiments on *periodically* perturbed chemical oscillators. The results, which show such features as entrainment, resonance, and chaos, are of considerable interest in the context of nonlinear dynamics, but shed little light on the question of mechanism.

A more mechanistically oriented approach was taken by Ruoff (1984), who attempted to test the FKN mechanism for the BZ reaction by adding various perturbing species to an oscillating BZ system and observing the phase shifts produced at a platinum electrode. The results obtained on adding solutions of KBr, AgNO$_3$, and HOBr were in good agreement with the predictions of the model, lending support particularly to the concept of bromide control of the oscillations, a key feature of the FKN mechanism.

A more generally applicable approach, known as *quenching*, has been developed by Hynne and Sørensen (1987). In most chemical oscillators, oscillation arises via a supercritical Hopf bifurcation that produces a saddle focus surrounded by a limit cycle. The unstable steady state that lies inside the oscillatory trajectory in phase space has an unstable manifold, that is, a surface along which trajectories move away from the steady state and toward the limit cycle, and a stable manifold, that is, a surface (possibly just a line) along which trajectories move toward the steady state. The situation is illustrated in Figure 5.9.

If the system is oscillating and one adds to (or subtracts from, e.g., by dilution) the concentration of one or more species, one may, if the perturbation is of the appropriate magnitude and occurs at the correct phase, be able to shift the system onto the stable manifold as shown in Figure 5.9. If this is accomplished, the

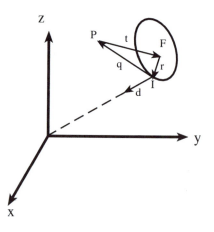

Figure 5.9 Schematic diagram of a quenching experiment in phase space. *F* is the steady state; the circle is the limit cycle, with *I* the point at which the system is perturbed. The quenching vector **q** shifts the state from *I* to *P*, which lies on the stable manifold along **t**. The system then moves to *F*. The vector **d** shows the result of a dilution. The "radius" **r** varies with the phase of oscillation. (Adapted from Hynne and Sørensen, 1987.)

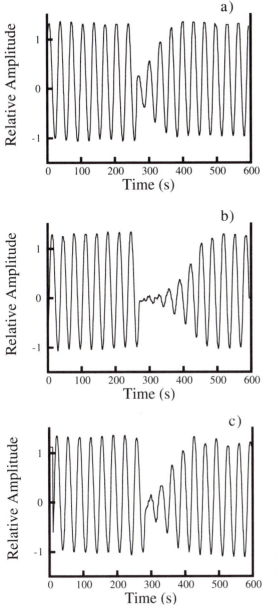

Figure 5.10 Three quenching experiments with equal additions of bromide ion to the BZ reaction at different phases of the oscillatory cycle: (a) −97°, (b) −104°, (c) −112°. (Adapted from Sørensen and Hynne, 1989.)

system will quickly move along the stable manifold to the steady state, causing the oscillations to cease, that is, to be quenched. Oscillations will resume when a random fluctuation moves the system from the steady state to the unstable manifold.

The results of an actual quenching experiment on the BZ system are shown in Figure 5.10. Note the importance of the timing of the perturbation. True quenching can be accomplished only if both the amplitude and the phase of the perturbation are correctly chosen, a process that can require a great deal of trial and error.

Quenching experiments yield information of considerable mechanistic value. One can obtain the relative amplitudes and phases of the species used in the quenchings and compare these with the predictions of a mechanism. For example, extensive quenching experiments on the BZ reaction (Sørensen and Hynne, 1989) reveal that the variables (Ce^{4+}), $(HBrO_2)$, and (Br^-) are not sufficient to account for all of the data, implying that the Oregonator model needs to be supplemented with at least one other variable. The quenching method is one that has been used only sparingly and is worthy of further development.

Finally, Arkin and Ross (1995) proposed a perturbation method, *correlation metric construction* (CMC), derived from electronic circuit and systems theory, which is generally applicable to the elucidation of complex reaction mechanisms. They subject a system near its steady state to random concentration changes in a set of input species. They measure as many concentrations as possible at time intervals no shorter than the slowest relaxation time in the network. The data are used to construct a *time-lagged correlation matrix* $r_{ij}(\tau)$, which measures the inter-species and time correlations within the network. Statistical techniques are used to extract information regarding the connections among species, essentially to see which species are involved in the same reaction. This information is then quantified by converting the matrix $r_{ij}(\tau)$ into a matrix that gives the "distance" between species in a multidimensional space, in which smaller distances are associated with mechanistic connections between species. When this set of distances is projected onto a two-dimensional space, it becomes possible to construct a picture of the reaction network. This last step involves a cluster analysis that uses the distance matrix to construct a hierarchy of interactions among the subsystems in the network. From these results and some chemical intuition, it should be possible to construct the reaction mechanism.

The CMC approach is a particularly powerful one, because it makes it possible, in principle, to deduce a complex reaction mechanism purely from a set of experimental measurements. Ideally, for the method to work, one should have a complete list of the species present in the system and be able to follow the concentrations of all of them. These practical requirements are severe, and it will be necessary to see to what extent the CMC technique can function in real situations, where one has only an incomplete set of information to work with.

6

Waves and Patterns

In this chapter, we will examine how diffusion can interact with the sort of nonlinear reaction kinetics that we have discussed in previous chapters to generate propagating waves of chemical reactivity and structured, sometimes spectacular, spatial patterns. One might argue, particularly if one were a physicist rather than a chemist, that the oscillatory phenomena that we have discussed thus far, and even the chaotic behavior that we will treat a bit later, can be just as easily generated by nonlinear electrical circuits, and that those circuits are much easier to work with than wet, smelly, and often toxic, chemicals. However, as far as chemistry goes, "spatial is special" in the sense that the spatial phenomena that we will discuss in this chapter are most naturally and conveniently generated in chemical systems. These same patterns also occur in a number of biological systems; the resemblance between a structure that develops in an inorganic chemical reaction and one found in a living organism can be startling.

We are inclined to think of diffusion as tending to make a system *more* homogeneous, with it acting to reduce and ultimately to obliterate local fluctuations in the concentration of a chemical species. The spontaneous appearance of concentration gradients might seem to violate the Great Law (i.e., the Second Law of Thermodynamics), since the free energy should be at a minimum or the entropy at a maximum when concentrations are uniform. However, as we have seen in the case of temporal oscillation, a system far from equilibrium can generate spontaneous transient spatial inhomogeneities and, in an open system, sustained spatial patterns as well.

The most common kind of chemical wave is the single propagating front, where, in an unstirred medium, there is a relatively sharp boundary between reacted and unreacted material, and this boundary or *wavefront* moves through the solution at an essentially constant speed. These waves are, of course, most noticeable when the reaction involves a color change. Many autocatalytic reactions exhibit this phenomenon of traveling waves. In some systems, the medium returns to its original state after passage of the front; such a phenomenon is often referred to as a *pulse*. It is necessary for the reaction to be somewhat difficult to initiate, so that it starts only at a single spot, and to be slow enough so that once it begins, the whole medium does not react so quickly that we are unable to see the movement of the wave.

Even more impressive than single-front or pulse propagation is the generation of repeated fronts, which, in a two-dimensional medium, leads to the sort of target or spiral patterns that can be seen in the BZ reaction. The phenomena that we will discuss in this chapter are dynamic—there is either a single traveling wave or a pattern that is continuously generated by new waves being born and moving through the medium. In Chapter 14, we shall discuss stationary patterns that can also be obtained in reaction–diffusion systems.

Most experiments on chemical waves and patterns to date have been carried out using essentially a one- or a two-dimensional geometry—a thin tube or a Petri dish. Typically, one examines the conditions (reactant concentrations, temperature) necessary to generate fronts or waves, perhaps some features of the wave initiation process, and the dependence of the propagation velocity on the conditions. The waves in such systems are necessarily a transient phenomenon, analogous to batch oscillations in a closed stirred system. More recently, people have begun to develop clever ways of studying spatial phenomena in open systems, so that the structures can persist for long periods of time. We shall look at some of these approaches toward the end of this chapter. First, we shall consider several approaches to understanding how waves arise and how we might calculate their velocities.

6.1 Stability Analysis

When we studied the emergence of temporal oscillations in Chapter 2, we found that it was useful to examine whether a small perturbation to a steady state would grow or decay. We now attempt a similar linear stability analysis of a system in which diffusion, as well as reaction, can occur. First, consider the general reaction–diffusion equation:

$$\frac{\partial c_i}{\partial t} = D_i \nabla^2 c_i + R_i(\{c\}) \qquad i = 1, 2, \ldots, n \tag{6.1}$$

where c_i is the concentration of the ith species, which has diffusion constant D_i, ∇^2 is the sum of second derivatives with respect to the spatial variables (the Laplacian), and R_i is the rate law, which depends on the set of concentrations $\{c\}$, for the net production of the ith species. Equation (6.1) accounts for the two ways that the concentration can change at any point in the solution. A chemical species can diffuse in or out, or it can be produced or destroyed by the chemical reaction.

In order to solve a partial differential equation like eq. (6.1), we need to know the boundary conditions. In most situations of interest—for example, in a Petri dish with an impermeable wall—the appropriate condition to use is the *zero-flux boundary condition*:

$$\mathbf{n} \cdot \nabla c_i = 0 \qquad i = 1, 2, \ldots, n \tag{6.2}$$

where $\mathbf{n}$ is a vector normal to the boundary of the system and ∇ is the concentration gradient at the boundary.

Just as in the case of purely temporal perturbations, it is useful to examine the behavior of a solution to eq. (6.1) composed of a steady-state solution plus an infinitesimal perturbation. Here, the steady-state solution is uniform in space, that is, $\{\mathbf{c}\}$ is independent of x, y, and z, and is constant in time, that is, $R(\{\mathbf{c}\}) = 0$. We chose our temporal perturbation to be $\alpha e^{\lambda t}$, an eigenfunction of the time derivative operator $\partial/\partial t$. Here, we choose the spatiotemporal perturbation to be a product of eigenfunctions of the time derivative and the Laplacian operators:

$$\alpha_i e^{\lambda t} u_m(\mathbf{r}) \tag{6.3}$$

where u_m is an eigenfunction of the Laplacian operator satisfying the boundary condition of eq. (6.2). Note that, because of the requirement that the boundary condition be satisfied, the actual form of u_m will be determined by the geometry of the boundary.

To make things more concrete (and somewhat simpler), consider a system that has only one spatial dimension, that is, a line of length L. If we put the boundaries of the system at $z = \pm L/2$, then condition (6.2) is simply the requirement that u_m/z vanish at the boundary points. It is not difficult to see that the function

$$u_m = \cos(m\pi z/L) \tag{6.4}$$

satisfies this criterion and is also an eigenfunction of the one-dimensional Laplacian operator:

$$\partial^2 u_m/\partial z^2 = -(m\pi/L)^2 u_m \tag{6.5}$$

where the integer m specifies the wavelength of the perturbation. The system may be stable to some wavelengths but unstable to others.

Let us now consider a specific example, the Brusselator model of eqs. (1.7)–(1.10). With A and B constant and all rate constants set to unity, the reaction–diffusion equations become:

$$\partial x/\partial t = A - (B+1)x + x^2 y + D_1 \partial^2 x/\partial z^2 \tag{6.6a}$$

$$\partial y/\partial t = Bx - x^2 y + D_2 \partial^2 y/\partial z^2 \tag{6.6b}$$

We set

$$x = x_{ss} + \alpha e^{\lambda t} \cos(m\pi z/L) \tag{6.7a}$$

$$y = y_{ss} + \beta e^{\lambda t} \cos(m\pi z/L) \tag{6.7b}$$

If we substitute eqs. (6.7) into eqs. (6.6), we obtain the same terms as when we did the stability analysis for the homogeneous case, except for an additional contribution from the diffusion terms. The spatial derivative gives $-(m\pi/L)^2$ times the perturbation. Therefore, the Jacobian now looks like

$$\mathbf{J}_m = \begin{bmatrix} B - 1 - D_1(m\pi/L)^2 & A^2 \\ -B & -A^2 - D_2(m\pi/L)^2 \end{bmatrix} \tag{6.8}$$

We can solve the characteristic equation $\det(\mathbf{J}_m - \lambda\mathbf{I}) = 0$ to obtain the eigenvalues λ and hence the stability. Clearly, we get different λ values for each value of m ($m = 0$ is the homogeneous case that we examined previously). Do we get anything new for $m \neq 0$? If we multiply out the determinant, we find

$$\lambda^2 + b\lambda + c = 0 \tag{6.9}$$

where

$$b = (\beta_m - \alpha_m) \tag{6.10a}$$
$$c = A^2 B - \alpha_m\beta_m \tag{6.10b}$$

and

$$\alpha_m = B - 1 - (m\pi/L)^2 D_1 \tag{6.11a}$$
$$\beta_m = A^2 + (m\pi/L)^2 D_2 \tag{6.11b}$$

The system is unstable if $b < 0$ or $c < 0$. The condition that $b < 0$ is equivalent to $\alpha_m > \beta_m$ or

$$B > 1 + A^2 + (m\pi/L)^2(D_1 + D_2) \tag{6.12}$$

The condition on c [eq. (6.10b)] is equivalent to $\alpha_m\beta_m > A^2 B$, or

$$B > 1 + (D_1/D_2)A^2 + A^2/D_2(L/m\pi)^2 + D_1(m\pi/L)^2 \tag{6.13}$$

As in the homogeneous case, large B promotes instability. For instability, we also want both diffusion constants not to be too large and the ratio D_1/D_2 to be relatively small. The conditions (6.12) and (6.13) tell us a number of other things about stability, which can be generalized to other reaction–diffusion systems, including systems in two and three spatial dimensions. We see that if the system is too small, that is, $L \to 0$, no instability is possible, since the right-hand sides of both inequalities grow without bound. In essence, the system cannot support waves if there is not enough room for at least one full wavelength. With $m = 0$, condition (6.12) is just the condition for instability of the homogeneous state. When m is not zero, the right-hand side increases with m, so that any instability resulting from this condition can persist only up to a finite value of m, which depends not only on the concentrations and diffusion constants, but also on the system length. If we consider what happens when m is varied in condition (6.13), we find that instability can occur only between upper and lower limits of the wavelength. Thus, concentration perturbations may grow and give rise to spatial structure, but only if they are of an appropriate wavelength.

As in the homogeneous case, linear stability analysis can tell us whether a small perturbation will grow, but it does not reveal what sort of state the system will evolve into. In Figure 6.1, we illustrate the range of m values over which instability occurs in the Brusselator.

Figure 6.2 shows an example of a traveling wave pattern in a one-dimensional Brusselator model.

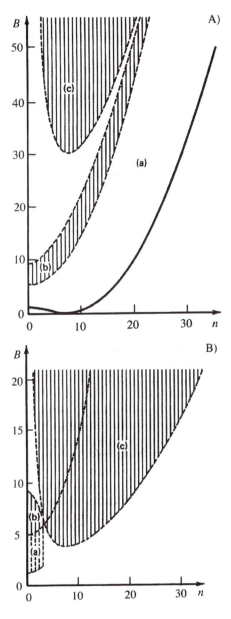

Figure 6.1 Linear stability diagrams from eqs. (6.12) and (6.13). The bifurcation parameter B is plotted against wavenumber n. Letters (a) and (b) indicate regions of complex eigenvalues, (b) indicates a region corresponding to an unstable focus, and (c) indicates an unstable region corresponding to a saddle point. Vertical lines indicate allowed discrete values of n for a system subject to zero flux or fixed boundary conditions. (A) $A = 2, l = 1, D_1 = 8.0 \times 10^{-3}$, $D_2 = 1.6 \times 10^{-3}$; (B) $A = 2, l = 1$, $D_1 = 1.6 \times 10^{-3}, D_2 = 8.0 \times 10^{-3}$. (Reprinted by permission of John Wiley & Sons, Inc. from Nicolis, G.; Prigogine, I. 1977. *Self-Organization in Nonequilibrium Systems.* Wiley: New York.)

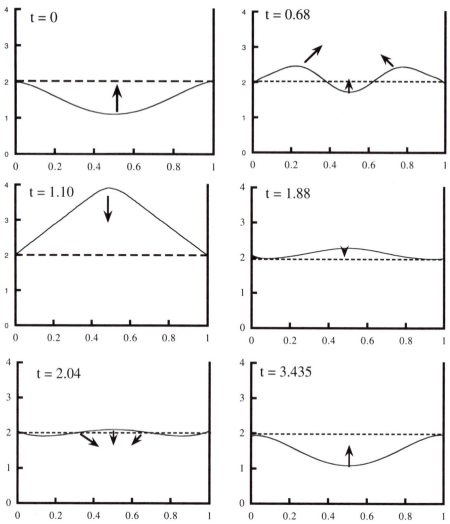

Figure 6.2 Evolution of the spatial profile of X in the Brusselator model. $L = 1$, $A = 2$, $B = 5.45$, $D_x = 8 \times 10^{-3}$, $D_y = 4 \times 10^{-3}$. (Adapted from Nicolis and Prigogine, 1977.)

6.2 An Exactly Soluble System—The Arsenous Acid–Iodate Reaction

The above analysis indicates that interesting things can happen in reaction–diffusion systems, but, except for the numerical integration that generated Figure 6.2, it does not actually tell us *what* will happen. In principle, if we know the kinetics of the homogeneous reaction and the diffusion coefficients for all the species, we should be able to predict not only the velocity of the wave, if one occurs, but also its shape. Nature, of course, is never as simple as we might hope, and, in practice, actually doing this turns out to be a considerable challenge. Because we are dealing with partial differential equations and nonlinear kinetics, even numerical integration of the reaction–diffusion equations is difficult, though recent improvements in computer hardware and software are making this a more feasible procedure.

Nevertheless, numerical integration of reaction–diffusion equations is time-consuming, especially if we have to redo the problem each time we change the initial concentrations. It would be nice if we had at least one system in which we could obtain an analytical solution for the velocity and wave profile in terms of the rate and diffusion constants and the initial concentrations.

In order to do this, we need to be clever and a little bit lucky. The most thorough analytical treatment of wave propagation in any system to date is the study by Showalter and coworkers of the arsenous acid–iodate reaction (Hanna et al., 1982). The key here is that the essential part of the kinetics can be simplified so that it is described by a single variable. If we treat one-dimensional front propagation, the problem can be solved exactly.

When arsenous acid is in stoichiometric excess ($[As(III)]_0 > 3[IO_3^-]_0$), the stoichiometry of the iodate oxidation of arsenous acid is given by

$$IO_3^- + 3H_3AsO_3 + 5I^- \rightarrow 6I^- + 3H_3AsO_4 \tag{6.14}$$

The rate law is

$$d[I^-]/dt = (k_a + k_b[I^-])[I^-][IO_3^-][H^+]^2 \tag{6.15}$$

Figure 6.3a shows the sharp change with time in the concentrations of the iodine-containing species in a stirred batch reaction. Using a more detailed model of the reaction (Hanna et al., 1982), Showalter calculated the concentrations of iodate, iodide, and iodine as functions of time. Figure 6.3b shows the results. Notice that the iodine concentration is amplified 350 times. This is the key to simplifying the reaction because we can now make the approximation that all the iodine is present as either iodide or iodate:

$$[I^-] + [IO_3^-] = [IO_3^-]_0 \tag{6.16}$$

Putting eq. (6.16) into eq. (6.15) enables us to eliminate the iodate concentration:

$$d[I^-]/dt = (k_a + k_b[I^-])[I^-]([IO_3^-]_0 - [I^-])[H^+]^2 \tag{6.17}$$

We substitute $[I^-] = C$, $[H^+] = h$, and $[IO_3^-]_0 = I_0$. We assume that h is a constant (no H^+ is produced or consumed in reaction (6.14)). Now we can write the full partial differential equation for the single dependent variable C:

$$\frac{\partial C}{\partial t} = D\frac{\partial^2 C}{\partial x^2} = (k_1 + k_2 C)(I_0 - C)h^2 C \tag{6.18}$$

We are looking for wave solutions, so we assume a solution of the form

$$C(x, t) = C(x - vt) = C(q), \qquad q = x - vt \tag{6.19}$$

Then,

$$\frac{\partial C}{\partial t} = \frac{dC}{dq}\frac{\partial q}{\partial t} = -vC' \tag{6.20}$$

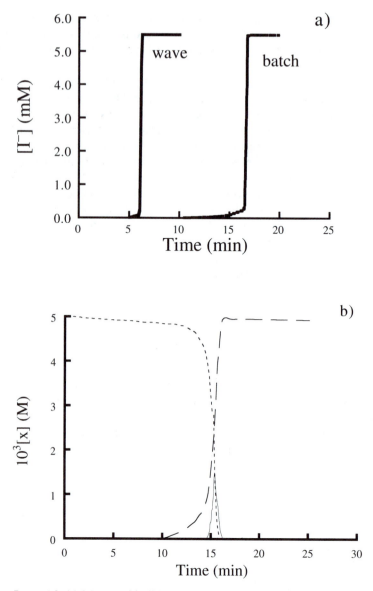

Figure 6.3 (a) Measured iodide concentration as a function of time in a reaction mixture containing excess arsenous acid. Solution composition: $[NaIO_3]_0 = 5.00 \times 10^{-3}$ M, $[H_3AsO_3]_0 = 5.43 \times 10^{-2}$ M, $[H^+]_0 = 7.1 \times 10^{-3}$ M. (b) Simulated concentrations of iodate (short dashes), iodide (long dashes), and 350 × iodine (solid line) under the same conditions. (Adapted from Hanna et al., 1982.)

where $'$ indicates d/dq. Similarly, we find

$$\frac{\partial^2 C}{\partial x^2} = C''$$

and our equation becomes

$$DC'' + vC' + (k_1 + k_2 C)(I_0 - C)h^2 C = 0$$

We now have a single nonlinear ordinary differential equation for the iodide concentration C involving the unknown velocity v. If we can find a solution, we should also get an expression for v in terms of other parameters in the problem. As usual, we have only trial and error or appeal to tradition to resort to in solving the equation. We know physically that at very large and very small x, that is, way behind or way before the wavefront, C should be constant, so we try a function that behaves that way. We guess that

$$C(q) = (\alpha + \beta e^{kq})/(1 + \gamma e^{kq}) \tag{6.21}$$

As $q \to -\infty$, $C \to \alpha$, while as $q \to +\infty$, $C \to \beta/\gamma$. We can already anticipate that one of these values should be zero, while the other should be I_0, but let us see how it comes out. What we will do is evaluate the derivatives, substitute them into the equation, set the coefficients of each power of e^{kq} equal to zero in the resulting equation, and then try to solve for the unknown parameters α, β, k, and v.

We have

$$C' = k(\beta - \gamma\alpha) e^{kq}/(1 + \gamma e^{kq})^2$$
$$C'' = k^2(\beta - \gamma\alpha)(e^{kq} - \gamma e^{2kq})/(1 + \gamma e^{kq})^3$$
$$k_1 + k_2 C = [(k_1 + k_2\alpha) + (k_1\gamma + k_2\beta) e^{kq}]/(1 + \gamma e^{kq})$$

Substituting into eq. (6.21), we obtain

$$\begin{aligned}
\{Dk^2(\beta - \gamma\alpha)(e^{kq} - \gamma e^{2kq}) &+ v(1 + \gamma e^{kq})k(\beta - \gamma\alpha) e^{kq} \\
&+ h^2(\alpha + \beta e^{kq})[(k_1 + k_2\alpha) \\
&+ (k_1\gamma + k_2\beta) e^{kq}][(I_0 - \alpha) \\
&+ (\gamma I_0 - \beta) e^{kq}]\}/(1 + \gamma e^{kq})^3 = 0
\end{aligned} \tag{6.22}$$

Equating like powers, we have

$$1: h^2\alpha(k_1 + k_2\alpha)(I_0 - \alpha) = 0 \tag{6.23a}$$

$$\begin{aligned}
e^{kq}: \ Dk^2(\beta - \alpha\gamma) + vk(\beta - \alpha\gamma) + h^2[\alpha(k_1 + k_2\alpha)(\gamma I_0 - \beta) \\
+ \alpha(k_1\gamma + k_2\beta)(I_0 - \alpha) + \beta(k_1 + k_2\alpha)(I_0 - \alpha)] = 0
\end{aligned} \tag{6.23b}$$

$$\begin{aligned}
e^{2kq}: \ -\gamma Dk^2(\beta - \alpha\gamma) + vk(\beta - \gamma^2\alpha) + h_2[\alpha(k_1\gamma + k_2\beta)(\gamma I_0 - \beta) \\
+ \beta(k_1 + k_2\alpha)(\gamma I_0 - \beta) + \beta(k_1\gamma + k_2\beta)(I_0 - \alpha)] = 0
\end{aligned} \tag{6.23c}$$

$$e^{3kq}: \ h^2\beta(k_1\gamma + k_2\beta)(\gamma I_0 - \beta) = 0 \tag{6.23d}$$

From eq. (6.23a), we conclude that $\alpha = 0$, $-k_1/k_2$, or I_0. From eq. (6.23d), we conclude that $\beta = 0$, $-\gamma k_1/k_2$, or γI_0. If either $\alpha = -k_1/k_2$ or $\beta = -\gamma k_1/k_2$, then C will become negative for large q (γ cannot be negative because then C would be infinite at $q = 0$). If both α and β are negative, then C is constant. If $\alpha = 0$, we cannot have $\beta = 0$ because then $C = 0$ everywhere, which is a solution but not a wave. Similarly, if $\alpha = I_0$, we cannot have $\beta = \gamma I_0$ because then we also have a homogeneous steady state, $C = I_0$. Thus, our two wave solutions correspond to $(\alpha, \beta) = (I_0, 0)$ or $(0, \gamma I_0)$. The first is the one that Showalter obtains. The other corresponds to a wave traveling in the opposite direction—that is, with the same magnitude but different sign of the velocity.

So we substitute in $\alpha = I_0$ and $\beta = 0$. The remaining equations, (6.23b) and (6.23c) become

$$e^{kq}: \quad -Dk^2\gamma I_0 - vk\gamma I_0 + h^2[I_0(k_1 + k^2 I_0)\gamma I_0] = 0 \tag{6.24a}$$

$$e^{2kq}: \quad \gamma^2 Dk^2 I_0 - vk\gamma^2 I_0 + h^2 I_0 k_1 \gamma^2 I_0 = 0 \tag{6.24b}$$

We cannot have $\gamma = 0$ because then $C = \alpha$, constant everywhere. Therefore, we divide eqs. (6.24a) and (6.24b) by γ and γ^2, respectively, to get

$$-Dk^2 I_0 - vk I_0 + h^2 I_0^2(k_1 + k_2 I_0) = 0 \tag{6.25a}$$

$$Dk^2 I_0 - vk I_0 + h^2 I_0^2 k_1 = 0 \tag{6.25b}$$

Subtracting the second equation from the first eliminates v and gives

$$-2Dk^2 I_0 + h^2 I_0^3 k_2 = 0 \tag{6.26a}$$

$$k^2 = (h^2 I_0^2 k_2)/2D \tag{6.26b}$$

Solving eq. (6.25b) for v and using eq. (6.26b) for k gives

$$v = Dk + h^2 I_0 k_1/k = DhI_0(k_2/2D)^{1/2} + h^2 I_0 k_1(2D/k_2)^{1/2}/hI_0$$
$$= (k_2 D/2)^{1/2} hI_0 + (2D/k_2)^{1/2} k_1 h \tag{6.27}$$

By putting in reasonable numbers, $k_1 = 4.5 \times 10^3 \text{ M}^{-3} \text{ s}^{-1}, k_2 = 1.0 \times 10^8 \text{ M}^{-4} \text{ s}^{-1}$, $D = 2 \times 10^{-5} \text{ cm}^2 \text{ s}^{-1}$ and the concentrations from Figure 6.3, we see that the second term on the right-hand side of eq. (6.27) is negligible (dividing by the large k_2). The expression predicts that $v = 1.15 \times 10^{-3} \text{ cm s}^{-1}$, compared with the experimental value of $4.12 \times 10^{-3} \text{ cm s}^{-1}$. This is an impressive result.

6.3 Propagator–Controller Systems

Linear stability analysis provides one, rather abstract, approach to seeing where spatial patterns and waves come from. Another way to look at the problem has been suggested by Fife (1984), whose method is a bit less general but applies to a number of real systems. In Chapter 4, we used phase-plane analysis to examine a general two variable model, eqs. (4.1), from the point of view of temporal oscillations and excitability. Here, we consider the same system, augmented with diffusion terms à la Fife, as the basis for chemical wave generation:

$$\partial u/\partial t = \varepsilon\nabla^2 u + (1/\varepsilon)f(u, v) \tag{6.28a}$$

$$\partial v/\partial t = \varepsilon\nabla^2 v + g(u, v) \tag{6.28b}$$

Fife refers to a set of partial differential equations like eqs. (6.28) as a *propagator–controller system*, for reasons that will become clear later. We recall that the small parameter $\varepsilon \ll 1$ makes u change much more rapidly than v. It also makes the time scale of the chemical change much shorter than that of the diffusion; that is, the diffusion terms are small compared with the chemical rates of change.[1] Note that all the chemistry is in the functions f and g, which are constructed so that neither concentration can grow to infinity. We will want to consider either a one-dimensional geometry (spatial variable x) or a radially symmetric two-dimensional system, like a Petri dish.

If ε is very small, eqs. (6.28) are well approximated nearly everywhere in time and space by:

$$f(u, v) = 0 \tag{6.29a}$$

$$dv/dt = g(u(v), v) \tag{6.29b}$$

where, in eq. (6.29b), we have expressed the fast variable u as an instantaneous function of the slow variable v. If, as we assumed in Chapter 4, the function $f(u, v)$ is s-shaped (i.e., it is something like a cubic equation), then the "function" $u(v)$, which is obtained by solving the algebraic equation (6.29a), is not, mathematically speaking, a function, but rather a relation, since it may be multivalued. A single value of v may, as shown in Figure 6.4, correspond to three values of u, since there are, in general, three solutions to a cubic equation.

As we saw in Chapter 4, the left-hand and right-hand branches of the curve in Figure 6.4 are stable, while the middle one is unstable. We define two functions, $h_+(v)$ and $h_-(v)$, that express, respectively, the functional dependence of u on v on the left and right branches of the curve. If we now imagine a one- or two-dimensional medium in which the chemistry and the diffusion are described by eqs. (6.28), we will find, after any transients have decayed, that either (1) the whole system is in one of the stable states described by $h_+(v)$ and $h_-(v)$, or (2) part of the system is in one of these states and part is in the other state. In case (2), the area that separates regions in the positive and negative states is called a *boundary layer*. For small ε, this layer is extremely narrow—in the limit as $\varepsilon \to 0$, it is a point or a curve in a one- or two-dimensional medium, respectively. In two dimensions, this layer is typically a very thin curved region over which the value of u jumps almost discontinuously between $h_+(v)$ and $h_-(v)$. This is the wavefront.

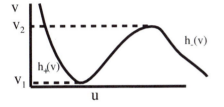

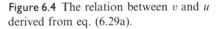

Figure 6.4 The relation between v and u derived from eq. (6.29a).

[1] We could have used different scaling factors for the reaction and diffusion parts, but since the numerical values are not critical, we avoid introducing another parameter.

Fife's analysis, formally called *singular perturbation analysis*, involves re-examining the behavior of the system in the boundary later by stretching the spatial variable there in the direction normal to the layer, by a factor of $1/\varepsilon$, and using the original eqs. (6.28). The analysis shows that, so long as the value of v in the layer lies between the limits v_1 and v_2 shown in Figure 6.4 (i.e., on the unstable branch of the curve), the wavefront will move with a normal velocity $c(v)$. Thus, the wave moves via a jump in the value of the *propagator*, variable u, with a velocity determined by the *controller*, variable v.

If we bear in mind that v, and therefore u as well, continues to evolve everywhere but in the tiny boundary layer according to eq. (6.30),

$$dv/dt = g_{\pm}(v) = g(h_{\pm}(v), v) \tag{6.30}$$

then we see that there are two mechanisms by which waves can arise in this system. Let us sit at a point x and wait for a wavefront to reach us. One way in which this can occur is that the wavefront, which propagates at velocity $c(v)$, arrives at x. This phenomenon, which clearly requires diffusion (since it is only by inclusion of diffusion terms in analyzing the behavior of the boundary layer that we find that the wavefront moves), is known as a *trigger wave*. Trigger waves move with velocities $c(v)$ that depend on both the kinetics of the system and the diffusion coefficients.

A second possibility is that since v at point x continues to change according to eq. (6.30), it reaches the critical value v_1 or v_2 at which transition from one branch to the other occurs. If the system is not uniform, then different points in the system will reach this value at different times and the wavefront will "move" from point to point. We have put the word "move" in quotation marks because this kind of wave, called a *phase wave*, does not require diffusion. In fact, if we were to separate the system into two halves with an impenetrable barrier, a phase wave *would pass right through the barrier*, since it depends only on the independent evolution of v at each point via eq. (6.30). An analog of this phenomenon is the appearance of waves of flashing light on a theater marquee. The velocity of a phase wave is inversely proportional to the magnitude of the gradient in v. The sharper the gradient, the slower the wave, because larger differences in concentration mean longer times between which the critical value of v will be reached at different points. Thus, as the system becomes more homogeneous as a result of diffusion, the phase wave moves faster and faster until it can no longer be detected. Trigger waves are extremely stable and will replace phase waves in a system, if they move rapidly enough to overtake the phase waves, because the trigger wave wipes out the gradient.

6.4 Wave Initiation

The above discussion provides an explanation for how a chemical wave can be generated and propagated in an excitable medium, given a sufficiently large initial perturbation. Where that initial perturbation comes from is a question that still needs to be addressed.

One possibility, of course, is that the experimenter can generate the initial perturbation deliberately. A system can be pushed beyond the threshold of excitability by chemical means (such as adding a drop of acid to a pH-sensitive reaction), by electrical stimulation (Hanna et al., 1982), or by thermally increasing the rate of reaction (e.g., by touching a hot wire to a point in the solution). Waves in photosensitive systems can be initiated by illumination at the appropriate wavelength. In a particularly impressive example of this approach, Kuhnert et al. (1989) used a laser to project various images onto a Petri dish containing a photosensitive ruthenium-catalyzed version of the BZ reaction. The resulting chemical waves preserved the images and even smoothed out defects in some of them. Yamaguchi's group developed this technique further as a means to perform image processing and contrast enhancement (Figure 6.5).

A more interesting question, of course, is how chemical waves can arise spontaneously. A great deal of effort has gone into trying to establish the route or routes by which waves arise in the BZ system. At least two possibilities exist. First, a spontaneous concentration fluctuation at a particular point in the medium may exceed the threshold of excitability and cause initiation of a wave. A second, inhomogeneous, mechanism involves the presence of a *pacemaker* or catalytic site, such as a dust particle. This site has the property that its kinetics are oscillatory, that is, the null clines cross on the middle segment of the s-shaped curve, as in Figure 4.4b, while away from the catalytic site the medium is excitable and the null clines cross on one of the outer branches of the s-shaped curve. When the solution in contact with the catalyst oscillates, it generates a phase wave, which ultimately turns into a trigger wave that can propagate through the excitable bulk

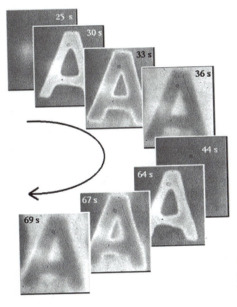

Figure 6.5 Contrast enhancement in a photosensitive BZ system. (Courtesy of T. Yamaguchi.)

of the medium. The next front starts after the oscillatory region has gone through its cycle and the excitable part has returned to its resting state after the refractory period. Thus, we see a series of repeated waves.

Several investigators have attempted to address the issue of whether hetero-geneous pacemakers are necessary for the initiation of waves in the BZ and other systems. By careful microfiltration of all solutions and maintenance of "dust-free" conditions, it is possible to increase greatly, but not to suppress completely, the average time required for the spontaneous appearance of waves in a BZ solution (Pagola and Vidal, 1987). The results suggest that dust particles do serve as pace-makers, but it remains an open question as to whether waves can arise from random concentration fluctuations or whether it was simply too difficult to elim-inate all heterogeneities from the system.

We can apply our knowledge of BZ chemistry to understand qualitatively how waves propagate in this system. If the bromide concentration is kept at a high enough value (e.g., by having a large initial concentration of bromomalonic acid), the system remains in a reduced steady state because the bromide shuts down the autocatalytic oxidation of the catalyst. If the local bromide concentration drops below a critical value, which can be accomplished electrochemically or through bromine adsorption on a dust particle, this part of the medium becomes oscilla-tory. As the medium becomes oxidized, a gradient in $[HBrO_2]$ develops; the autocatalyst will then diffuse into the neighboring excitable medium. The $HBrO_2$ excites the medium, causing it to oxidize and produce more $HBrO_2$ that diffuses into and excites the next region, and so on. After some time, the system slowly becomes reduced as the bromide concentration builds until the reduced steady state is achieved. As Tyson and Keener (1988) point out, the oxidizing wavefront in the BZ system is a trigger wave propagated via the reaction and diffusion of bromous acid, while the reducing waveback is actually a phase wave that follows behind the gradient created by the wavefront. Ferroin is typically used as the catalyst in place of cerium because the oxidized form of ferroin is blue and the reduced form is red, so waves can be readily observed. A recipe for preparing BZ waves is presented in Appendix 1.

6.5 Waves in Two Dimensions

Chemical waves are easiest to think about, and to simulate numerically, in a medium that has one spatial dimension, such as a very narrow tube. The wave-front consists of a single point or a very narrow interval where concentrations jump from one nearly constant level to another. The wave profile in space resem-bles the concentration vs. time profile for the homogeneous reaction. Mathematical treatment of such systems is relatively straightforward, either ana-lytically or numerically. Waves in three dimensions, which we shall discuss briefly later in this chapter, are extremely difficult to obtain experimentally, and far harder to model computationally. The intermediate case, waves in a two-dimen-sional system, is one that occurs frequently in nature, can easily be created in the laboratory by studying thin layers of solution in a Petri dish, and allows a certain amount of mathematical analysis. Moreover, two-dimensional waves provide some of the most beautiful and thought-provoking phenomena in chemistry.

6.5.1 Curvature Effects

One feature that arises in two- (and three-)dimensional waves, but is absent in a single spatial dimension, is the possibility that the wavefront may be curved. We can quantitatively characterize the curvature of any point on a wavefront in terms of its *radius of curvature*—the radius of the circle that best fits the front in the neighborhood of the point in question. If the radius of curvature is R, the *curvature K* is defined as $1/R$, with the further convention that K is positive if the front curves toward the direction of propagation (e.g., an arc of a circle moving toward the center of the circle) and negative if the front curves in the opposite direction. Thus, a plane wave (which has an infinite radius of curvature) has $K = 0$, while uniformly contracting or expanding circular waves of radius r have curvatures $K = 1/r$ and $-1/r$, respectively.

Several investigators have shown theoretically (Zykov, 1980; Keener and Tyson, 1986) and experimentally (Foerster et al., 1988) that the velocity of a curved wavefront depends on the curvature. This dependence is given by the *eikonal equation*:

$$N = c + DK \tag{6.31}$$

where N is the normal velocity of the wavefront, c is the velocity of a plane wave as determined by the controller concentration v at the wavefront, and D is the diffusion coefficient of the propagator species u. For typical chemical systems, the curvature has a significant effect on the wave velocity only when the radius of curvature is smaller than about 100 μm (Tyson and Keener, 1988). The general result of the curvature effect in eq. (6.31) is to smooth out kinks in the wavefront and to turn "wiggly" fronts into smoother ones that resemble planar or circular waves.

6.5.2 Target Patterns and Spirals

In a uniform two-dimensional medium, such as a Petri dish, a wave emanating from a point produces a circular front, since it travels at the same velocity in all directions. If, as described in section 6.4, we have a system that generates repeated waves, we obtain a pattern of concentric circles, known as a *target pattern*. When two or more initiation sites are present, waves can collide, which results in annihilation of the colliding waves and can lead to patterns of considerable complexity. Typical examples in the BZ reaction are shown in Figure 6.6.

If a target pattern is broken by physical disruption of the medium, say, by pushing a pipette through the solution, the broken ends of the wave will curl in and form a pair of steadily rotating spirals as shown in Figure 6.7. With more elaborate manipulation, it is possible to obtain multiarmed spirals (Agladze and Krinsky, 1982) like those in Figure 6.8.

Although target patterns are easier to obtain in a Petri dish, spiral patterns are far more common in complex media, such as living systems, because irregularities

Figure 6.6 Target patterns in the BZ reaction.
(Courtesy of T. Yamaguchi.)

in the medium cause a break in the circular waves that would otherwise form target patterns. Spiral waves have been observed not only in aqueous solution but also in a wide range of excitable systems, such as carbon monoxide oxidation on single crystals of platinum (Jakubith et al., 1990), aggregating slime molds (Siegert and Weijer, 1989), retinal tissue (Gorelova and Bures, 1983), developing frog eggs (Camacho and Lechleiter, 1993) and heart muscle (Davidenko et al., 1992).

A great deal is known about the behavior of spiral waves in excitable media from both the mathematical and the experimental points of view. One feature of particular interest is the center or core from which the spiral waves emanate. The eikonal equation (6.31) allows us to obtain a rough estimate of the size of a spiral core. Chemical waves typically travel at speeds of millimeters per minute, or perhaps $(4\text{--}10) \times 10^{-3}$ cm s^{-1}. Diffusion constants of monomeric species in aqueous solution tend to be around 2×10^{-5} cm^2 s^{-1}. If we consider an expanding circle (which has a negative curvature) on the perimeter of the core, and plug these values for c and D into eq. (6.31), we find that $N = 0$ when

Figure 6.7 Spiral patterns in the BZ system.
(Courtesy of T. Yamaguchi.)

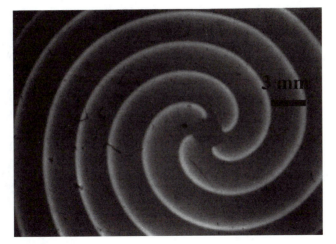

Figure 6.8 A multiarmed spiral in the BZ system. (Courtesy of S. Müller.)

$K = -c/D = -(2\text{–}5) \times 10^2 \text{ cm}^{-1}$, or at a radius $r = 1/K$ of 20–50 μm. By using sophisticated spectroscopic imaging techniques, it is possible to measure the concentration profile of ferroin in the BZ reaction with a spatial resolution of a few micrometers every few seconds (Müller et al., 1985). By superimposing a set of these images, it is possible to construct a picture of the spiral wave's dynamics. What one finds, in agreement with our estimate from the eikonal equation, is a circular core—with a radius of about 30 μm—in which the concentration changes very little with time and from which the spiral waves emerge (Figure 6.9).

If you watch a single spiral wave carefully over a long period of time, you may see something surprising. Under certain circumstances, the core of the spiral moves around! But it does not just move randomly, as it might if the core were a dust particle executing Brownian motion; instead the core, and the spiral it carries with it, can undergo a variety of complex but regular motions, ranging from circles to flowerlike patterns to epicycles resembling those in Ptolemaic astronomy. This motion of the spiral core is known as *meander*, and was first discovered in numerical experiments on the Oregonator model (Winfree, 1972; Jahnke et al., 1989) and then confirmed by experiments on the ferroin-catalyzed BZ reaction (Plesser et al., 1990). Further theoretical work (see Winfree, 1991, for a review) shows that meander is a general property of spiral waves in excitable media, and that one can understand the progression from one meander pattern to another in much the same way that one can view bifurcations among temporal states of a system as parameters are varied. Figure 6.10 shows a "meander diagram" depicting the various patterns that occur as two parameters, ε and f, are varied in a two-variable version of the Oregonator (Keener and Tyson, 1986).

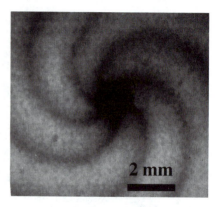

Figure 6.9 Spectroscopic imaging of a spiral wave core. (Courtesy of S. Müller.)

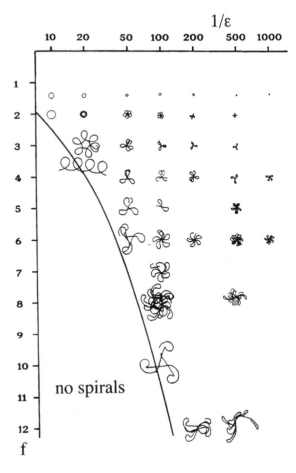

Figure 6.10 Meander patterns in the Keener–Tyson Oregonator model. Below the solid line, spiral waves fail to propagate. (Reprinted with permission from Jahnke, W.; Winfree, A. T. 1991. "A Survey of the Spiral-Wave Behaviors in the Oregonator Model," *Int. J. Bifurc. Chaos 1*, 445–466. © 1991 World Scientific Publishing.)

6.6 Three-Dimensional Waves

Just as moving from one to two spatial dimensions greatly increases the variety of patterns and phenomena possible in reaction–diffusion systems, an expansion into the third dimension opens up a vast range of potential new behavior. Experimentally, this behavior has thus far proved difficult to observe in chemical systems like the BZ reaction, largely because of a variety of irksome but important practical issues—bubble formation, oxygen effects, convection, etc. Some success has been achieved by running the BZ reaction in a sealed vessel (Welsh and Gomatam, 1990) or in a gel (Jahnke et al., 1988), but the most promising media for studying three-dimensional waves experimentally may well be biological tissues. Three-dimensional waves of electrical and chemical activity appear to play crucial roles in the functioning of such organs as the heart and the brain.

In contrast to the rather primitive experimental state of knowledge about three-dimensional waves, considerably more is known from a theoretical point of view. The simplest kinds of waves, other than plane waves, are the three-dimensional extensions of target patterns and spiral waves: spherical waves and scroll waves, respectively. One can imagine drawing out the point at the center of a spiral wave into a filament that serves as the axis of a scroll wave. If the filament remains straight, we have a simple scroll wave that looks like a jelly roll. We can imagine, however, bending the filament into various shapes. A u-shaped filament would give a pair of counter-rotating spirals. Far more elaborate patterns are possible if we are willing to twist the filament or even tie it into knots. An example of a scroll ring, where the filament forms a circle, is shown in Figure 6.11. Winfree and Strogatz (1983a, 1983b, 1983c, 1984) performed an extensive geometric and topological analysis of the sorts of waves that are possible in three dimensions. They derived rules, relating such quantities as the number of linked rings and the number of twists in a scroll wave pattern, that must be satisfied if the wave is to be compatible with the requirements of physical chemistry. Whether some of the exotic patterns discussed in the theoretical literature can be realized experimentally is open to question, though one might imagine being able to create the requisite initial conditions by laser perturbation of a photosensitive excitable chemical system.

6.7 Kinetics with a Ruler

A model that reproduces the homogeneous dynamics of a chemical reaction should, when combined with the appropriate diffusion coefficients, also correctly predict front velocities and front profiles as functions of concentrations. The ideal case is a system like the arsenous acid–iodate reaction described in section 6.2, where we have exact expressions for the velocity and concentration profile of the wave. However, one can use experiments on wave behavior to measure rate constants and test mechanisms even in cases where the complexity of the kinetics permits only numerical integration of the rate equations.

The BZ reaction provides an intermediate case, where one can approximate the kinetics of wave propagation by focusing on the autocatalytic reaction that produces bromous acid. This step has the rate equation

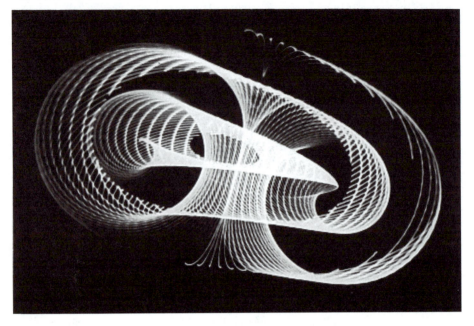

Figure 6.11 Simulation of a scroll ring. (Reprinted from Winfree, A. T.; Strogatz, S. H. 1983. "Singular Filaments Organize Chemical Waves in Three Dimensions. II. Twisted Waves," *Physica 9D*, 65–80, with kind permission from Elsevier Science, The Netherlands.)

$$\frac{d[\text{HBrO}_2]}{dt} = k[\text{H}^+][\text{BrO}_3^-][\text{HBrO}_2] \tag{6.32}$$

We would like to determine the rate constant k in eq. (6.32), but measuring it directly is difficult because HBrO_2 is not a stable species that we can simply mix with bromate. In the BZ system, formation of HBrO_2 is accompanied by oxidation of the metal ion catalyst. Therefore, as a traveling wave of oxidation propagates, we can follow the color change in the ferroin. Tyson (1985) suggests that the wave velocity in the BZ reaction is given in terms of the diffusion coefficient of HBrO_2, and the rate constant of eq. (6.32) by

$$\text{Velocity} = 2(k[\text{H}^+][\text{BrO}_3^-]D)^{1/2} \tag{6.33}$$

A plot of the square of the velocity vs. $[\text{H}^+][\text{BrO}_3^-]$ (Figure 6.12) should give a straight line whose slope equals $4kD$.

We see that although the velocity is quite constant in a given experiment, the straight line in the part b plot in Figure 6.12, from which k is determined, is less than perfect, which means that the rate constant determined in this way has considerable uncertainty. The error results, in part, from experimental difficulties in making the measurements and, in part, from the approximation inherent in

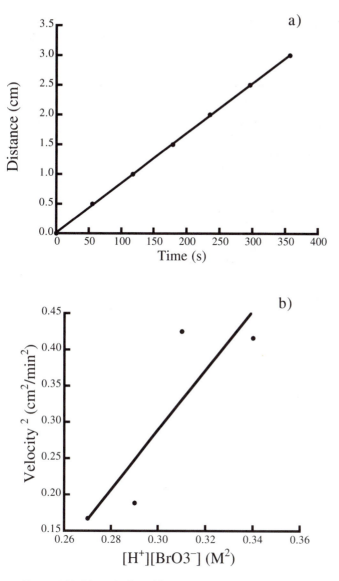

Figure 6.12 The velocity of ferroin-catalyzed bromate oxidation waves as a function of the acidic bromate concentration. (a) Experimental data in a single experiment; the velocity is the slope of the straight line. (b) The slope, 4.1 cm^2 min^{-2} M^{-2}, equals $4kD$, according to eq. (6.33). Using 2×10^{-5} cm^2 s^{-1} as the diffusion coefficient of HBrO$_2$ yields a rate constant of 14 M^{-2} s^{-1}.

using eq. (6.33). (For more information on carrying out this experiment in an undergraduate lab, see Appendix 2.) More accurate expressions for the velocity dependence of the wave can be obtained by utilizing the full mechanism instead of the single rate-determining step of eq. (6.33) (Eager et al., 1994).

Another system that has been studied in this manner is the oxidation of iron(II) by nitric acid. In Chapter 3, we discussed a model that reproduces well the clock reaction experiments. Pojman et al. (1991a) numerically integrated the partial differential equations describing front propagation in that system. Figure 6.13 shows the experimental and calculated dependence of the front velocity on the initial iron(II) and nitrate concentrations using two variations of the model. Both

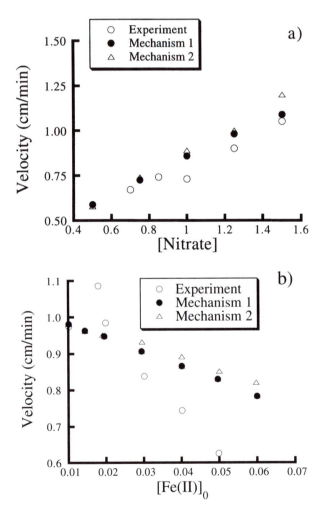

Figure 6.13 Comparison of the velocity of front propagation in the iron(II)–nitric acid system on (a) the initial nitrate concentration and (b) the initial Fe(II) concentration. (Adapted from Pojman et al., 1991a.)

give good agreement with the nitrate dependence but fail for the iron(II) dependence. We thus conclude that the model requires further refinement.

If a reaction is highly exothermic, it may be difficult to perform isothermal homogeneous experiments because the heat cannot be removed fast enough. By running the front in a very narrow capillary tube, which maximizes the surface area through which heat can escape, it may be possible to obtain an isothermal front whose velocity can be compared with numerical simulations.

In general, front experiments provide another tool, often a very sensitive one, for testing proposed mechanisms.

6.8 Chemical Waves and Patterns in Open Systems

Most of the wave experiments that we have been describing were carried out in relatively simple closed systems: the one-dimensional (i.e., thin tube) and two-dimensional (i.e., Petri dish) equivalents of the beaker used in the earliest experiments on temporal (zero-dimensional) oscillations. We have seen, in Chapter 3, the enormous impact that the introduction of open flow reactors had on the study of temporal behavior. Similar considerations apply to the study of spatial waves and patterns. The patterns that we have discussed, however fascinating, can only be transiently present in a closed system like a Petri dish. As in the case of a stirred beaker, an unstirred closed system is destined by thermodynamics to reach a stable, homogeneous, unchanging state of equilibrium. In this section, we describe briefly some recent experimental advances that have made possible the study of spatial waves and patterns in open systems.

The most straightforward approach is to try to build a Petri dish that is simultaneously fed with fresh reactants, so that it is an open system, and free of stirring or other fluid motion, so that patterns and waves can develop. A moment's thought should convince the reader that this is not going to be easy! Any kind of flow into a liquid is going to disperse the contents of the dish in such a way as to break up any patterns that might tend to form. Swinney and collaborators (Tam et al., 1988a) developed a very clever solution to this problem. First, they carried out the reaction in a chemically inert polyacrylamide gel to prevent fluid motion. Then, they introduced the reactants into a CSTR whose contents reached one surface of the gel through an array of glass capillary tubes. The reactants then spread through the gel by diffusion. Because of the experimental arrangement, dubbed a continuously fed unstirred reactor (CFUR) and shown in Figure 3.9, all the mass transport between the CSTR and the gel occurs perpendicular to the gel surface and, hence, cannot influence patterns forming parallel to that surface.

A similar reactor, in which the reaction occurs in an annular ring of gel, gives rise to a "chemical pinwheel," a stable array of waves that rotate around the ring (Nosztziczius et al., 1987). The pattern is initiated by introducing a concentrated solution of acidic bromate at one point in the gel. This pacemaker generates pairs of counter-rotating waves. If, say, one annihilates the clockwise-rotating waves by introducing in the neighborhood of the pacemaker a flow of distilled water to dilute the reagents, then, after a few waves are generated, the pacemaker and annihilator flows can be removed and we are left with a series of counterclockwise-rotating waves.

A rather different approach to generating spatiotemporal patterns utilizes a Couette reactor (Grutzner et al., 1988; Tam et al., 1988a; Ouyang et al., 1989). A Couette reactor, illustrated schematically in Figure 3.10, consists of two coaxial cylinders. The reactants are introduced into the annular gap between the cylinders from two CSTRs situated at the ends of the cylinders. The inner cylinder rotates, creating a turbulent flow that mixes the solution and gives rise to a one-dimensional diffusion along the axis. All species diffuse in this direction with a diffusion constant D that is determined by the rotation rate and can be much higher than typical molecular diffusion rates (as much as 1 cm^2 s^{-1}). As D is tuned by varying the rotation rate or as the composition of the reactants is varied, the system goes through a series of bifurcations among a set of spatiotemporal states. A sequence of states in the chlorite–iodide–malonic acid reaction is shown in Figure 6.14.

Finally, we observe that waves and patterns may be obtained in an effectively open system if the reaction occurs in one medium (usually solid) that is immersed in another medium (usually liquid) that provides, via transport, a continuous supply of fresh reactants. Several attractive systems of this type have been developed. Maselko and Showalter studied the BZ reaction on the surface of beads of cation-exchange resin on which the ferroin catalyst had been immobilized (Maselko and Showalter, 1989; Maselko et al., 1989). When the beads were placed in a solution containing sulfuric acid, bromate, and malonic acid, spiral waves formed. With small (100–1000 μm), closely packed beads, the wave moved from bead to bead. With larger, isolated beads, it was possible to observe the movement of a spiral wave on the surface of a single bead.

Two other media that offer promise for further studies of this sort are membranes and sol–gel glasses. Showalter and colleagues (Winston et al., 1991) studied the behavior of waves on the surface of a Nafion membrane impregnated

single steady front   single oscillating front   three steady fronts

Figure 6.14 Patterns observed in the chlorite–iodide–malonic acid reaction in a Couette reactor. The CSTR composition, flow rate, and rotation rate are held fixed, except for chlorite composition in one CSTR, which serves as the bifurcation parameter. In each frame, the abscissa represents the position along the reactor and the ordinate represents time. The dark color results from the presence of the starch–triiodide complex. (Adapted from Ouyang et al., 1991.)

with ferroin and bathed in a solution of BZ reagent, while Epstein et al. (1992) observed target and spiral patterns at the surface of a silicate glass that had been prepared by incorporating ferroin and was then bathed in a BZ solution without catalyst. Figure 6.15 shows an example of the patterns observed. Note how the wave can "jump across" a narrow break in the glass via the diffusion of material through the solution.

6.9 Chemical Waves in Nonuniform Media

We conclude this chapter with some examples of more "exotic" behaviors of which chemical waves are capable, behaviors that may suggest new and important applications for these fascinating phenomena. The phenomena in which we are interested involve media that are not uniform, but whose properties are different in different regions of space. Catalytic, biological, and geological systems are examples of nonuniform media that may give rise to chemical waves.

In Figure 6.15, we observe a circular wave moving from a glass through a region of aqueous solution and then into another piece of glass. How, if at all, might the properties of the wave, particularly the direction and the speed, be expected to change as it moves from one region to another? If we were talking about an electromagnetic or optical wave, the answer would be clear. Snell's law

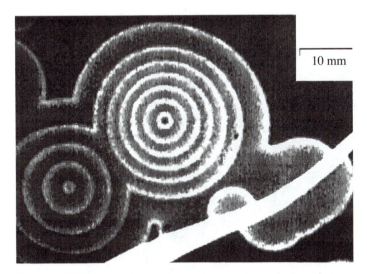

Figure 6.15 Target patterns observed in a piece of glass that was formed by heating a gel of $Si(OCH_3)_4$ containing ferroin and then immersed in a solution containing sulfuric acid, malonic acid, and sodium bromate. (Reprinted from Epstein, I. R.; Lengyel, I.; Kádár, S.; Kagan, M.; Yokoyama, M. 1992. "New Systems for Pattern Formation Studies," *Physica A, 188*, 26–33, with kind permission from Elsevier Science, The Netherlands.)

tells us how the speed and the direction of a wave passing from one medium to another are related:

$$\sin \varphi_i / \sin \varphi_r = v_i/v_r \tag{6.34}$$

where φ_i and φ_r are the angles of incidence and refraction, respectively, and v_i and v_r are the speeds of the incident and transmitted (refracted) waves, respectively. Elementary geometric optics also tells us that a plane wave cannot move from one medium to another when $v_i < v_r$ unless the angle of incidence is less than the critical angle φ_{cr}:

$$\varphi < \varphi_{cr} = \sin^{-1}(v_i/v_r) \tag{6.35}$$

If no refraction is possible, then specular reflection occurs, with the angle of incidence equal to the angle of reflection.

Do chemical waves obey these laws of optics? Zhabotinsky et al. (1993) analyzed the behavior of waves in the ferroin-catalyzed BZ reaction in a medium consisting of rectangular pieces of gel having different thicknesses and oxygen concentrations, and, hence, different propagation speeds for the chemical wave. They found, both theoretically and experimentally, that Snell's law and the condition (6.35) for refraction to take place when the wave moves into a medium with higher speed both hold for chemical waves. Reflection, however, behaves quite differently in chemical and electromagnetic waves. Reflection of chemical waves can occur only if the incident wave originates in the medium with lower speed and if $\varphi_i > \varphi_{cr}$. In this case, reflection is not specular; for chemical waves, the angle of reflection always equals the critical angle. Figures 6.16 and 6.17 show examples of refracted and reflected waves in the BZ reaction.

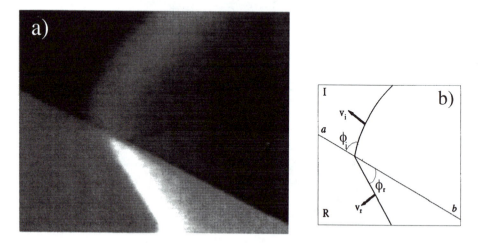

Figure 6.16 Refraction of a chemical wave. (a) Experimental observation in a silica gel layer open to air, with stepwise variation in thickness. (b) Schematic representation of part a: ab, boundary between regions I (thickness = 2.15 mm) and R (thickness = 1 mm); v_i and v_r, wave vectors of incident and refracted waves, respectively; φ_i and φ_r, angles of incidence and refraction, respectively. Measured speeds: $v_r/v_i = 0.43 \pm 0.002$, measured angles: $\sin \varphi_r / \sin \varphi_i = 0.45 \pm 0.05$. (Adapted from Zhabotinsky et al., 1993.)

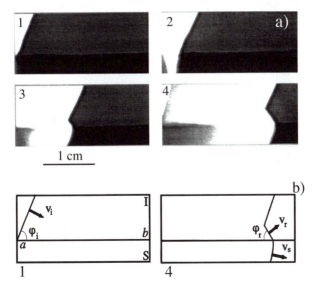

Figure 6.17 Reflection of a chemical wave. Light-colored wave is incident from the upper left. (a) Experimental observation in a polyacrylamide gel layer open to air, with stepwise variation in thickness. Interval between frames 1 and 2, 75 s; 2 and 3, 130 s; 3 and 4, 200 s. (b) Schematic representation of frames 1 and 4 of part a. ab, boundary between regions I (thickness $= 0.45$ mm) and S (thickness $= 0.75$ mm); v_i, v_s, and v_r, wave vectors of incident, secondary circular, and reflected waves, respectively; φ_i and φ_r, angles of incidence and reflection, respectively. Measured speeds: $v_i/v_s = 0.76 \pm 0.03$ ($\varphi_{cr} = 49.5 \pm 2°$); measured angles: $\varphi_i = 67.4 \pm 3°$ ($\varphi_r = 54.0 \pm 2°$). (Adapted from Zhabotinsky et al., 1993.)

Showalter and coworkers have utilized the properties of chemical wave propagation in two very clever ways. First, they use waves in the BZ reaction to solve a classic problem in mathematics and computer science—finding the shortest path from one point to another in a complex geometrical arrangement (Steinbock et al., 1995b). They produce a series of labyrinths, like that shown in Figure 6.18, by taking vinyl–acrylic membranes saturated with BZ reaction mixture and cutting out sections of the membrane to make a set of barriers. Under conditions where spontaneous initiation of waves is exceedingly rare, they initiate a wave at one point on the membrane by contact with a silver wire. They then follow the progress of the wave by image analysis of monochromatic light reflected from the membrane. At each junction in the labyrinth, the wave splits, ultimately reaching all points in the maze that are connected to the point of initiation. By superimposing successive snapshots, it is possible to calculate the velocity field of the

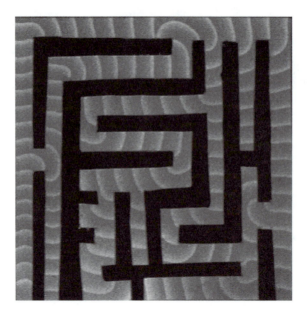

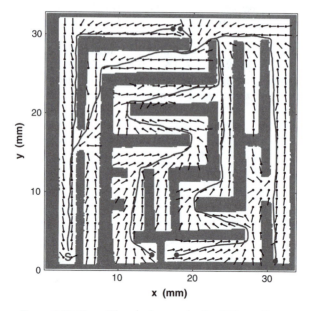

Figure 6.18 Top: Chemical wave in the BZ reaction propagating through a membrane labyrinth. The wave was initiated at the lower left corner of the 3.2 × 3.2 cm maze. Bottom: Velocity field showing local propagation direction obtained from analysis of fifty images obtained at 50-s intervals. Lines show shortest paths between five points and the starting point S. (Reprinted with permission from Steinbock, O.; Tóth, A.; Showalter, K. 1995. "Navigating Complex Labyrinths: Optimal Paths from Chemical Waves," *Science 267*, 868–871. © 1995 American Association for the Advancement of Science.)

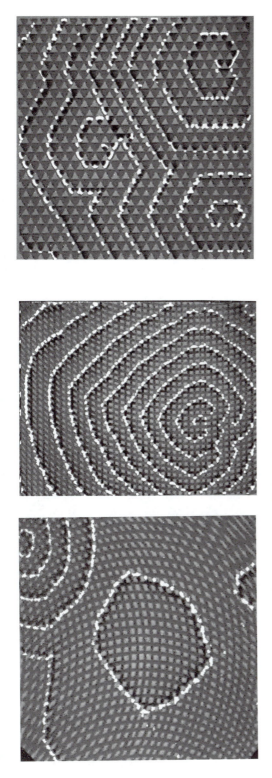

Figure 6.19 Propagating waves on BZ membranes with predetermined geometric patterns of catalyst. Catalyst loaded and unloaded regions are dark and light gray, respectively. Positions of waves, shown with white for wavefront and black for waveback, are obtained by subtraction of successive video images. Top: Triangular cells giving rise to hexagonal pattern. Middle: Superimposed linear and circular grids giving rise to diamond-shaped patterns. Bottom: Super-imposed circular grids giving rise to pentagonal patterns. (Reprinted with permission from Steinbock, O.; Kettunen, P.; Showalter, K. 1995. "Anisotropy and Spiral Organizing Centers in Patterned Excitable Media," *Science 269*, 1857–1860. © 1995 American Association for the Advancement of Science.)

wave and, by following the direction of the field, to calculate the shortest path between the starting point and any other point in the medium.

In a second experimental tour de force, Steinbock et al. (1995a) use an ink-jet printer to lay down patterns of bathoferroin catalyst on a polysulfone membrane. Exposure to a gel containing BZ solution results in a variety of intricate propagating wave patterns whose geometry is determined, though not always in obvious ways, by the geometric arrangement of the catalyst. Some examples are shown in Figure 6.19.

7

Computational Tools

It is fair to say that the field of nonlinear chemical dynamics would not be where it is today, and perhaps it would not exist at all, without fast digital computers. As we saw in Chapter 1, the numerical simulation of the essential behavior of the BZ reaction (Edelson et al., 1975) did much both to support the FKN mechanism and to make credible the idea that chemical oscillators could be understood without invoking any new principles of chemical kinetics. In 1975, solving those differential equations challenged the most advanced machines of the day, yet the computers used then were less powerful than many of today's home computers! Despite the present widespread availability of computing power, there remain many challenging computational problems in nonlinear dynamics, and even seemingly simple equations can be difficult to solve or maybe even lead to spurious results. In this chapter, we will look at some of the most widely used computational techniques, try to provide a rudimentary understanding of how the methods work (and how they can fail!), and list some of the tools that are available.

There are several reasons for utilizing the techniques described in this chapter:

1. For a complicated system, it is generally not possible to measure all of the rate constants in a proposed mechanism. One way to estimate the remaining parameters is to simulate numerically the behavior of the system, varying the unknown rate constants until the model satisfactorily reproduces the experimental behavior.
2. If a mechanism, which may consist of dozens of elementary chemical reactions, is valid, then it should reproduce the observed dynamical behavior.[1] Proposed

---

[1] Of course, giving agreement with a finite set of observed data does not guarantee the validity of a mechanism. Several mechanisms or sets of rate constants may yield satisfactory agreement with a given set of observations.

mechanisms are most commonly tested by integrating the corresponding rate equations numerically and comparing the results with the experimental time series, or by comparing the results of many such simulations with different initial conditions (or of a numerical continuation study) to the experimental phase diagram.

3. Numerical results can act as a guide to further experiments. The real reason for developing models is not to interpolate between our experimental observations but to extrapolate into unknown realms. A model can direct the experimentalist into the proper region of parameter space, saving time and money.

We shall not attempt to provide an expert-level description of numerical methods, but rather try to convey to readers enough information so that they can appreciate the key principles, understand the literature, and make rational choices as to what methods to employ in their own work.

7.1 Types of Modeling

Numerical integration (sometimes referred to as solving or simulation) of differential equations, ordinary or partial, involves using a computer to obtain an approximate and discrete (in time and/or space) solution. In chemical kinetics, these differential equations are typically the rate laws that describe the time evolution of the system. One obtains results for the mean concentrations, without any information about the (typically very small) fluctuations that are inevitably present. Continuation and sensitivity analysis techniques enable one to extrapolate from a numerically obtained solution at one set of parameters (e.g., rate constants or initial concentrations) to the behavior of the system at other parameter values, without having to carry out a full numerical integration each time the parameters are changed. Other approaches, sometimes referred to collectively as stochastic methods (Gardiner, 1990), can provide data about fluctuations, but these require considerably more computational labor and are often impractical for models that include more than a few variables.

Two other approaches treat a spatially distributed system as consisting of a grid or lattice. The cellular automaton technique looks at the numbers of particles, or values of some other variables, in small regions of space that interact by set rules that specify the chemistry. It is a deterministic and essentially macroscopic approach that is especially useful for studying excitable media. Lattice gas automata are mesoscopic (between microscopic and macroscopic). Like their cousins, the cellular automata, they use a fixed grid, but differ in that individual particles can move and react through probabilistic rules, making it possible to study fluctuations.

7.2 Stochastic Methods

In the *master equation* approach (Gardiner, 1990), the quantity whose time evolution one seeks to calculate is not the concentration vector $C(t)$, but rather the probability distribution $P(C, t)$ that the system will have concentrations C at time *t*. The kinetics are represented by a birth and death process that takes into account the change in the number of particles of each species in each elementary process. For example, if the reaction of interest is $A + B \rightarrow C$, then the master equation for the numbers of particles, a, b, and c, looks like

$$P(a, b, c, t + 1) = q[(a + 1)(b + 1)P(a + 1, b + 1, c - 1, t) - abP(a, b, c, t)] \quad (7.1)$$

where q is proportional to the rate constant of the reaction, the first term on the right-hand side of eq. (7.1) gives the probability per unit time that A and B react to form the cth particle of C when there are $a + 1$ and $b + 1$ particles of each, and the second term gives the probability of reaction when there are a, b, and c particles, respectively. If we can solve for the distribution P (a formidable task), then we can use it to calculate not only the mean concentrations, such as

$$\langle a(t) \rangle = \int aP(a, b, c, t) \, da \, db \, dc \quad (7.2)$$

but also the moments of the distribution, which give us the fluctuations; for example, $(\langle a(t)^2 \rangle - \langle a(t) \rangle^2)$ is the mean-square fluctuation in the number of particles of A. When the total number of particles is very large, the distribution function is sharply peaked about the mean, and the relative size of the fluctuations is nearly always very small.

Another approach, in which the stochastic nature of chemical systems plays a major role, is the *Monte Carlo method* (Gillespie, 1977; Broeck and Nicolis, 1993; Erneux and Nicolis, 1993; Nicolis and Nicolis, 1993; Nicolis and Gaspard, 1994; Vlachos, 1995). Here, the rate constants and numbers of particles are used to calculate the probabilities that each of the elementary steps will occur during some time interval, which is chosen small enough so that no more than one elementary reaction is likely to occur. To simulate the time evolution, one chooses a set of initial particle numbers. A random number between 0 and 1 is then selected and used, together with the resulting probabilities, to decide which reaction, if any, takes place during the time step. For example, suppose that step 1 has probability 0.1 and step 2 has probability 0.2 during the given time interval. Then, if the random number is less than 0.1, step 1 occurs; if it lies between 0.1 and 0.3, step 2 takes place; and if it is greater than 0.3, there is no reaction during this time step. The particle numbers and probabilities are updated after each time step, and a new random number is chosen. A simulation of this sort gives rise to very large fluctuations, which can be reduced by repeating the simulation with the same initial conditions but a different sequence of random numbers and then averaging the results over the set of runs.

In the *molecular dynamics* approach, one represents the system by a set of particles randomly distributed in space and then solves Newton's equations for the motion of each particle in the system. Particles react according to the rules assumed for the kinetics when they approach each other sufficiently closely. While the particle motion is deterministic, a stochastic aspect is present in the choice of the initial particle positions and velocities. Again, an average over "replicate" runs will reduce the statistical fluctuations. By necessity, only a very small system over a very short time (picoseconds) can be simulated (Kawczynski and Gorecki, 1992). We will not treat stochastic methods in any further detail here, but, instead, we will emphasize approaches that are readily available in "canned" packages or are likely to become so in the near future.

7.3 Solving Ordinary Differential Equations

The most frequently encountered numerical problem in nonlinear chemical dynamics is that of solving a set of ordinary, nonlinear, first-order, coupled, autonomous differential equations, such as those describing the BZ reaction. We hope you understand by now what nonlinear means, but let us comment on the other modifiers. The equations are ordinary because they do not contain partial derivatives (we consider partial differential equations in the next section), first order because the highest derivative is the first derivative, and coupled because the time derivative of one species depends on the concentrations of other species. In the absence of time-dependent external forcing, rate equations are autonomous, meaning that time does not appear explicitly on the right-hand side.

We have seen earlier that only the most trivial of chemically relevant differential equations can be solved analytically. Even seemingly simple expressions derived from mass action kinetics can be challenging to solve numerically, and a great deal of effort has gone into developing reliable methods. Here, we treat the case of a single dependent variable, but the techniques described are easily extended to treat arbitrarily many coupled equations by treating the variable A as a vector rather than a scalar. To illustrate some of the issues, we shall use Euler's method, probably the simplest technique for numerical integration, to obtain an approximate solution of the equation:

$$dA/dt = f(A) \tag{7.3}$$

where $f(A)$ might be, for example, $-kA$.

7.3.1 Euler's Method

Numerical integration is essentially an exercise in curve fitting. One uses information about the function to be integrated—its value and possibly the values of its derivatives—at certain points, to approximate the function over some interval by functions that we know how to integrate, usually polynomials. In Euler's method, we utilize only the value of the function and its derivative at a single point and make the simplest approximation, namely, that the function can be represented by a straight line through the point, with slope equal to the derivative. Thus, in eq. (7.3), if we are given the initial value $A = A_0$ at $t = 0$, the value at $t = h$ is

$$A(h) = A_0 + hf(A_0) \tag{7.4}$$

This amounts to approximating the instantaneous derivative as the slope of a line [in this case, the tangent to the curve $A(t)$ at $t = 0$] over a finite interval, as shown in Figure 7.1.

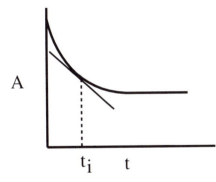

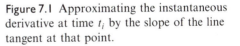

Figure 7.1 Approximating the instantaneous derivative at time t_i by the slope of the line tangent at that point.

If we divide the entire interval over which we seek to solve eq. (7.3) into equal subintervals of length h, we can "march along," using the value of A obtained at one step to give us an approximation of A at the next step:

$$A(nh) = A((n-1)h) + hf(A((n-1)h)) \tag{7.5}$$

As Figure 7.1 suggests, this approach is less than perfect. It gives the correct answer only if $f(A)$ is a constant—that is, only if $A(t)$ is a straight line—but this, of course, is just what does not happen in nonlinear chemical dynamics. For a complex, nonmonotonic function like the one depicted in Figure 7.2, this problem can be quite serious. Even the qualitative, much less the quantitative, features of the solution may not be approximated correctly.

An obvious way to deal with this problem is to make the time increment h smaller. If one looks at a small enough interval, any reasonably behaved function is well approximated by a straight line. One can, in fact, show that as h approaches zero, the error due to approximating the function by a straight line also approaches zero, and that this error is proportional to h. This fact, and its simplicity, make Euler's method seem attractive, and it is, *if* the functions to be integrated change relatively slowly. However, if we look at realistic examples in nonlinear chemical dynamics, we soon discover that choosing h small enough to give good accuracy means that the amount of computer time consumed becomes very large. This problem is exacerbated by the fact that we are typically integrating not a single equation but a set of many coupled equations.

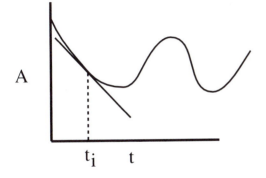

Figure 7.2 Overshooting becomes a very serious problem for complex functions unless a very small Δt is used.

7.3.2 Multistep Methods

What can we do to improve the efficiency and accuracy of our numerical integration technique? Perhaps the most important advance we can make is to use more of the information available to us about the function that we want to integrate. If we knew its value and maybe its derivative, not just at one point but at several, we could fit something more flexible than a line, perhaps a quadratic or a cubic, that would pass through the known points with the correct slopes and give us a better shot at getting the ups and downs of the function right. Methods that use information from several previous points or steps in the integration are known as *multistep methods*. They play a key role in the numerical solution of nonlinear, ordinary differential equations (Gear, 1971). A general formula for a k-step method has the form

$$A(nh) = \sum_{i=1}^{k} [a_i A((n-i)h) + hb_i A'((n-i)h)] \qquad (7.6)$$

where the prime denotes differentiation with respect to t and the coefficients a_i and b_i are chosen to give the best fit to the data (Gear, 1971). In the Euler method, for example, we have $k = 1$ and $a_1 = b_1 = 1$.

We can see that taking $k > 1$ should give us considerably more flexibility in fitting the function to be integrated, and therefore allow us to obtain the same accuracy with larger values of h and, hence, less computing time. There is a potential problem, however. If our method requires data at two or more points and we only have information at a single point initially, how can we ever get started? One solution is to use a single-step method, like Euler, to get started, until we have calculated the function and its derivative at the first k points. We can then use the more sophisticated k-step method for the remainder of the integration.

7.3.3 Predictor–Corrector Techniques

Methods such as those in eqs. (7.5) and (7.6) are known as *explicit* methods because they enable us to calculate the value of the function at the next point from values that are already known. They are really extrapolations. We utilize information about a function on an interval to estimate its value outside the interval. Extrapolation is inherently more error-prone than interpolation. There is no assurance that $A(nh)$ calculated solely from values of the function for $t \le (n-1)h$ will bear any resemblance to the true value that satisfies the differential equation (7.3). Another group of techniques are known as *implicit* methods. They give the value of the function not only in terms of what has happened at previous points, but also in terms of the behavior at the point of interest. For example, if we knew the derivative f at $t = nh$, we should be able to get a better approximation to $A(nh)$ by using the average derivative in the interval $(n-1)h < t < nh$ rather than simply the derivative at $t = (n-1)h$. Thus a better formula than eq. (7.5) might be

$$A(nh) = A((n-1)h) + h[f(A((n-1)h)) + f(A(nh))]/2 \qquad (7.7)$$

Since we do not actually know $f(A(nh))$, except as a function of $A(nh)$, eq. (7.7) constitutes a nonlinear equation for $A(nh)$ that we need to solve numerically for

$A(nh)$. This extra effort may well be rewarded by the increased accuracy of the solution obtained. A more effective approach, which is frequently used in modern software packages, is to combine an explicit method with an implicit one to give a *predictor–corrector* technique. An explicit formula like eq. (7.5) or eq. (7.6) is used, not to give a final result, but to provide an initial guess of the solution, $A_p(nh)$, at $t = nh$. We can then insert this value of A into the differential equation (7.4) to obtain the predicted value of the derivative $A'_p(nh) = f(A_p(nh))$. These values of A and A' are then used in the implicit method to yield the final, corrected solution $A_c(nh)$. A secondary advantage of predictor–corrector methods is that the difference between A_p and A_c provides a reliable estimate of the error, typically (Acton, 1970),

$$e \approx |A_c - A_p|/20 \qquad (7.8)$$

7.3.4 Sources of Error

As eq. (7.8) implies, any numerical integration method has errors associated with it. These arise from several sources. For a particular differential equation, one kind of error may be more significant than the others, but a satisfactory method will ensure that none of these gets too large. We list here the sources of error in numerical integration. The reader may wish to consult more specialized texts, (e.g., Gear, 1971; Celia and Gray, 1992) for more details.

1. *Discretization error* arises from treating the problem on a grid of points ($t = 0, h, 2h, \ldots$) rather than as a continuous problem on an interval. In effect, we have replaced the original differential equation by a difference equation whose exact solution at the grid points may not coincide with that of the differential equation, even if we are able to solve the difference equation to perfect accuracy. In most methods, this error is proportional to the step size h to some positive integer power. Thus, the discretization error can be reduced by decreasing the step size.

2. *Truncation error* results from using a function like a polynomial to approximate the true solution. To reduce truncation error, we could use more sophisticated functions—for example, higher order polynomials—which require more information. Multistep methods do this, but the additional accuracy gained by increasing the order of the method drops off very quickly beyond about $k = 4$.

3. *Roundoff error* occurs because all computers have a finite precision. At each step of the computation, a small error is introduced. While these errors would tend to average out if the solution were linear, generally they accumulate in an unpredictable fashion, though one can make statistical estimates of their magnitude. Decreasing h to lower the discretization error means that there are more steps and hence a larger roundoff error. There is no way around this problem except to use a machine with a greater precision. For most machines and nonlinear problems, double precision is the minimum requirement to obtain satisfactory results.

The above factors contribute a *local error* at each step of the calculation. As we move from point to point, the local errors combine to produce a *global error*. Good integration packages estimate the amount of error as the calculation moves along. The user specifies a tolerance that defines the magnitude of local error that is acceptable. If this tolerance cannot be met, h is automatically decreased; if satisfactory results still cannot be obtained, another method (e.g., one with more steps) may be tried.

It is also important that a method be *stable*. The concept here is very similar to the notion of stability for the steady state and other asymptotic solutions of our differential equations. If we introduce a small change (error) into the numerical solution at a grid point, we would like that error to decay rather than to grow. Since this behavior will depend on the equation being solved, stability is typically defined with respect to a standard equation like $dA/dt = -kA$. Unstable methods are generally unsuitable for numerical computation, because even small roundoff errors can cause the calculated solution to "explode," that is, to get so large that a floating overflow point error occurs.

7.3.5 Stiffness

One problem that is encountered in solving the equations that describe real chemical oscillators is *stiffness*. Stiffness is a property of differential equations whose solutions contain two or more very different time scales.

The concept can be illustrated with a simple one-variable example (Byrne and Hindmarsh, 1987):

$$dy/dt = -10^3[y - e^{-t}] - e^{-t} \qquad y(0) = 0 \qquad (7.9)$$

The solution of eq. (7.9) is

$$y(t) = \exp(-t) - \exp(-10^3 t) \qquad (7.10)$$

We see that the solution—eq. (7.10)—consists of two parts: one that changes on a time scale of about one time unit, the other varying on a time scale of about 0.001 time units. The initial sharp rise in the function, seen in Figure 7.3, results from the rapid decay of the second term on the right-hand side of eq. (7.10); during this time, the first term remains nearly constant. To follow this behavior accurately, a numerical method must use a step size considerably smaller than the 0.001-unit time scale on which the second component of the solution decays. After

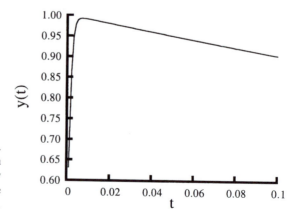

Figure 7.3 The solution of eq. (7.9)—that is, eq. (7.10)—in which y changes very rapidly and then changes much more slowly.

a few thousandths of a time unit, the fast component has decayed essentially to zero, and the behavior of the solution is basically that of the slow component, $\exp(-t)$. In order to follow the solution until it approaches its final value of zero, we must run our numerical integration out to several time units. Employing the small step size that we needed to describe the fast component during the period dominated by the slow component is exceedingly inefficient computationally. It is this situation of having a component that changes very rapidly, necessitating a small time step, and a second component that varies much more slowly, requiring a lengthy interval of integration, that constitutes stiffness.

Again, let us consider the Oregonator. Although the equations are deceptively simple, they are difficult to integrate. Figure 7.4 shows the numerical solution obtained on Matlab, a commercial simulation package. Notice that intervals in which the bromous acid concentration is changing only slightly alternate with very rapid increases and decreases. In order to capture the fast changes, a very short time step must be used. If this step were employed throughout the calculation, the computer time required would be prohibitive.

Getting around the problem of stiffness requires rather sophisticated numerical methods. These techniques typically employ multistep methods and, most importantly, a way of dynamically varying the time step, and in some cases the integration method, according to the rate of change of the solution. Thus, small step sizes are used when the solution is changing rapidly, but these can be increased significantly when the solution is changing relatively slowly. In the latter case, additional savings can be obtained, without loss of accuracy, by invoking the corrector step in a predictor–corrector scheme once every few time steps rather than at each step. The major determinant of the integration time is the number of evaluations of the function to be integrated that must be made during the calculation. One of the first, and still among the most frequently employed, software packages for

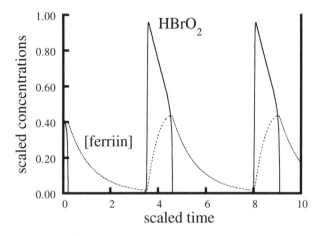

Figure 7.4 Simulation of the Oregonator, using the formulation of Keener and Tyson (1986). Notice how rapidly the $HBrO_2$ concentration changes after long periods of very low concentration.

integrating stiff ordinary differential equations (ODEs) was developed by Hindmarsh (1974) and is called GEAR. There are now many such codes available, and it is certainly not necessary for the chemist to write his or her own. Some of the features of these programs and others have been reviewed by several authors (Byrne and Hindmarsh, 1987; Brenan et al., 1996; Hairer et al., 1996), who point out that different packages have different strengths and weaknesses, so that for particularly difficult computational problems, it may be worth trying more than one program.

7.3.6 A Final Caution

Errors in integration can lead not only to quantitative changes in the details of the solution, but also to spurious, qualitatively incorrect, dynamical behavior. Figure 7.5 shows an example (Citri and Epstein, 1988) in which chaotic behavior is found for loose error tolerances in numerical integration of a model for the bromate–chlorite–iodide reaction, but periodic behavior occurs when the tolerance is more stringent. Inaccurate numerical integration can turn steady states into oscillatory ones and vice versa. Chaotic systems are particularly tricky, and there is considerable interest among mathematicians and numerical analysts, at present, in the question of "shadowing," that is, to what extent numerical solutions to chaotic

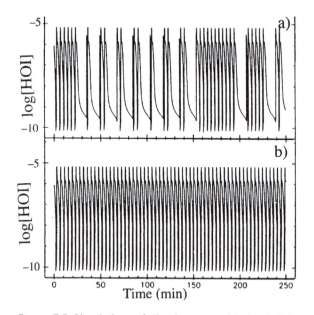

Figure 7.5 Simulation of the bromate–chlorite–iodide reaction. (a) With an error tolerance of 10^{-4}, chaotic behavior is obtained. (b) Tightening the tolerance to 10^{-8} leads to the correct result, periodic behavior. (Adapted from Citri and Epstein, 1988.)

differential equations approach, or "shadow," the true solutions as the numerically introduced error is decreased (Hammel et al., 1987).

7.4 Solving Partial Differential Equations

Solving the ordinary differential equations that describe homogeneous systems can be tough, but treating chemical waves and other spatial phenomena requires us to go still further and to solve partial differential equations, specifically the reaction–diffusion equations we saw in the previous chapter. Partial differential equations present a far more formidable challenge. While considerable advances have been made during the last decade in hardware and software for solving partial differential equations, their numerical solution is by no means routine, particularly if the problem involves more than one spatial dimension. If we can be satisfied with solving the problem on a line, then most solvers on a good personal computer or workstation can do the job for a model with as many as five to ten dependent variables. Two- and three-dimensional problems require more powerful computers and are difficult to solve for realistic models that comprise more than two or three concentration variables. The state of the art continues to advance rapidly, and implementing numerical techniques for partial differential equations on parallel computers offers considerable promise.

To get some insight into how such systems are handled numerically, we will consider the one-dimensional diffusion equation:

$$\partial C / \partial t = \partial^2 C / \partial x^2 \tag{7.11}$$

In addition to discretizing the time, as we did for ordinary differential equations, we divide the space into cells of length δx, labeled $i + 1$, etc., as shown in Figure 7.6. We convert the differential equation into a difference equation:

$$C(i, t + \delta t) = C(i, t) + [C(i - 1, t) - 2C(i, t) + C(i + 1, t)] \, \delta t / \delta x^2 \tag{7.12}$$

The general approach of representing both the time and the space derivatives by difference quotients is called the *finite difference method*. The particular scheme in eq. (7.12) is an example of the *method of lines*, in which one discretizes the diffusion term and then uses methods of the sort we have just discussed to solve

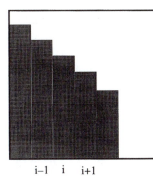

i–1   i   i+1

Figure 7.6 The space discretization used in the finite difference method, eq. (7.12). Heights of bars represent magnitude of concentration C.

the ODE problem across a whole line in space at each time step. The method works well for the diffusion equation. It can be applied to reaction–diffusion problems by decoupling the chemistry from the mass transport, that is, the equations describing the chemical reactions can be integrated with respect to time in each cell between diffusion steps. Although this simple approach works well for problems with small gradients (e.g., simulating fronts in the iodate–arsenous acid system), it fails for more complicated problems. Stiffness, in particular, is an even more serious problem in partial than in ordinary differential equations.

Stiffness can occur not only from the rate terms but also because of the spatial gradients. Figure 7.7 shows a simulation of a thermal front with a sharp temperature gradient using an adaptive grid, meaning that the space discretization δx changes with the gradient of the dependent variable(s). If a uniform grid were used on this problem, with the resolution necessary to represent accurately the gradient in the front, then an enormous amount of computing time would be wasted on the regions ahead of and behind the front. Ahead of the front, no significant reaction occurs, and behind the front, the reaction is complete.

There are many other approaches with different advantages, and several are available in standard packages like IMSL (Visual Numerics, Houston, Tex.). Their consideration is beyond the scope of this book. We refer the reader to works on numerical methods, such as Celia and Gray (1992).

7.5 Continuation Methods

Numerically integrating the differential equations in a model is analogous to performing an experiment. Just like an experiment, the system has to be run long enough to be sure that asymptotic behavior has been reached. Then, the initial or feed concentrations or the flow rate can be varied and the integration performed again. In this way, a numerical bifurcation diagram showing the parameter regions in which different types of behavior occur can be constructed and compared with the experimental one. The amplitude and period of oscillations for different parameter values can also be compared with the experimental results. Whether this process of obtaining results over a range of parameters is carried out experimentally or numerically, it is likely to be very time-consuming.

In many cases, the most important information we seek is the location of the bifurcations—the points, lines or surfaces in the parameter space where qualitative changes in behavior occur. For example, to construct a bifurcation diagram, all we really need to know is where the bifurcations occur. Also, away from the bifurcations, interpolation of values for such quantities as steady-state concentrations or periods of oscillation is usually quite reliable. Numerical continuation methods provide a particularly efficient way to obtain a great deal of information about a system of differential equations at relatively small computational cost. Continuation methods (Parker, 1989; Seydel, 1994) start from a known solution, such as a steady state, and essentially extrapolate that solution as a function of a parameter. By keeping track of the eigenvalues of the Jacobian matrix during this procedure, one is able to determine where changes in sign, which signal steady-

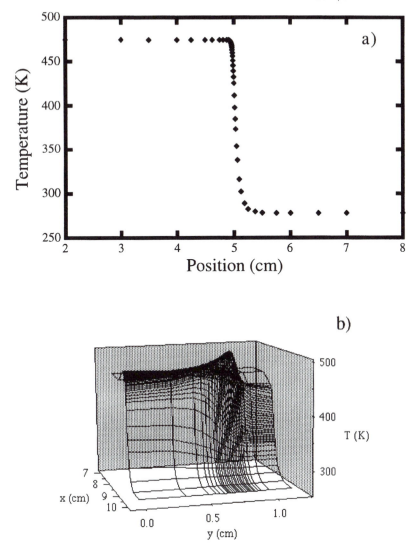

Figure 7.7 (a) One-dimensional simulation of a thermal front with adaptive grid. (b) Two-dimensional simulation of a thermal front. (Plots courtesy of S. Solovyov.)

state bifurcations, arise. Two widely used packages that implement numerical continuation methods are AUTO (Doedel et al., 1991) and CONT (Marek and Schreiber, 1991). These programs trace out curves of steady states and follow the properties of limit cycles by using techniques similar to the predictor–corrector approach described above to improve their computational efficiency. An example is shown in Figure 7.8.

To afford the reader some insight into how numerical continuation methods work, we offer two examples. The problems we tackle are trivial; they are easily

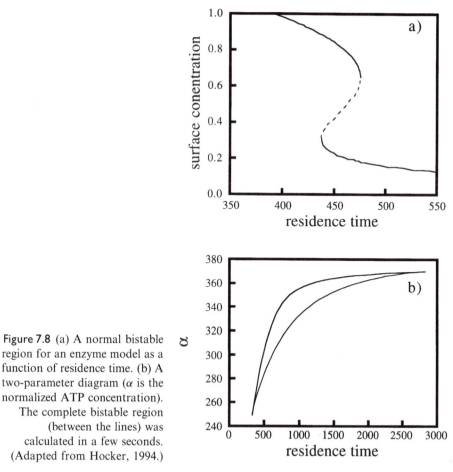

Figure 7.8 (a) A normal bistable region for an enzyme model as a function of residence time. (b) A two-parameter diagram (α is the normalized ATP concentration). The complete bistable region (between the lines) was calculated in a few seconds. (Adapted from Hocker, 1994.)

solved exactly by analytical methods. Nonetheless, they should provide some intuitive feel for the essential features of the continuation approach.

Imagine that you have been in an automobile accident and have sustained a head injury that has wiped out your knowledge of the quadratic formula but has left your mathematical abilities otherwise unimpaired.[2] You are interested in the properties of the equation

$$x^2 + bx + 4 = 0 \tag{7.13}$$

over a wide range of the parameter b. You would particularly like to know whether there are any bifurcation points at which the number of solutions changes. Of course, you could simply put in values of b and solve eq. (7.13) by trial and error (or, since you still know calculus, by something more sophisticated, like Newton's method). An alternative approach, which reflects the philosophy of

---

[2]Stranger results of brain lesions are well documented in the neurological literature.

the continuation method, is first to solve the equation for one value of b, for example, by trial and error. Next, we differentiate eq. (7.13) with respect to the parameter of interest. We will then use our knowledge of dx/db to extrapolate from our solution for a single value of b—we will take $b = 6$, which has the solutions $x = -5.23607$ and $x = -0.76393$—to other values of b, including, we hope, a prediction of any bifurcations that occur. Differentiating eq. (7.13) gives

$$2x\, dx/db + x + b\, dx/db = 0 \tag{7.14}$$

Solving for dx/db, we have

$$dx/db = -x/(b + 2x) \tag{7.15}$$

Equation (7.15) predicts that the derivative of the solution with respect to the parameter will approach infinity as $b + 2x$ approaches zero. Intuitively, as illustrated in Figure 7.9, it makes sense that an infinitely rapid change in the derivative would correspond to a qualitative change in the character of the solution. It is indeed the case that an infinite derivative indicates a bifurcation. We shall see shortly how to follow x as b changes, but a bit of algebra will tell us immediately what we want to know. We can rearrange eq. (7.13) in the form

$$b = -(4 + x^2)/x \tag{7.16}$$

so that

$$b + 2x = (x^2 - 4)/x \tag{7.17}$$

Setting $b + 2x = 0$ in eq. (7.17), yields our bifurcation condition, $x^2 - 4 = 0$, or $x = \pm 2$, which corresponds to $b = \pm 4$. This result is, of course, easily obtained from the amnesia-blocked quadratic formula, which gives the discriminant of eq. (7.13) as $b^2 - 16$. When $|b| > 4$, the discriminant is positive, and there are two real roots. When $b = \pm 4$, the two roots merge, that is, the equation can be written as $(x \pm 2)^2 = 0$, If $|b| < 4$, the discriminant is negative and there are no real roots. In other words, for $|b| > 4$, there are two real roots; at the bifurcation points, $b = \pm 4$, there is a single real root, $x = \mp 2$; and between the bifurcation points, no real roots exist.

The continuation approach thus reveals the existence and location of the two bifurcation points. It can also enable us to follow the roots as we vary b away from its initial value of 6. We view x as a function of b and write

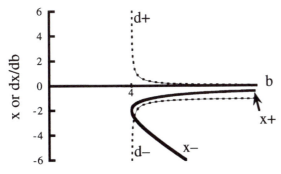

Figure 7.9 Plots of the roots, x_+ and x_-, of eq. (7.13) and of the derivatives $d_\pm = dx_\pm/db$ of eq. (7.15) for $b \ge 0$. The plot for $b \le 0$ is a mirror image.

Table 7.1 Exact Roots x_e and Numerical Continuation Approximations x_a for eq. (7.13)

b	x_{1e}	x_{1a}	x_{2e}	x_{2a}
6	−5.23607	−5.23607	−0.76393	−0.76393
5.9	−5.11852	−5.11899	−0.78101	−0.78148
5.8	−5.00000	−5.00097	−0.80000	−0.79950
5.7	−4.88039	−4.88190	−0.81961	−0.81854
5.6	−4.75959	−4.76168	−0.84041	−0.83870
5.5	−4.63746	−4.64016	−0.86254	−0.86010

$$x(b + \Delta b) = x(b) + dx/db|_b \, \Delta b \tag{7.18}$$

In Table 7.1, we compare the exact roots of eq. (7.13) calculated from the quadratic formula with the values obtained by repeated application of eqs. (7.15) and (7.18) with a Δb of 0.1. Even better results can be obtained by using a smaller increment in the parameter.

Our other example consists of a differential equation. Again, we choose an example whose exact solution we know:

$$dx/dt = -kx \tag{7.19}$$

Rather than obtain the exact solution $x(t) = x(0) \exp(-kt)$, let us assume that we have solved eq. (7.19) numerically for some specified value of k, and now we want to know what the steady-state solution looks like for other values of k and, in particular, whether bifurcations occur.

Although this problem is trivial for eq. (7.19), it is still instructive, since we can easily imagine extrapolating our approach to more interesting equations or systems. The algebraic equation for the steady state(s) of eq. (7.19) is

$$-kx = 0 \tag{7.20}$$

Differentiating eq. (7.20) with respect to the parameter k gives

$$-x - k \, dx/dk = 0 \tag{7.21}$$

or

$$dx/dk = -x/k \tag{7.22}$$

Since the solution to eq. (7.20) is $x_{ss} = 0$, the equation that determines how the steady state changes with k becomes $dx/dk = 0$; that is, the steady-state value is independent of k, except possibly at $k = 0$, where eq. (7.22) becomes indeterminate. In fact, at $k = 0$, a bifurcation occurs, because the solution changes from a decaying to a growing exponential, and the steady state $x = 0$ loses its stability. This fact can also be derived by following the Jacobian, $\partial(dx/dt)/\partial x = -k$, which changes sign at $k = 0$, again suggesting that a bifurcation occurs at this parameter value.

The procedures outlined here are, in a more sophisticated form, those that go into a numerical continuation package. Starting from a reference solution at some chosen set of parameters, the derivatives of the steady-state or oscillatory solutions and of the Jacobian are estimated with respect to the parameters of interest (analytically if possible, numerically if necessary), and these solutions and the Jacobian are then extrapolated until a bifurcation is located. If the derivatives become very large, it may be advisable to calculate a new reference solution along the way to decrease the numerical error.

7.6 Sensitivity Analysis

Another valuable technique, which is similar to numerical bifurcation analysis in that it attempts to estimate how solutions change with variations in the parameters, is *sensitivity analysis*. Sensitivity analysis (Hwang et al., 1978), however, is directed not at locating bifurcations or qualitative changes in the behavior of asymptotic solutions, but rather at determining the smaller, smooth changes that occur as parameters are varied far from bifurcation points. In evaluating a model with many parameters, it is useful to know what aspects of the system's behavior are most sensitive to each of the parameters. In this way, we can design experiments that will enable us to fit parameters in the most effective fashion. If one or more parameters have no measurable effect on any observable quantity, then Occam's razor should be applied to shave those parameters out of the model.

More precisely, sensitivity analysis concerns itself with quantities of the form

$$s_{ij} = \partial x_i / \partial p_j \qquad (7.23)$$

where x_i is a dependent variable, like a concentration or some quantity that depends on the concentrations, and p_j is a parameter, like a rate constant or an initial concentration.

To see how the analysis might go, we return to our trivial but illustrative example, eq. (7.19). This model has only a single variable, x, and two parameters, the rate constant k and the initial concentration x_0. There are thus two sensitivities that we could calculate:

$$y = \partial x / \partial k, \qquad z = \partial x / \partial x_0 \qquad (7.24)$$

Differentiating eq. (7.19) with respect to the parameters gives us two equations for the time evolution of the sensitivities:

$$dy/dt = -x - ky \qquad (7.25)$$
$$dz/dt = -kz \qquad (7.26)$$

To solve eq. (7.25), we substitute the solution $x(t) = x_0 \exp(-kt)$ for x, and obtain

$$y(t) = (y_0 - x_0 t) \exp(-kt) \qquad (7.27)$$

What is y_0? At $t = 0$, with fixed x_0, the sensitivity y is zero, since changing the parameter k has no effect on the value of x. Thus, $y_0 = 0$, and the sensitivity to the parameter k is given by

$$y(t) = -x_0 t \exp(-kt) \qquad (7.28)$$

What does eq. (7.28) tell us? First, it shows that varying the parameter k has negligible effect on the solution at either very short or very long times

$(y(0) = y(\infty) = 0)$. By differentiating eq. (7.28) with respect to t, we can find that the maximum sensitivity occurs at $t = 1/k$, so that in a set of experiments starting from the same initial concentration, the rate constant is most accurately determined by making careful measurements in the neighborhood of the relaxation time $\tau = 1/k$.

Solving eq. (7.26) is somewhat easier. The definition of z [eq. (7.24)] implies that $z_0 = 1$, and the differential equation has the solution

$$z(t) = \exp(-kt) \qquad (7.29)$$

This equation simply states the obvious result that changing the initial concentration rescales the entire solution, preserving its exponential form, with the maximum effect occurring at $t = 0$ and the effect disappearing as $t \to \infty$.

While the results obtained for our example may seem self-evident, sensitivity analysis of multivariable, multiparameter models can be quite illuminating. It is possible, for example, to compute the dependence of the period of oscillation, one of the most easily measured experimental quantities, on the parameters of a model like the Oregonator (Edelson and Thomas, 1981). The equations for the sensitivities, like eqs. (7.25) and (7.26), have the same form as the equations for the concentrations and can easily be appended to the original model, yielding a set of equations that can be solved simultaneously, even for a model with as many as ninety-eight elementary steps (Edelson and Allara, 1980).

Another promising approach (Vajda et al., 1985; Turányi et al., 1993) employs sensitivity analysis to answer such questions as when the steady-state approximation is valid and what steps in a complex mechanism can safely be eliminated or considered to be at equilibrium in attempting to construct a simpler model. In general, sensitivity analysis shows great potential for facilitating the intelligent study of larger systems, a potential that has largely remained unexploited.

7.7 Cellular Automata

Partial differential equations represent one approach (a computationally intensive one) to simulating reaction–diffusion phenomena. An alternative, more approximate, but often less expensive and more intuitive technique employs *cellular automata*. A cellular automaton consists of an array of cells and a set of rules by which the state of a cell changes at each discrete time step. The state of the cell can, for example, represent the numbers of particles or concentrations of species in that cell, and the rules, which depend on the current state of the cell and its neighbors, can be chosen to mimic diffusion and chemical reaction.

We illustrate here how cellular automata may be used to model excitable media. Recall that an excitable medium has two important properties. First, there is an *excitation threshold*, below which the system remains unexcited. Second, after it has been excited, the medium cannot be excited again for a fixed time, called the *refractory period*. We consider a two-dimensional lattice, or grid, in which the state of each cell in the next time step depends on its current state and on the states of its neighbors. For simplicity, we choose a rule for which only nearest neighbors have an effect.

This simplest grid is rectangular and operates as follows: If a cell is quiescent (in the off state), it remains so unless one or more of its nearest neighbors is excited (in the on state). Then, in the next time step, it becomes excited. A cell remains excited for only one time period. It then enters the refractory state and remains there for a specified number of time periods. As we see in Figure 7.10, if a

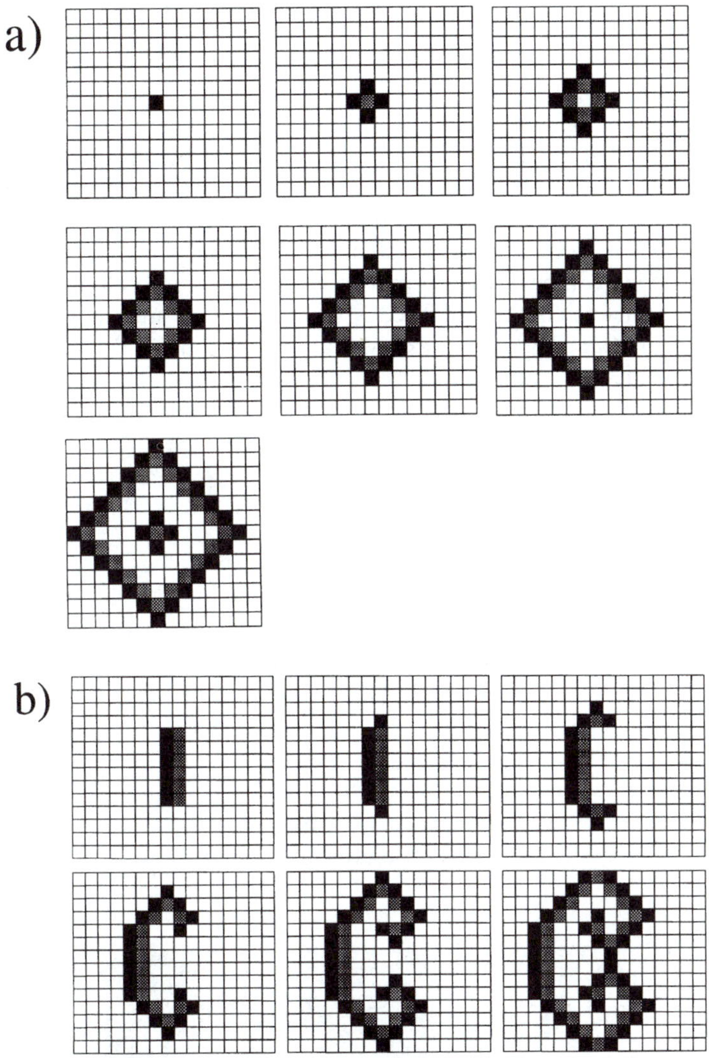

Figure 7.10 Evolution of a simple cellular automaton with a rectangular grid. White indicates excitable cells, black indicates excited cells, and gray indicates a cell in its refractory period. (a) Appearance of target patterns when the central cell is a pacemaker. (b) Appearance of a spiral when a line of cells is excited adjacent to cells that are refractory.

single cell starts in the excited state, then a ring of excitation expands out from it. If the initial excited cell is made to oscillate, then target patterns will develop. If a line of excited cells is initially adjacent to refractory cells, then a spiral pattern forms.

More sophisticated versions can be implemented that have many states and/or many variables, corresponding to increasingly more detailed specification of the actual reaction–diffusion system (Figure 7.11) (Markus and Hess, 1990). How does this method capture the essence of an excitable reaction–diffusion system? Recall the BZ system, where a perturbation in $HBrO_2$ concentration propagates because diffusion carries it into a neighboring region. When the concentration exceeds the excitation threshold, the system rapidly oxidizes and produces more $HBrO_2$, which can diffuse and repeat the process. Once excited, the solution cannot be re-excited until the bromide concentration builds up and the metal ion catalyst is reduced. The automaton can be made more realistic by making a cell's evolution depend on more than its nearest neighbors; the range of interaction can be chosen according to the actual diffusion constant. The excitation and recovery processes can be described by two variables in each cell (Gerhardt et al., 1990) and parameters chosen that specify the number of time steps spent in each state so as to imitate the kinetics and the known thresholds for propagation of excitation and de-excitation waves. The amazing fact is this simple, gamelike approach captures the essence of a very complex excitable medium and qualitatively reproduces target patterns and spiral waves.

7.8 Lattice Gas Automata

An interesting variant of the cellular automaton approach is the *lattice gas automaton*. Despite the name, lattice gas automata can be used to simulate reactions in condensed as well as gaseous media. They were introduced initially as an alternative to partial differential equations (PDE) for modeling complex problems in fluid flow (Hardy et al., 1976) and have been adapted to include the effects of chemical reaction as well (Dab et al., 1990). Lattice gas automata are similar to cellular automata in that they employ a lattice, but differ because they focus on the motions of individual particles along the lattice and because they can account

Figure 7.11 (a) A plane wave and spiral in two dimensions. (b) Cross section of a three-dimensional medium. (Reprinted with permission from Markus, M.; Hess, B. 1990. "Isotropic Cellular Automaton for Modelling Excitable Media," *Nature 347*, 56–58. © 1990 Macmillan Magazines Limited.)

for fluctuations. Although they can, in principle, be applied to systems with any number of variables (Kapral et al., 1992), they are most effective in dealing with systems in which the number of variable concentrations is small and the effects of the solvent and the constant concentrations can be incorporated into the lattice.

Each lattice site is occupied by a certain (small) number of particles. The particles represent the chemical species of interest and the lattice represents the solvent. Time is discrete, and each particle has a velocity that causes it to move, at each time step, onto one of the neighboring lattice points. The magnitude of the velocity is same for all the particles, but they differ in direction. Figure 7.12 shows the time evolution for particles. The particles are initially randomly distributed on the lattice with a random distribution of velocities (directions). At each time step, particles move to neighboring points according to their velocities. After each move, the velocity distribution is randomized, that is, the particle velocities undergo rotation, simulating the effect of elastic collisions with the solvent.

Chemical reactions are treated by a rule that reflects the mass action kinetics. For example, for a bimolecular reaction of $A + B \rightarrow C$, when an A particle and a B particle occupy the same lattice point, they have a probability of being converted to a C particle that depends on the rate constant of the reaction and the time scale. The procedure used to determine what reactions occur in a given interval is identical to that described in section 7.2 for the Monte Carlo method.

The lattice gas approach has two advantages over the traditional method of integrating PDEs. The simulations can be performed very rapidly because the only floating point operations are the choices of the random numbers. The actual integration only involves updating integer matrices by a simple rule. Second, through the randomization processes, the stochastic nature of the chemical reactions is accounted for, allowing the study of fluctuations in pattern formation. Figure 7.12 shows particles on a lattice gas automaton. The results of a lattice gas simulation that gives rise to Turing patterns are illustrated in Figure 7.13.

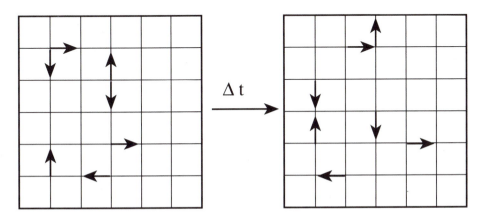

Figure 7.12 The dynamics of particles in a lattice gas automaton on a square lattice. Particles are randomly distributed on the lattice with a random distribution of velocities. After a time step Δt, particles move to neighboring lattice points. (Adapted from Boon et al., 1996.)

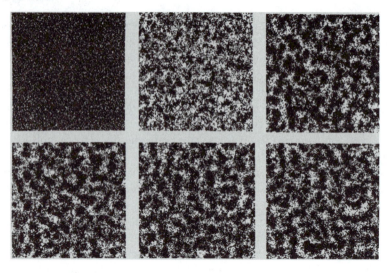

Figure 7.13 Evolution of Turing patterns using a lattice gas automaton. (Adapted from Boon et al., 1996).

7.9 Conclusions

Numerical methods for solving ordinary and partial differential equations constitute an entire discipline. We have only scratched the surface here, concentrating on general principles and on some pitfalls that may await the unwary user. In recent years, a number of powerful and user-friendly packages for performing these calculations have become widely available, so a grasp of the broad outlines of the subject should enable readers to perform these calculations with some confidence.

The numerical bifurcation and sensitivity analysis techniques have not been as widely employed, but they offer valuable adjuncts to more conventional numerical simulation methods. They make it possible both to explore efficiently large ranges of parameter space and to test and condense models by examining how parameter variations affect experimentally measurable quantities. Both techniques implement, in a systematic fashion, approaches that one would naturally want to employ even if only by trial and error. Testing models and designing new experiments clearly benefit from a knowledge of the effects of changing model parameters; these techniques simply ask and answer those questions in very efficient ways.

Cellular automata and the related lattice gas automaton models provide less quantitative, more cost-effective, and often more intuitive alternatives to differential equation models. Although they can be constructed to include considerable detail, they are best used to provide qualitative insights into the behavior of carefully designed models of complex reaction–diffusion systems.

PART II

SPECIAL TOPICS

8

Complex Oscillations and Chaos

After studying the first seven chapters of this book, the reader may have come to the conclusion that a chemical reaction that exhibits periodic oscillation with a single maximum and a single minimum must be at or near the apex of the pyramid of dynamical complexity. In the words of the song that is sung at the Jewish Passover celebration, the Seder, "Dayenu" (It would have been enough). But nature always has more to offer, and simple periodic oscillation is only the beginning of the story. In this chapter, we will investigate more complex modes of temporal oscillation, including both periodic behavior (in which each cycle can have several maxima and minima in the concentrations) and aperiodic behavior, or chaos (in which no set of concentrations is ever exactly repeated, but the system nonetheless behaves deterministically).

8.1 Complex Oscillations

Most people who study periodic behavior deal with linear oscillators and there-fore tend to think of oscillations as sinusoidal. Chemical oscillators are, as we have seen, decidedly nonlinear, and their waveforms can depart quite drastically from being sinusoidal. Even after accepting that chemical oscillations can look as nonsinusoidal as the relaxation oscillations shown in Figure 4.4, our intuition may still resist the notion that a single period of oscillation might contain two, three, or perhaps twenty-three, maxima and minima. As an example, consider the behavior shown in Figure 8.1, where the potential of a bromide-selective electrode

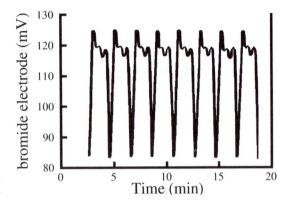

Figure 8.1 Mixed-mode oscillations in the BZ reaction in a CSTR. (Adapted from Hudson et al., 1979.)

in the BZ reaction in a CSTR shows one large and two small extrema in each cycle of oscillation.

8.1.1 Mixed-Mode Oscillations and the Slow-Manifold Picture

The oscillations shown in Figure 8.1 are of the *mixed-mode* type, in which each period contains a mixture of large-amplitude and small-amplitude peaks. Mixed-mode oscillations are perhaps the most commonly occurring form of complex oscillations in chemical systems. In order to develop some intuitive feel for how such behavior might arise, we employ a picture based on *slow manifolds* and utilized by a variety of authors (Boissonade, 1976; Rössler, 1976; Rinzel, 1987; Barkley, 1988) to analyze mixed-mode oscillations and other forms of complex dynamical behavior. The analysis rests on the schematic diagram shown in Figure 8.2.

The system must have at least three independent variables. The folded surface labeled as the slow manifold is a region on which the system moves relatively slowly through the concentration space. If the system is in a state that does not lie on this manifold, rapid processes cause it to move swiftly onto that surface. The steady state, which lies on the lower sheet of the manifold, is stable to perturbations transverse to the manifold, but is unstable along the manifold. Thus, if the

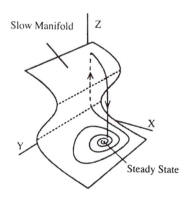

Figure 8.2 Schematic representation of a slow manifold in the concentration space of a chemical reaction that exhibits mixed-mode oscillations. The trajectory shown has one large and three small extrema in X and Y for each cycle of oscillation. (Reprinted with permission from Barkley, D. 1988. "Slow Manifolds and Mixed-Mode Oscillations in the Belousov–Zhabotinskii Reaction," *J. Chem. Phys.* **89**, 5547–5559. © 1988 American Institute of Physics.)

system is initially off the manifold, it tends to approach the steady state, but if it is perturbed away from that state along the manifold, it begins to spiral away from the steady state, as shown in Figure 8.2. When the system spirals out far enough, it crosses the edge of the lower sheet of the manifold and jumps rapidly to the upper sheet (dashed line in the figure). It now moves along the upper sheet until it reaches the edge of that surface and falls back to the lower sheet, where it begins to spiral away from the steady state again. The excursions between the two sheets correspond to the large-amplitude oscillations, while the spiraling process gives rise to the small-amplitude peaks. Note that if the slow manifold is oriented as shown, only the variables X and Y will show both the large- and small-amplitude oscillations, while Z will exhibit single-peaked, large-amplitude periodic behavior.

8.1.2 Model A

There are several variants of such slow-manifold models, which are discussed by Barkley (1988). Here, we shall look here only at a single model, dubbed model A by Barkley, who argues persuasively that it best characterizes the sort of mixed-mode oscillations seen in the BZ system.

Model A is constructed from two two-variable subsystems, so it actually contains four variables. Each subsystem exhibits relatively familiar, simple behavior. Subsystem I undergoes Hopf bifurcation, with a single, stable steady state becoming unstable via a supercritical Hopf bifurcation as a control parameter β is varied beyond a critical value. This behavior is shown schematically in the top part of Figure 8.3. When β falls below its critical value, the steady state gives way to a nearly sinusoidal periodic oscillation. There are many mathematical models that

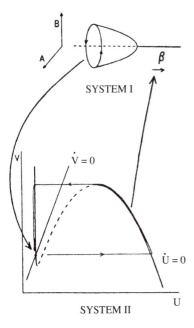

Figure 8.3 Schematic diagram of model A. System I undergoes a Hopf bifurcation to nearly sinusoidal periodic oscillation in A and B when β falls below a critical value. System II has a slow manifold (the $dU/dt = 0$ null cline) and is excitable. Changes in B in system I perturb the steady-state point in system II, while changes in V in system II alter the effective value of β in system I. (Reprinted with permission from Barkley, D. 1988. "Slow Manifolds and Mixed-Mode Oscillations in the Belousov–Zhabotinskii Reaction," *J. Chem. Phys. 89*, 5547–5559. © 1988 American Institute of Physics.)

yield such behavior. One chemically relevant model has been suggested by Gray and Scott (1983) to describe the reaction system in a CSTR:

$$A + 2B \rightarrow 3B \tag{8.1}$$

$$B \rightarrow C \tag{8.2}$$

The resulting differential equations for the concentrations of A and B are

$$dA/dt = -AB^2 - k_0(A - A_0) \tag{8.3}$$

$$dB/dt = AB^2 - k_1 B - k_0(B - B_0) \tag{8.4}$$

Subsystem II, whose phase-plane behavior is shown in the lower portion of Figure 8.3, describes systems like the BZ reaction in which the null cline corresponding to the rapidly changing variable is s-shaped. Such systems give rise to either excitability or relaxation oscillations, depending upon whether the steady state, that is, the intersection of the two null clines, occurs on one of the outer, stable branches or on the middle, unstable branch of the s shape. The system is attracted to the trajectory shown, along which it moves slowly on the segments for which it coincides with the s-shaped null cline, and rapidly when it jumps between these segments (i.e., along the horizontal lines marked with arrows). A simple model that behaves in this fashion is one proposed by Tyson and Fife (1980) to describe certain aspects of the BZ reaction:

$$dU/dt = (1/\varepsilon)[U(1 - U) - bV(U - a)/(U + a)] \tag{8.5}$$

$$dV/dt = U - V \tag{8.6}$$

In the figure, the $dV/dt = 0$ null cline in system II is located so that this subsystem is excitable rather than oscillatory. The steady state is stable, but if V is lowered beyond the turn in the s-shaped null cline, the system will undergo a large-scale excursion, as shown by the arrows, before returning to the steady state.

To create model A, we must link the two subsystems in an appropriate fashion. We introduce two coupling parameters, g_1 and g_2, and a third parameter, k, that adjusts the relative time scales of the two subsystems. The first coupling parameter causes changes in the variable V of system II to affect the control parameter β in system I. To accomplish this, we first identify β with the constant term $k_0 A_0$ in eq. (8.3) and then replace β by $\beta + g_1 V$. We also introduce a coupling in the other direction by replacing V by $V + g_2 B$ on the right-hand sides of eqs. (8.5) and (8.6). This modification causes oscillations of sufficient magnitude in system I to push system II beyond the threshold of excitability.

With the definitions $\alpha = k_0$, $\gamma = k_1 + k_0$, and $\delta = k_0 B_0$, we can add the appropriate coupling terms to eqs. (8.3)–(8.6) to obtain the full model A:

$$dA/dt = -AB^2 - \alpha A + \beta + g_1 V \tag{8.7}$$

$$dB/dt = AB^2 - \gamma B + \delta \tag{8.8}$$

$$dU/dt = (k/\varepsilon)[U(1 - U) - b(V + g_2 B)(U - a)/(U + a)] \tag{8.9}$$

$$dV/dt = k(U - V - g_2 B) \tag{8.10}$$

If the timing parameter k is chosen appropriately, the s-shaped null cline in the two-dimensional subsystem II becomes the slow manifold in the fully coupled model A, as shown in Figure 8.4.

To obtain oscillations, we need only set the parameters so that when V is near the steady-state value of system II, $g_1 V + \beta$ lies below the oscillatory threshold of system I. A small decrease in V will decrease the effective value of β, causing the amplitude of the oscillations in B to grow. Eventually, these oscillations in B will carry system II past the point of excitability, resulting in a large increase in U, followed by an increase in V. When V becomes large enough, it moves the effective value of β beyond the critical value, and the oscillations in system I cease. As V relaxes back toward the steady-state value in system II, the oscillations in system I begin again. This sequence is shown in Figure 8.5, where the small amplitude oscillations in A correspond to the nearly sinusoidal oscillation of system I and the spiral in Figure 8.4, while the large amplitude peaks in both variables correspond to the trajectory in system II when the system is excited beyond its threshold.

We have spent a great deal of time discussing model A because we believe that the scenario that it presents—a fast oscillatory subsystem that is driven in and out of oscillation by a slower, coupled subsystem that moves between two states—is both intuitively comprehensible and chemically relevant. Moreover, it can be used to derive insight into other sorts of complex dynamical behavior, such as quasi-periodicity or chaos, as well. The slow-manifold picture is, of course, not the only way in which mixed-mode oscillation can arise. Another route to this form of behavior is discussed by Petrov et al. (1992).

8.1.3 Experimental Observations

We turn now to a few examples of experimentally observed complex oscillations. As is so often the case, the BZ reaction provides the earliest instances and the widest variety of the phenomenon of interest.

Probably the first examples of complex periodic oscillations in chemistry are found in studies of the BZ reaction in a CSTR in the mid-1970s. In these experiments, a number of investigators (Zaikin and Zhabotinsky, 1973; Sørensen, 1974; Marek and Svobodova, 1975; De Kepper et al., 1976; Graziani et al., 1976) observed *bursting*, a form of oscillation commonly seen in neurons, but previously unobserved in simple chemical reactions. Bursting consists of periods of relatively

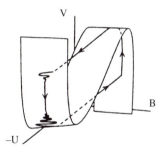

Figure 8.4 Three-dimensional projection of the slow manifold and a representative trajectory for model A. (Reprinted with permission from Barkley, D. 1988. "Slow Manifolds and Mixed-Mode Oscillations in the Belousov–Zhabotinskii Reaction," *J. Chem. Phys. 89,* 5547–5559. © 1988 American Institute of Physics.)

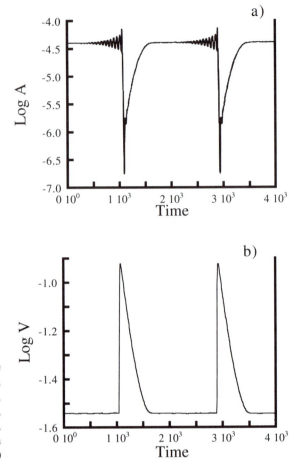

Figure 8.5 Mixed-mode oscillations in model A with $\alpha = 0.001$, $\beta = 0.03$, $\gamma = 0.12$, $\delta = 0.002$, $\varepsilon = 0.001$, $a = 0.01$, $b = 2.5$, $k = 0.005$, $g_1 = 0.2$, $g_2 = -0.02$. (Adapted from Barkley, 1988.)

quiescent behavior, in which concentrations change very little, alternating with periods of large-amplitude oscillations, as illustrated in Figure 8.6. As a control parameter is varied, the number of oscillations per burst increases or decreases until a bifurcation to the steady state or some other form of oscillatory behavior occurs. Bursting can be understood in terms of a slow-manifold picture (Rinzel, 1987) much like the one we employed to illuminate the phenomenon of mixed-mode oscillation. In Chapter 12, we will discuss another example of bursting produced when two CSTRs containing the components of the chlorine dioxide–iodide reaction are coupled.

A second form of complex oscillation seen in the BZ reaction in a CSTR (Richetti, 1987) is *quasiperiodicity*. Here, the oscillatory behavior contains two different, incommensurate frequencies, a fact easily discerned from consideration of the Fourier spectrum, which shows two fundamental peaks in addition to linear combinations of the fundamentals. Quasiperiodicity is commonly observed when two nonlinear oscillators are coupled or when a single oscillator is periodically forced, but it is less often found in a single oscillator. Its presence may indicate the

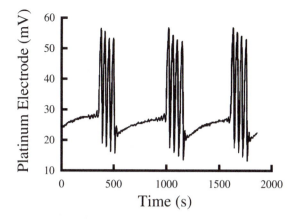

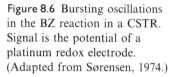

Figure 8.6 Bursting oscillations in the BZ reaction in a CSTR. Signal is the potential of a platinum redox electrode. (Adapted from Sørensen, 1974.)

existence of two independent oscillatory pathways. An experimental example is shown in Figure 8.7. Another system that gives rise to quasiperiodicity is the electrodissolution of copper (Bassett and Hudson, 1989).

The mixed-mode oscillations discussed in the previous section constitute probably the most commonly observed form of complex periodic oscillation in chemical systems. Again, the BZ reaction provides the richest source, but there are other examples as well. Zhabotinsky (1964b) was the first to report multipeaked periodic states in the BZ system, but the most thorough study is that of Maselko and Swinney (1986), who report an amazing wealth of behavior. To characterize the patterns that they observe, these authors define basic patterns, denoted L^S, consisting of L large-amplitude followed by S small-amplitude oscillations. Figure 8.8 depicts two such basic patterns, as well as a compound state in which each period contains a sequence of seven basic patterns, some of which repeat, totaling eleven large and fourteen small oscillations! In fact, all of the states observed in this set of experiments, which were conducted on the manganese-catalyzed BZ reaction in a CSTR, can be characterized as concatenations of basic L^S states.

Most of the states found consist of one or two basic patterns. By fixing malonic acid concentration $(MA)_0$ and varying the residence time, it was possible to study

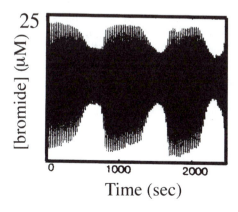

Figure 8.7 Quasiperiodicity in the BZ reaction. (Adapted from Richetti, 1987.)

Figure 8.8 Mixed-mode oscillations in the manganese-catalyzed BZ reaction. Signal is the potential of a bromide-selective electrode: (a) a 1^3 state, (b) a 5^6 state, (c) a compound state containing four different basic patterns. Dots above the traces are separated by one period. Experimental conditions: $[MnSO_4]_0 = 0.00416$ M, $[H_2SO_4]_0 = 1.5$ M in all cases. (a) Residence time $\tau = 660$ s, $[MA]_0 = 0.018$ M, $[KBrO_3]_0 = 0.0275$ M; (b) $\tau = 480$ s, $[MA]_0 = 0.30$ M, $[KBrO_3]_0 = 0.0275$ M; (c) $\tau = 435$ s, $[MA]_0 = 0.056$ M, $[KBrO_3]_0 = 0.033$ M. (Adapted from Maselko and Swinney, 1986.)

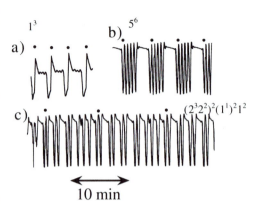

10 min

the sequence of basic states $1^5 \rightarrow 1^4 \rightarrow \ldots \rightarrow 1^1 \rightarrow 2^1 \rightarrow \ldots \rightarrow 6^1$. For some conditions, bistability between different periodic states was observed. In a narrow region of behavior, sequences composed of three and even four basic patterns, as in Figure 8.8c, were found. We shall consider such states in more detail below. Maselko and Swinney find it convenient to characterize the complex periodic states by a *firing number*, the ratio of small-amplitude oscillations to total oscillations per period. While the firing number does not uniquely specify each state (two or more states, e.g., 1^1 and 2^2, may have the same firing number), when all other control parameters are fixed, the firing number is a monotonically increasing function of the flow rate for the states observed in these experiments.

Other mixed-mode oscillations of this type have been seen in the Bray reaction (Chopin-Dumas, 1978) and several chlorite-based oscillators (Orbán and Epstein, 1982; Alamgir and Epstein, 1985a, 1985b). In Figure 8.9, we present a phase diagram that shows the progression of 1^S oscillations in the chlorite–thiosulfate system.

A sequence of states like that shown in Figure 8.9 is sometimes referred to as a *period-adding* sequence, since the period, measured in terms of the number of oscillations per cycle, increases by 1 as a control parameter is varied. Complex oscillations often arise via another scenario, known as *period-doubling*. In this case, instead of observing 1, 2, 3, 4, . . . oscillations per cycle as the control parameter is varied, one sees, 1, 2, 4, 8, ... At each bifurcation point, as shown in Figure 8.10, neighboring pairs of peaks that were identical before the bifurcation become slightly different in amplitude beyond the bifurcation. Again, period-doubling in chemical systems is best characterized in the BZ reaction. This behavior has received a great deal of attention in nonchemical contexts, since it is intimately associated with the phenomenon of chaos, which we shall encounter a bit later in this chapter.

8.1.4 Farey Sequences

At this point, we will make a brief mathematical digression in order to look at a fascinating aspect of the mixed-mode oscillations found in the BZ reaction.

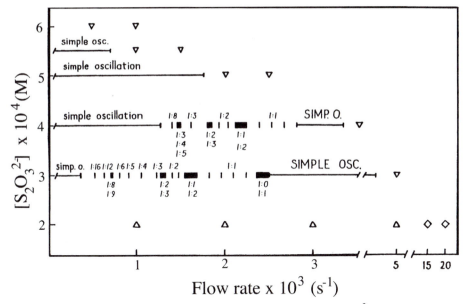

Figure 8.9 Phase diagram of the chlorite–thiosulfate system in the $[S_2O_3^{2-}]_0 - k_0$ plane with $[ClO_2^-]_0 = 5 \times 10^{-4}$ M and pH 4. "SIMPLE" and "simple" denote pure large- and pure small-amplitude oscillations, respectively. 1:N denotes 1^N multipeak oscillations. Vertical segments | show flow rate at which new 1:N state appears. Filled-in boxes denote regions in which basic states are mixed to form aperiodic states. ∇, Low-potential steady state (SSI); $\triangle$, high-potential steady state (SSII); $\diamond$, bistability (SSI/SSII). (Reprinted with permission from Orbán, M.; Epstein, I. R. 1982. "Complex Periodic and Aperiodic Oscillation in the Chlorite–Thiosulfate Reaction," *J. Phys. Chem.* **86**, 3907–3910. © 1982 American Chemical Society.)

Consider two rational numbers reduced to lowest terms, p/q and r/s. There is a kind of arithmetic (you may have used it as a child, without realizing its profound significance, when you were first introduced to fractions!) in which the sum of two such numbers is given by

$$p/q \oplus r/s = (p+r)/(q+s) \qquad (8.11)$$

This form of addition, known as *Farey addition* (Farey, 1816), plays a significant role in number theory (Hardy and Wright, 1989). It turns out, for reasons beyond

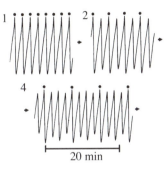

Figure 8.10 A period-doubling sequence $(1 \rightarrow 2 \rightarrow 4)$ in the BZ reaction. (Adapted from Coffman et al., 1987.)

the scope of this chapter, that Farey arithmetic provides a useful description of the mixed-mode states observed in the BZ reaction (Maselko and Swinney, 1987).

We noted above that mixed-mode states like those in Figure 8.8 consist of concatenations of a few basic L^S patterns. Maselko and Swinney observed that the firing numbers, $F = S/(L + S)$, of the states they found in their experiments form a *Farey sequence*, that is, a sequence of rational numbers such that (1) between each pair of states with firing numbers F_1 and F_2 (the parents), there lies a state (the daughter) with firing number $F_1 \oplus F_2$; and (2) if the states are arranged in a *Farey tree*, with parents on one level and daughters on the next, as in Figure 8.11, states that are related as parents or as parent and daughter obey the relationship $|p_1q_2 - p_2q_1| = 1$, where $F_i = p_i/q_i$. This relationship is demonstrated in Table 8.1 for the Farey sequence illustrated in Figure 8.11.

Obviously, one cannot expect to observe an infinite number of generations on the Farey tree, but Maselko and Swinney did find that when they were able to adjust their residence time with sufficient precision, they saw the intermediate states predicted by the Farey arithmetic, though after a few cycles the system would drift off to another, higher level state on the tree, presumably because their pump could not maintain the precise flow rate corresponding to the intermediate state. An even more complex and remarkable Farey arithmetic can be formulated for states consisting of sequences of three basic patterns (Maselko and Swinney, 1987). The fact that the mixed-mode oscillations in the BZ system form a Farey sequence places significant constraints on any molecular mechanism or dynamical model formulated to explain this behavior.

8.2 Chemical Chaos

The notion of chaos in science has been seen by many as one of the fundamental conceptual breakthroughs of the second half of the twentieth century (Gleick,

Table 8.1 Farey Tree Arithmetic for the Sequence of States shown in Figure 8.11. States are Listed in Order of Decreasing Residence Time

Farey Sum	$p_1q_2 - p_2q_1$
Parent state: 2/3	
	$20 - 21 = -1$
$2/3 \oplus 5/7 = 7/10$	
	$49 - 50 = -1$
$2/3 \oplus 3/4 = 5/7$	
	$55 - 56 = -1$
$5/7 \oplus 3/4 = 8/11$	
	$120 - 121 = -1$
$8/11 \oplus 3/4 = 11/15$	
	$209 - 210 = -1$
$11/15 \oplus 3/4 = 14/19$	
	$322 - 323 = -1$
$14/19 \oplus 3/4 = 17/23$	
	$68 - 69 = -1$
Parent state: 3/4	

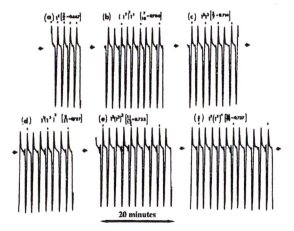

Figure 8.11 Several states in the Farey tree for a sequence of states observed as the residence time decreases in the Mn-catalyzed BZ reaction. Each state consists of a concatenation of the basic 1^2 and 1^3 patterns. Firing numbers are given in brackets. The 4/3 and 17/23 states in Table 8.1 are not shown here. (Reprinted with permission from Maselko, J.; Swinney, H.L. 1986. "Complex Periodic Oscillations and Farey Arithmetic in the Belousov–Zhabotinskii Reaction," *J. Chem. Phys.*, 85, 6430–6441. © 1986 American Institute of Physics.)

1987; Stewart, 1989). In many respects, the idea that systems with a deterministic dynamics can behave in ways that we normally associate with systems subject to random forces—that "identical" experiments on macroscopic systems can lead to very different results because of tiny, unavoidable differences in initial conditions—is truly revolutionary. Chaos has become a more central part of mathematics and physics than of chemistry, but it is now clear that chemical systems also exhibit chaotic behavior (Scott, 1991). In this section, we define chaos and some related concepts. We also give some examples of chaotic chemical systems. In the next section, we discuss the intriguing notion that understanding chaos may enable us to control chaotic behavior and tailor it to our needs.

Although there are many definitions of chaos (Gleick, 1987), for our purposes a chaotic system may be defined as one having three properties: deterministic dynamics, aperiodicity, and sensitivity to initial conditions. Our first requirement implies that there exists a set of laws, in the case of homogeneous chemical reactions, rate laws, that is, first-order ordinary differential equations, that govern the time evolution of the system. It is not necessary that we be able to write down these laws, but they must be specifiable, at least in principle, and they must be complete, that is, the system cannot be subject to hidden and/or random influences. The requirement of aperiodicity means that the behavior of a chaotic system in time never repeats. A truly chaotic system neither reaches a stationary state nor behaves periodically; in its phase space, it traverses an infinite path, never passing more than once through the same point.

Related to, but distinct from the property of aperiodicity is the sensitivity of a chaotic system to its initial conditions. This property is perhaps the most striking and surprising aspect of chaotic systems. In the systems with which chemists are most familiar, a tiny difference in the initial conditions—say, 0.0001% in the concentrations of reactants introduced into a beaker—leads to similarly small variations in the evolution and the final state of the system. In chaotic systems, in contrast, initial differences, however small, ultimately grow to macroscopic proportions. This phenomenon is sometimes referred to as the "butterfly catastrophe," in reference to one of the earliest recognized chaotic systems, a model of the atmosphere devised by meteorologist Edward Lorenz (1964). If the weather is governed by a set of equations with chaotic solutions, then, even if we program our most powerful computers with those equations and even if we have sensors distributed all over the world to feed information into our weather-forecasting programs, the infinitesimal, but unaccounted for, difference in initial wind velocity resulting from a butterfly flapping its wings somewhere in the wilds of Borneo will ultimately invalidate any long-term forecast generated by our weather-prediction algorithms. The data we put in may lead to our expectation of a beautiful fall day in Mississippi six months from now, but addition of the butterfly's perturbation may result in a hurricane instead.

Another remarkable property of chaotic systems is *universality*. Chaotic systems as different as a dripping water faucet, a vibrating strip of metal, or a chemical reaction in a flow reactor share key features that are predictable from a stunningly simple characterization of their dynamics.

8.2.1 One-Dimensional Maps

The universal aspect of chaos is best appreciated by studying a set of mathematical objects considerably simpler than the rate equations that describe the BZ or comparable chemical reactions. Consider a rule for generating a sequence of numbers. Given one number, the rule tells us how to calculate the next one:

$$x_{n+1} = f(x_n) \tag{8.12}$$

Such a rule is called a one-dimensional (1-D for short) map. At first glance, it would seem unlikely that a recipe like this, particularly if the function f is not very complicated, could generate anything very interesting. But wait!

Let us take a simple-looking example, the so-called *logistic map*,

$$x_{n+1} = \lambda x_n(1 - x_n) \tag{8.13}$$

where λ is a parameter to be chosen. We note that if λ lies between 0 and 4, then an initial choice of x_0 in the range (0,1) will cause all subsequent x_n values to stay within this range.[1] Now let us pick a few values for λ and a starting value for x_0, and see what happens. We shall be particularly interested in the long-term behavior of the sequence, that is, to what number, if any, it converges.

The clever reader may say, "Aha, there's no need to plug numbers in. If I just set x_{n+1} equal to x_n, then all subsequent numbers in the sequence will equal x_n, and the sequence will converge to that value." A little algebra gives

---

[1]Since $x(1 - x)$ has its maximum value of 1/4 at $x = 1/2$, x_{n+1} must always be less than 1 if $\lambda < 4$. Choosing an initial x value between 0 and 1 ensures that $x(1 - x)$ will always be positive.

$$x_{n+1} = x_n = \lambda x_n (1 - x_n)$$
$$x_n = 0 \text{ or } 1 - 1/\lambda$$
$$(8.14)$$

If $\lambda < 1$, the only "steady-state" solution in $(0,1)$ is $x = 0$, while if $\lambda > 1$, there are two such solutions. What does this mean? Does the sequence actually converge to the solution of eq. (8.13)? To which solution does it converge when there is more than one? Does the initial value x_0 matter, or is the asymptotic behavior independent of the starting point?

It turns out that steady states (more commonly referred to as *fixed points* or *equilibria*) of 1-D maps have stability properties like the steady states of ordinary differential equations. We can ask whether, if we choose x_n infinitesimally close to, but not exactly at, the fixed point, the sequence will approach the fixed point or move away from it. If subsequent numbers in the sequence converge toward the fixed point, then it is stable. If not, then it is unstable. If there is more than one stable fixed point, then each of these points will have its own basin of attraction, just as we found for ordinary differential equations with multiple stable states. Without attempting to derive it (it is not difficult), we simply present the result that a fixed point x of a 1-D map $f(x_n)$ is stable if the derivative df/dx_n at the fixed point x lies between -1 and 1, that is,

$$|df(x_n)/dx_n|_{x_n=x}| < 1 \qquad (8.15)$$

Since the derivative of the 1-D map in eq. (8.13) is just $\lambda(1 - 2x)$, the fixed point at 0 is stable for $\lambda < 1$, that is, when it is the unique fixed point, and it is unstable for all larger values of λ. Thus, for small values of λ, the sequence should approach zero. A bit more algebra shows that the second fixed point, at $x = 1 - 1/\lambda$, is stable for $1 \leq \lambda \leq 3$. What happens if we pick a value of λ greater than 3? At this point, it seems to be time to plug in some numbers, so let us choose a few representative values of λ and a starting point x_0 (the choice of x_0 will not affect the asymptotic solution—try it yourself!), say, 0.25, and see what happens. Some results are shown in Table 8.2.

As expected, the sequences for $\lambda = 0.5$ and $\lambda = 1.5$ converge to $x = 0$ and $x = 1 - 1/\lambda = 1/3$, respectively. The more interesting results occur for the larger values of λ, where both fixed points are unstable. For $\lambda = 3.1$, the sequence converges to a period-2 oscillation between $x = 0.5580$ and $x = 0.7646$. If we increase λ to 3.50, the asymptotic solution has period 4. A further increase to $\lambda = 3.54$ yields a sequence of period 8. Note how the period-8 sequence is nearly a period-4 sequence; compare, for example x_{146} and x_{150}, or x_{144} and x_{148}. If we continue to increase λ in small increments, we can find sequences that converge to repetitive patterns with period 16, 32, 64, ... At each bifurcation point λ_j, the asymptotic solution goes from having period 2^j to having period 2^{j+1}; the period doubles. Just past the bifurcation point, as in the case when $\lambda = 3.54$, terms that were equal because they were a distance 2^j apart in the period-2^j sequence become slightly different, generating the period-2^{j+1} sequence. This behavior is seen clearly in the bifurcation diagram of Figure 8.12, where the asymptotic values of x_n are plotted as a function of λ. Thus, for example, if the asymptotic solution for a particular value of λ has period 4, there will be four lines above that value of λ.

Table 8.2 Sequences Derived from Equation (8.13) for Various Values of λ

	λ					
n	0.5	1.5	3.1	3.50	3.54	3.7
	x_n	x_n	x_n	x_n	x_n	x_n
0	0.2500	0.2500	0.2500	0.2500	0.2500	0.2500
1	0.0938	0.2813	0.58125	0.6563	0.6638	0.6938
2	0.0425	0.3032	0.7545	0.7896	0.7901	0.7861
3	0.0203	0.3169	0.5742	0.5816	0.5871	0.6221
4	0.0100	0.3247	0.7580	0.8517	0.8581	0.8698
5	0.0049	0.3289	0.5687	0.4420	0.4310	0.4190
6	0.0025	0.3311	0.7604	0.8632	0.8681	0.9007
7	0.0012	0.3322	0.5649	0.4132	0.4053	0.3309
8	0.0006	0.3328	0.7620	0.8486	0.8532	0.8192
9	0.0003	0.3331	0.5623	0.4496	0.4433	0.5481
10	0.0002	0.3333	0.7630	0.8661	0.8736	0.9164
⋮						
134	0.0000	0.3333	0.7646	0.8750	0.8833	0.6071
135	0.0000	0.3333	0.5580	0.3828	0.3650	0.8826
136	0.0000	0.3333	0.7646	0.8269	0.8205	0.3835
137	0.0000	0.3333	0.5580	0.5009	0.5214	0.8748
138	0.0000	0.3333	0.7646	0.8750	0.8834	0.4053
139	0.0000	0.3333	0.5580	0.3828	0.3645	0.8918
140	0.0000	0.3333	0.7646	0.8269	0.8202	0.3569
141	0.0000	0.3333	0.5580	0.5009	0.5221	0.8493
142	0.0000	0.3333	0.7646	0.8750	0.8833	0.4736
143	0.0000	0.3333	0.5580	0.3828	0.3650	0.9224
144	0.0000	0.3333	0.7646	0.8269	0.8205	0.2648
145	0.0000	0.3333	0.5580	0.5009	0.5214	0.7203
146	0.0000	0.3333	0.7646	0.8750	0.8834	0.7454
147	0.0000	0.3333	0.5580	0.3828	0.3645	0.7021
148	0.0000	0.3333	0.7646	0.8269	0.8202	0.7738
149	0.0000	0.3333	0.5580	0.5009	0.5221	0.6476
150	0.0000	0.3333	0.7646	0.8750	0.8833	0.8444

What about the sequence for λ = 3.7? It does not seem to repeat. In fact, no matter how far we extend the sequence [assuming that our calculator computes eq. (8.13) to infinite precision], the numbers will never repeat; the sequence is chaotic. One can demonstrate, either numerically or analytically, that the set of bifurcation points λ_j converges to the value λ = 3.569946, and that at higher values of λ, the sequence is chaotic. In Figure 8.12, chaotic solutions appear as continuous, filled regions; the asymptotic solution eventually comes arbitrarily close to all values of x within a certain range. You may have noticed that within the chaotic solutions are embedded narrow "windows" of periodic solutions. We shall not discuss these windows in any detail here, but we point out that not only the existence but also the order of the periods of these regions is predicted by theory (Metropolis et al., 1973) and found in experiments on chemical systems (Coffman et al., 1987). Because the order in which the windows occur is predicted

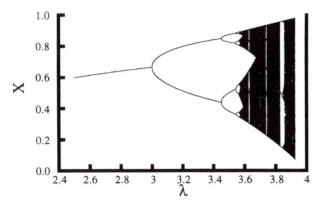

Figure 8.12 Bifurcation diagram for eq. (8.13) showing asymptotic solution as a function of the parameter λ. The first 100 iterations were discarded, and the next 500 are plotted. (Data courtesy of P. Strizhak.)

to be the same for a large class of 1-D maps independent of the details of the function f in eq. (8.12), the sequence of windows is referred to as the universal or U-sequence.

Period-doubling generates another remarkable form of universality, which we wish to consider in a bit more detail. We have noted that the bifurcation points of the period-doubling sequence for the logistic map get closer together and converge to a limit point λ*, beyond which chaos begins. It can be shown that for *any* map of the form of eq. (8.12), in which the function f has a single, quadratic maximum,[2] (i.e., $d^2f(x)/dx^2 \neq 0$ at the maximum), if there is a period-doubling sequence, the bifurcation points will converge to a point λ* in a very special way. As the limit point is approached, the distance between successive bifurcation points and λ* decreases geometrically, and *the convergence factor is the same for all such functions:*

$$(\lambda_j - \lambda^*)/(\lambda_{j+1} - \lambda^*) \to 4.69920\ldots \qquad \text{as } \lambda \to \infty \qquad (8.16)$$

The universal convergence factor 4.69920 . . . is known as the Feigenbaum number in honor of its discoverer (Feigenbaum, 1979). It is a transcendental number like π or e—the first to be found in modern times!

Is the sequence generated in Table 8.2 for $\lambda = 3.7$ really chaotic? In particular, is it sensitive to the initial condition x_0? In Table 8.3, we show the result of starting the sequence with two different initial values, 0.25000 and 0.25001. We observe that the two sequences slowly diverge from one another, staying reasonably close together for the first ten or so iterates. Once the map has been applied a few dozen times, there is no way to tell that the two sequences ever started off a mere 0.00001 away from each other.

---

[2]Instead of $x(1 - x)$, we could choose, for example, $\sin x$ (with $x \varepsilon [0, \pi]$) or $x \exp(-x)$ as $f(x)$.

Table 8.3 Sequences Derived from Equation (8.15) with $\lambda = 3.7$

n	x_n	x_n
0	0.25000	0.25001
1	0.69375	0.69377
2	0.78611	0.78608
3	0.62213	0.62219
4	0.86981	0.86976
5	0.41899	0.41913
6	0.90072	0.90080
7	0.33087	0.33063
8	0.81917	0.81886
9	0.54809	0.54882
10	0.91644	0.91618
⋮		
21	0.45120	0.36162
22	0.91619	0.85415
23	0.28412	0.46093
24	0.75256	0.91935
25	0.68899	0.27433
26	0.79285	0.27433
27	0.60769	0.71792
28	0.88209	0.74930
29	0.38483	0.69506
30	0.87593	0.78423
⋮		
41	0.651988	0.456366
42	0.839544	0.917955
43	0.498444	0.278677
44	0.924999	0.743755
45	0.256722	0.705177
46	0.706011	0.769244
47	0.767988	0.656788
48	0.659300	0.834066
49	0.831111	0.512100
50	0.519366	0.924466

The results in Tables 8.2 and 8.3 were calculated by iterating eq. (8.13) on a calculator (actually on a spreadsheet), but there is another, more intuitive approach to analyzing the behavior of a 1-D map. This technique is a graphical one and is illustrated in Figure 8.13. We construct a graph whose axes are x_n and x_{n+1}. We then plot the function $f(x_n)$ on this graph and also draw the line $x_n = x_{n+1}$. Any point where the graph of the function crosses the line is a fixed point of the map, because at this point $x_n = x_{n+1} = f(x_n)$. To follow the evolution of the system, find x_0 on the abscissa. If we now move vertically toward the graph of f, the value of the ordinate at which we intersect the graph is x_1. To obtain x_2, we move horizontally to the diagonal line, so that the abscissa is now x_1. Moving vertically from here to the graph of f will

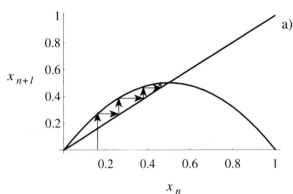

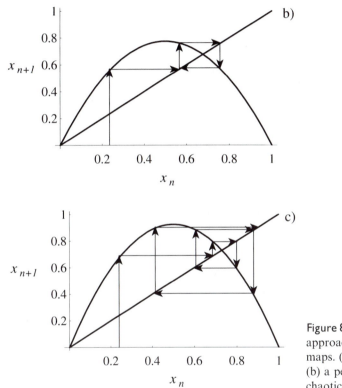

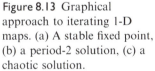

Figure 8.13 Graphical approach to iterating 1-D maps. (a) A stable fixed point, (b) a period-2 solution, (c) a chaotic solution.

produce x_2. Continuation of this process of moving vertically to the graph of f and then horizontally to the diagonal will generate the entire sequence of x_n values.

If the asymptotic solution is a stable fixed point, the path will approach that point as shown in Figure 8.13a. If the solution is periodic, the ultimate trajectory will jump back and forth among a finite set of points, as shown in the period-2 solution of Figure 8.13b. A chaotic system will have a trajectory that continues to circulate forever, as shown in Figure 8.13c.

8.2.2 Characterizing Chaos

Sequences generated from 1-D maps like the ones we have been looking at provide nice, neat, easily analyzable examples of chaos. Chemists, however, deal with real experimental data. A somewhat benighted congressman is reputed to have said during hearings on legislation regarding pornography, "I can't tell you exactly what it is, but I know it when I see it." Does a chemist, confronted with a time series in which the signal increases and decreases repeatedly, but never periodically, have to take the same view of chaos? Are there quantitative ways of distinguishing genuine, deterministic chaos from stochastic, noisy data? In fact, there are several methods by which one can attempt to determine whether a set of data is truly chaotic. We describe some of these techniques here and illustrate them in the next section using experimental examples from the BZ reaction.

One approach is to examine the behavior of the system for a range of parameter values. If, as a parameter is varied, we observe a period-doubling sequence that culminates in aperiodic behavior, then the aperiodic data are almost surely chaotic. Other scenarios leading to chaos are also possible. Two examples worth noting, since they have been observed in chemical systems, are periodic–chaotic sequences and intermittency. Periodic–chaotic sequences can arise when, as a parameter is varied, the number of oscillations changes, as in the period-adding sequence of mixed-mode oscillations described in section 8.1.2. Instead of observing a discrete transition from periodic state A to periodic state B as the control parameter is varied, one finds a region of chaotic oscillations containing irregular mixtures of A- and B-type oscillations separating the A region and the B region. Intermittency is a situation in which periodic behavior is interrupted by occasional episodes of (usually large amplitude) oscillations. As the control parameter is varied, these turbulent episodes become more prevalent and full-blown chaos takes over.

Often, we do not have available a sequence of time series from which the approach to chaos can be discerned. It is then necessary to analyze a single data set. The most useful way to do this is to look at the phase portrait of the system in a multidimensional phase space. Unfortunately, in real chemical systems we almost always have data on just a single variable, such as a concentration. Nature is not always cruel, however, especially when assisted by mathematics. If we have a series of measurements at regular time intervals, $\{x(t), x(t + T), x(t + 2T), \ldots\}$, we can construct an n-dimensional phase space in which the coordinates are $x(t), x(t + T), \ldots, x(t + nT)$. Amazingly, several embedding theorems (Whitney, 1935) guarantee that, for almost every choice of the delay time T, the attractor in the space constructed in this fashion will be topologically equivalent[3] to the attractor in the actual phase space.

This delay method (Takens, 1981) enables us to construct, for example, two- and three-dimensional portraits of the attractor. A first test to which the attractor can be subjected is to observe whether the attractor has a clear structure. If the

---

[3]Equivalent to within rotations or distortions that would change, for example, a circle into an ellipse, but not into a line or a parabola, that is, the two attractors will have "the same essential shape."

source of the data is stochastic noise rather than deterministic chaos, successive points will be randomly distributed, and no coherent structure will emerge. Periodic attractors form closed curves whose phase portraits may be difficult to distinguish from truly chaotic attractors if there are significant amounts of noise present.

More sensitive analyses can be applied to multidimensional phase portraits. A *Poincaré section* is obtained by cutting the *n*-dimensional attractor with an $(n - 1)$-dimensional hyperplane (the Poincaré plane). The points at which the trajectory of the attractor intersect the hyperplane constitute the Poincaré section. If the attractor is periodic, the section will consist of a finite number of points: one in the case of a simple periodic orbit or a cluster if the data contain some noise. A chaotic attractor should give rise to a continuous curve as a Poincaré section. Another convenient way to analyze the data is to create from the Poincaré section a *next-return map* by plotting the value of one of the components for the *j*th intersection of the trajectory with the Poincaré plane against the value of that component for the $j + 1$th intersection. A variation of this approach that is considerably easier to implement in practice is to construct a *next-amplitude map* by plotting the amplitude of the *j*th peak in the time series against the value of the $j + 1$th peak. If the data arise from a deterministic system, the points should form a smooth curve that permits one to predict the amplitude of each peak from the amplitude of the previous one by iterating this 1-D map in the manner described above. Again, periodic attractors yield maps that consist of a finite number of points (or clusters of noisy points).

A final measure of chaos that has proved useful in characterizing experimental data is the calculation of *Lyapunov exponents*. These quantities are the extensions to more general attracting trajectories of the Jacobian eigenvalues that characterize steady states. Like the eigenvalues that we calculated earlier, these exponents measure the rate of growth or decay of infinitesimal perturbations in various directions, starting from a point on the attractor. Since the attractor now consists of an infinite number of points, a single exponent will measure only the *average* rate at which perturbations grow or decay, where the averaging is done over the points of the attractor. Alternatively, one may think of the Lyapunov exponents as characterizing the average rate of growth or decay of an *n*-dimensional volume around the evolving attractor in the *n*-dimensional phase space (Scott, 1991).

As in the case of steady states, it is the most positive eigenvalue or exponent that determines the long-time behavior of the system. For a point attractor (steady state), if this exponent is negative, the attractor is stable, and all neighboring points evolve toward it. For a periodic orbit, there is one direction—motion along the orbit—for which the corresponding exponent is zero; perturbations along the orbit neither grow nor decay, since the trajectory remains on the orbit. Stability is then determined by whether or not there is a positive exponent in addition to the zero exponent. Exponents for periodic orbits are also known as *Floquet exponents*. A quasiperiodic orbit will have two zero exponents, since this orbit is a two-dimensional surface in phase space; if the other $n - 2$ exponents are negative, the torus will be an attractor.

A chaotic attractor, often referred to as a *strange attractor*, is different in that it not only has one zero exponent associated with motion along the attractor, but also has a positive exponent that leads to divergence of neighboring trajectories *within the attractor*. It is this divergence that is responsible for the sensitivity of chaotic systems to their initial conditions. Chaotic attractors, which can only occur in a phase space with at least three dimensions, also possess $n - 2$ negative exponents, corresponding to the directions along which the system approaches the attractor. Attractors with more than one positive Lyapunov exponent are often referred to as *hyperchaotic*. They will not be discussed further here.

Calculating Lyapunov exponents from experimental data is a numerically challenging task. Probably the most successful approach was developed by Wolf et al. (1985). The basic idea is that if we choose a point on the attractor and surround it by an n-dimensional sphere of infinitesimal radius $r(0)$, the sphere will evolve over time (as the initial point traverses the attractor) into an ellipse. The Lyapunov exponents can be calculated from the lengths of the principal axes of the ellipse, $r_i(t)$, as

$$L_i = \lim_{t \to \infty} (1/t) \, \log_2 \, [r_i(t)/r(0)] \tag{8.17}$$

In practice, calculating the entire spectrum of n Lyapunov exponents is a difficult prospect when dealing with real, noisy experimental data. We are often willing to settle for the one or two most positive exponents. To calculate the most positive exponent, the procedure consists of choosing an initial point on the attractor and the closest neighboring point off the attractor. We call the distance between them $L(t_0)$. We move along the trajectory until t_1, at which time the distance $L'(t_1)$ between the points, and the angle between the initial and current vectors that connect them, are still within preset limits. We now choose a "replacement point" on the trajectory that is close to the evolved points, and repeat the process. The process of stopping the evolution periodically and choosing a new pair of points ensures that we are probing only the dynamics near the attractor and avoiding the effects of trajectories that fold back upon themselves far from the attractor. The largest Lyapunov exponent is given by an average over all points chosen:

$$L_1 = (1/t_M - 1/t_0) \sum_{k=1}^{M} \log_2 \, [L'(t_k)/L(t_{k-1})] \tag{8.18}$$

A similar procedure can be used to calculate the sum of the two most positive Lyapunov exponents by starting from an initial area rather than a line segment. Both approaches are illustrated in Figure 8.14

8.2.3 The BZ Reaction—Experiments and Models

Chaos has been observed in a number of chemical systems, but by far the most thoroughly studied is the BZ reaction. In this section, we illustrate some features of chaos in the BZ reaction in a CSTR. There appear to be two different regions of chaos in the BZ system, generally referred to as the high-flow-rate (Hudson et al., 1979; Roux et al., 1980) and the low-flow-rate (Turner et al. 1981; Roux et al., 1983) regions. The next section lists examples of chaotic behavior in other chemical systems.

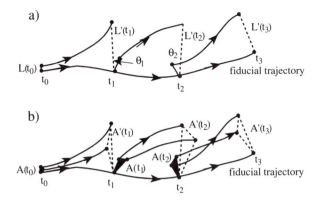

Figure 8.14 Schematic representation of the algorithm for calculating Lyapunov exponents. (a) Most positive exponent L_1 is calculated from the growth rate of an initial line segment. Replacement points are chosen when the length of the line segment or the change in its orientation reaches a preset limit. (b) Procedure for calculating sum of L_1 and L_2 from the area of a triangle consisting of one point on the trajectory and two neighboring points off it. (Adapted from Wolf et al., 1985.)

Three different routes to chaos—period-doubling, periodic–chaotic sequences, and intermittency—have been observed in the BZ reaction. Figure 8.15 shows time series for three periodic states of a period-doubling sequence, while Figure 8.16 illustrates both periodic and chaotic states from a periodic–chaotic sequence, and Figure 8.17 gives an example of intermittency.

The construction of two- and three-dimensional phase portraits from experimental data by use of the delay technique is now routine. Figure 8.18 shows two-dimensional phase portraits of periodic and chaotic attractors for the BZ reaction.

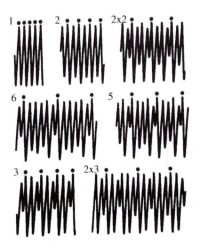

Figure 8.15 Period-doubling sequence in the BZ reaction. Signal is the potential of a bromide-selective electrode. Dots above the time series are separated by one period. Simple oscillation has a period of 115 s. (Adapted from Simoyi et al., 1982.)

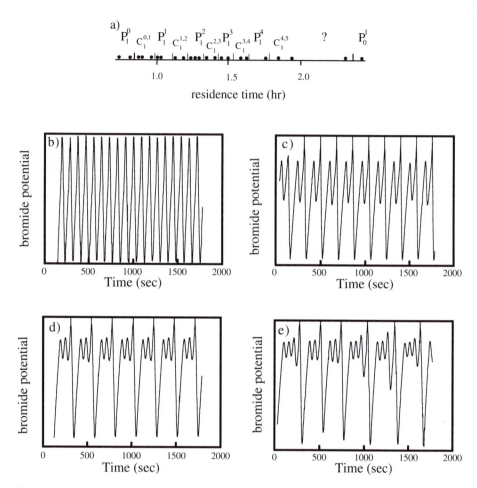

Figure 8.16 Periodic–chaotic sequence in the BZ reaction. (a) Bifurcation diagram as a function of flow rate, (b) simple periodic state L, (c) mixed-mode periodic state 1^1, (d) mixed-mode periodic state 1^2, (e) chaotic mixture of 1^3 and 1^2 patterns. (Adapted from Swinney, 1983.)

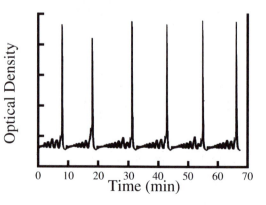

Figure 8.17 Intermittency in the BZ reaction. (Adapted from Pomeau et al., 1981.)

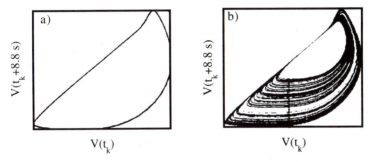

Figure 8.18 Two-dimensional phase portraits in the BZ reaction constructed by the delay technique. (a) A periodic state, (b) two-dimensional projection of a three-dimensional phase portrait of a chaotic attractor. (Adapted from Swinney, 1983.)

The phase portrait in Fig 8.18b is actually a two-dimensional projection of a three-dimensional phase portrait, where the third axis $V(t_k + 17.6 \text{ s})$ is normal to the plane of the page.

The dashed line in Figure 8.18b denotes the location of the two-dimensional Poincaré plane that was passed through the three-dimensional attractor to create the Poincaré section shown in Figure 8.19a. Figure 8.19b shows a next amplitude map created from the same data. Note the well-defined structures of both the section and the 1-D map, and the characteristic single-humped shape of the map, which gives rise to period-doubling and chaos according to the Feigenbaum scenario.

For a number of years, there was considerable controversy over whether true chaos existed in homogeneous chemical systems, and in the BZ reaction in particular. Interestingly, some of the skeptics included those who had fought hardest to convince the chemical community that periodic oscillation was a genuine property of chemical systems and not an artifact (Ganapathisubramanian and Noyes, 1982a). It now seems safe to declare that the argument has been clearly settled in favor of those who hold that chaos can arise from pure chemical processes. On the

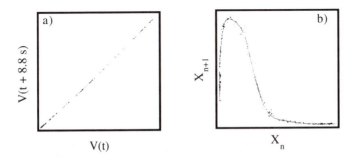

Figure 8.19 (a) Poincaré section formed by the intersection of trajectories in the three-dimensional phase space with a plane normal to the page and containing the dashed line in Figure 8.18b, (b) next amplitude map constructed from the data in Figure 8.18b. (Adapted from Swinney, 1983.)

experimental side, the decisive factors have been improved experimental techniques that make it possible to obtain high-quality data that demonstrate, for example, several states in a period-doubling sequence and permit the accurate computation of Lyapunov exponents and other relevant quantities. A second key achievement has been the development of chemically realistic models for the BZ reaction that give good agreement with the observed chaotic behavior and approach to chaos in the high-flow-rate case (Györgyi and Field, 1991; Györgyi and Field, 1992). Nonetheless, in any particular example of an aperiodic time series, one must proceed with caution and utilize the techniques described above before deciding that the data arise from chaos rather than from uncontrolled stochastic influences.

8.2.4 Other Examples

Although the BZ reaction is by far the best characterized chaotic chemical system, other examples of chaos abound. Homogeneous reactions in a CSTR in which evidence for chaos has been found include the chlorite–thiosulfate (Orbán and Epstein, 1982), bromate–chlorite–iodide (Maselko et al., 1986) and chlorite–thiourea (Alamgir and Epstein, 1985a) reactions. Some chaotic time series from the periodic–chaotic sequence in the chlorite–thiosulfate reaction are shown in Figure 8.20. The bromate–chlorite–iodide system exemplifies two principles: that chaos can be generated by coupling periodic oscillators, and that one must exercise caution before declaring a set of aperiodic oscillations to be chaotic. The bromate–chlorite–iodide system consists of two chemically coupled oscillators (see Chapter 12 for details), neither of which has been found to behave chaotically by itself. As we noted in Chapter 7, initial efforts at modeling this reaction (Citri and Epstein, 1988) gave time series that looked very much like the experimentally observed chaos. However, further simulations with a tighter error parameter showed that accurate numerical simulation of the model under study gave only complex periodic behavior in the region of interest; the chaos was a numerical artifact.

Chaotic behavior has been observed in an enzyme-catalyzed reaction, the peroxidase-catalyzed oxidation of the reduced form of nicotinamide adenine dinucleotide (NADH) (Olsen and Degn, 1977), in the gas-phase reaction between carbon monoxide and oxygen (Johnson and Scott, 1990; Johnson et al., 1991), and in the delay-controlled photodissociation of peroxydisulfuryl difluoride, $S_2O_6F_2$ (Zimmerman et al., 1984). Heterogeneous systems also give rise to chaotic behavior. We note here the oxidation of carbon monoxide catalyzed by a platinum single crystal (Eiswirth et al., 1988) and the potentiostatic electrodissolution of copper (Bassett and Hudson, 1987) as being particularly rich sources of interesting dynamical behavior.

We have alluded to the philosophical implications of chaotic systems. Before moving on, we invite the reader to consider one more. Until quite recently, the experimental systems studied by the pioneers of every major field of science were decidedly deterministic and reproducible. In fact, one of the first things that budding scientists are taught is that any worthwhile experiment can be duplicated, both by the original experimenter and by those wishing to pursue the work. If

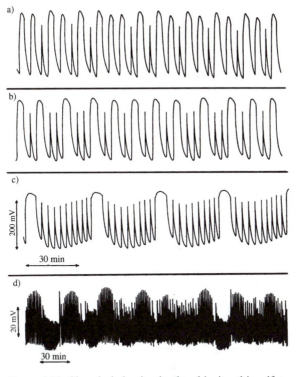

Figure 8.20 Chaotic behavior in the chlorite–thiosulfate reaction. Reprinted with permission from Orbán, M.; Epstein, I. R. 1982. "Complex Periodic and Aperiodic Oscillation in the Chlorite–Thiosulfate Reaction," *J. Phys. Chem. 86*, 3907–3910. © 1982 American Chemical Society.)

early scientists and philosophers had encountered and studied chaotic systems, would what we refer to as "the scientific method" look quite the way it looks today?

8.3 Controlling Chaos

There is no doubt that the concept of chaos has captured the imagination of the nonscientific community in a way that no other aspect of nonlinear dynamics has. An anecdotal history of the field (Gleick, 1987) was on the bestseller lists for weeks. One of us recently attended a conference on ethics and was astounded at being besieged by philosophers and policy makers seeking insights into their disciplines from the subject of chaos. Certainly, scientific ideas like relativity, the uncertainty principle, or chaos can provide helpful analogies for people in other fields, but it may be dangerous to take these analogies too literally. In the case of chaos, it can be particularly perilous to encourage those considering social problems either to throw up their hands and say that there are systems that we simply cannot control or, on the other hand, encouraged by papers like the one entitled "Exploiting Chaos to Predict the Future and Reduce Noise" (Farmer and Sidorowich, 1989), to believe that a solution to any problem is just a matter of finding the correct algorithm.

The notion of controlling chaos is obviously an attractive one, and, remarkably, chaos can be controlled in a certain sense. That sense, however, is a rather narrow one. In this section, we describe recent work aimed at controlling chaos by stabilizing a single, unstable periodic orbit embedded within a chaotic attractor. The basic idea, developed by Ott et al. (1990), is to maintain the system on the stable manifold of a particular unstable periodic orbit by introducing a carefully chosen, time-dependent perturbation of a control parameter. Their approach was successfully applied to an experimental system consisting of a chaotically vibrating magnetoelastic ribbon (Ditto et al., 1990), as well as to a diode resonator circuit (Hunt, 1991) and a chaotic laser (Roy et al., 1992). A variation of the method was employed by Garfinkel et al. (1992) to convert a drug-induced chaotic cardiac arrhythmia in a rabbit heart into periodic beating by the application of appropriate electrical stimuli.

We describe here an approach to controlling chaos in chemical systems pioneered by Showalter and collaborators (Peng et al., 1991). The algorithm does not require knowledge of the underlying differential equations, but rather works from an experimentally determined 1-D map—for example, the next amplitude map for the system. We consider the problem of stabilizing an unstable period-1 orbit, that is, an unstable fixed point in the 1-D map, of a chaotic system. We assume that we have measured the 1-D map and express that map as

$$x_{n+1}(p) = F(x_n, p) \qquad (8.19)$$

where p is a control parameter—for example, the flow rate through a CSTR. Now suppose that we change the parameter p by a small amount δp. The map will be changed, too, but if δp is small enough and we confine our attention to a narrow region of x—say, in the neighborhood of the fixed point x_s—the new map will be parallel to the old one.

Under these circumstances, the shift in x when we change p can be written as

$$\Delta x = g \, \delta p \qquad (8.20)$$

where g is a constant that depends on the partial derivatives of the map F, but operationally can be calculated by measuring the horizontal distance between the two maps F obtained at two nearby values of p. The trick, illustrated in Figure 8.21, is to choose δp so that a point $x_n(p)$ on the original map is shifted to a point $x_n(p + \delta p)$ on the new map and that will evolve into the fixed point $x_s(p)$ of the original map on the next iteration.

To carry out the scheme, we must wait until the system comes close to the fixed point.[4] If the value of x at this time is x_n, we must perturb the parameter p by an amount δp such that

---

[4]In the case of a chaotic attractor, this will always happen if we are willing to wait long enough, no matter how closely we insist that the system approach the fixed point.

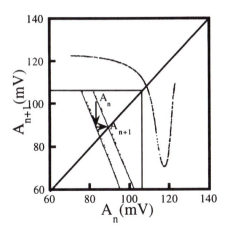

Figure 8.21 Example of the control algorithm where the next amplitude map is obtained from successive maxima of a bromide-sensitive electrode in the BZ reaction. Inset shows original map (left) and shifted map (right) in the neighborhood of the unstable steady state A_s. When the flow rate is shifted by 0.2%, the point A_n is shifted down to a point that evolves to the steady state A_s. (Adapted from Petrov et al., 1993.)

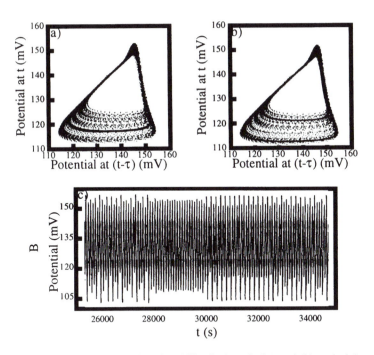

Figure 8.22 Phase portraits of stabilized (a) period-1 and (b) period-2 orbits embedded in a chaotic attractor in the BZ reaction. Scattered points show chaotic trajectory (delay time $\tau = 1.3$ s) before stabilization. (c) Time series showing potential of bromide-sensitive electrode. Control via change in input flow rate of cerium and bromate solutions was switched on from 27,800 s to 29,500 s to stabilize period-1 and from 30,000 s to 32,100 s to stabilize period-2. (Adapted from Petrov et al., 1993.)

$$\delta p = (x_n - x_s)/g \tag{8.21}$$

This perturbation will send the system toward x_s. Since the parameter value is no longer exactly p, we will not be exactly at the steady state, and we will need to make another correction using eq. (8.21) at the next return, that is, at a time T later, where T is the time interval used to discretize the time series. In practice, the corrections soon become small and the fixed point is stabilized. Higher period orbits can also be stabilized using the same algorithm. Figure 8.22 shows phase portraits and time series of period-1 and period-2 orbits selected from a chaotic attractor in the BZ reaction.

Controlling chaos, even in the rather limited fashion described here, is a powerful and attractive notion. Showalter (1995) compares the process to the instinctive, apparently random motions of a clown as he makes precise corrections aimed at stabilizing his precarious perch on the seat of a unicycle. A small, but carefully chosen, variation in the parameters of a system can convert that system's aperiodic behavior not only into periodic behavior, but also into any of a large (in principle, infinite) number of possible choices. In view of the appearance of chaos in systems ranging from chemical reactions in the laboratory, to lasers, to flames, to water faucets, to hearts, to brains, our ability first to understand and then to control this ubiquitous phenomenon is likely to have major consequences.

9

Transport and External Field Effects

Thus far, we have implicitly assumed that chemical species move only by diffusion. In fact, a number of external forces can affect mass transport, with significant and interesting effects on chemical waves. In this chapter, we consider three types of fields: gravitational, electric, and magnetic. These always exist, though their magnitudes are usually very small. As we shall see, small fields can have surprisingly large effects.

9.1 Gravitational Fields

9.1.1 Convection

Gravity is a ubiquitous force that all living and chemical systems experience. People largely ignored the profound effect that living with gravity has upon us until humans spent significant time in space. Bone loss and changes to the vascular systems of astronauts (Nicogossian et al., 1994) are still not well understood.

Eliminating the effects of gravity is not easy. Enormous cost and effort have been expended to simulate gravity-free conditions in drop towers, parabolic airplane flights, or in Earth orbit. A simple calculation seems to suggest that gravity should have negligible influence on chemical reactions. The mass of a molecule is on the order of 10^{-26} kg, which translates into a gravitational force of about 10^{-25} N. We can compare this with the force of attraction between the electron and the proton in a hydrogen atom, which is of the order 10^{-8} N. Even allowing for shielding effects, the electrostatic forces that cause chemical bonds to be made and broken will always be many orders of magnitude stronger than gravitational forces. So gravity does not affect the fundamental atomic and molecular interactions, but it can drastically alter the macroscopic transport of heat and matter through *convection*, or macroscopic fluid motion. Natural convection is the movement of fluid as the result of differences in density, so that denser fluid sinks and less dense fluid rises. This motion is resisted by the viscosity of the medium, which acts like friction does in slowing the motion of solids. The study of convection is an entire area of physics, and we will touch only on a few aspects. The reader is referred to some excellent texts on the subject (Tritton, 1988; Turner, 1979).

Convection is a much more efficient process than diffusion for transporting heat and matter. To appreciate that convection is a more effective mode of material transport, one need only consider what would happen if smoke in a fireplace were removed solely by diffusion. In a short time, the room would fill as the smoke particles dispersed randomly. Instead, if things work properly, the smoke goes up the chimney as the exothermic combustion reactions in the fire produce heat, which decreases the density of the gases and allows them to "float" up the flue because of buoyancy. We understand this idea of buoyancy intuitively when we say that "heat rises."

Bazsa and Epstein (1985) found that fronts in the nitric acid oxidation of iron(II) propagated as much as six times faster moving down a vertical tube than traveling up the same tube. This front velocity anisotropy, which we shall discuss in more detail later in this chapter, was a function of the tube diameter and could be eliminated by adding silica gel to the solution.

What causes this dramatic variation in the wavefront velocity? Is gravity really "pulling the wave down" in some sense? During an exothermic reaction[1] in a wavefront, the density can change either because of an increase in temperature or because of the change in chemical composition. This change in density can initiate the macroscopic transport of fluid. The fact that the exothermicity of a reaction can affect the solution density will probably not surprise most readers, but the effect of a change in chemical composition may. We know, though, that salty water is denser than fresh water and that a helium balloon rises in air because of its lower molecular mass. Composition changes in chemical reactions may be more subtle, but it is not unreasonable that a more highly charged ferric ion might hold polar molecules (e.g., the solvent water) more closely than the reactant ferrous ion. Changes in volume associated with chemical reactions are key pieces of data for thermodynamic analysis (Millero, 1972).

Consider a common reaction–diffusion system that exhibits sensitivity to orientation with respect to the force of gravity—a burning cigarette. The natural convection resulting from the exothermicity of the chemical wavefront in a cigarette has a large effect on the rate of propagation. The heat produced expands the gas mixture, decreasing its density, and subjecting it to a buoyant force in the upward direction. This is how the smoke rises from a lit cigarette—the combustion products are lighter than the surrounding air and rise like a hot-air balloon. A

---

[1] No endothermic reactions are known to propagate waves.

cigarette that is held filter up will burn as much as 30% faster than if it burns downward (Pojman, 1990).

9.1.2 Factors Affecting Solution Density

The density of a solution is a function of its temperature and composition. We can define coefficients that relate the density ρ to changes in temperature T and composition C.

$$\alpha = -\frac{1}{\rho}\frac{\partial \rho}{\partial T} \tag{9.1}$$

$$\beta_i = \frac{1}{\rho}\frac{\partial \rho}{\partial C_i} \tag{9.2}$$

where C_i is the molar concentration of the ith species. To a first approximation, the total density of the solution perturbed from a set of reference conditions ρ_0, T_0, C_{0i} is:

$$\rho = \rho_0[1 - \alpha(T - T_0) + \Sigma\beta_i(C - C_0)_i] \tag{9.3}$$

As the wavefront proceeds, the region into which the wave has not yet propagated will be at a different temperature than the region through which it has already passed. This can cause a density gradient, which may result in convection. Even if the reaction neither consumes nor generates heat, density gradients can arise if the partial molal volumes of the products differ from those of the reactants, and these gradients may be sufficient to cause convection.

If V_0 is the initial volume of the reaction mixture, the volume of reaction ΔV can be related to the change in density $\Delta\rho_c$ caused by the change in chemical composition by

$$\Delta\rho_c = \frac{1}{1 + \Delta V/V_0}\rho_0 - \rho_0 \equiv \beta\Delta C\rho_0 \tag{9.4}$$

where the middle equality holds if $\Delta V \ll V_0$, ρ_0 is the initial density of the solution, ΔC is the change in concentration of the limiting reagent, and β is the mean molar density coefficient. A positive ΔV corresponds to a decrease in density.

From α and the heat capacity C_p of the reaction solution, we can calculate the density change $\Delta\rho_T$ caused by the enthalpy change of an adiabatic reaction:

$$\Delta\rho_T = \Delta H \cdot \Delta C \cdot \frac{1}{C_p} \cdot \alpha \cdot \rho_0 \tag{9.5}$$

Alternatively, the change in temperature, $(-\Delta H \cdot \Delta C)/C_p$, can be measured directly.

9.1.3 Simple Convection

We now examine a reaction in which ΔV and ΔH are of opposite sign. In other words, the reaction is exothermic with an isothermal increase in volume during the reaction. These effects will add to cause a net density decrease:

$$\Delta\rho = \Delta\rho_c + \Delta\rho_T \qquad (9.6)$$

Consider an adiabatic vertical tube containing a solution in which a chemical wave may propagate. The wave can be initiated from either the top or the bottom. If the front is initiated at the top, a descending wavefront will ensue. Because $\Delta\rho$ is negative, the density of the solution will be lower above the front than below it—a stable situation. Figure 9.1 shows an ascending front, which results in a density gradient in which the less dense fluid is under the more dense one. This configuration may be unstable. Several factors will affect whether free convection will occur. The greater the density gradient, the less stable is the column of fluid, since the gravitational force that tends to cause fluid motion will be larger. A narrow tube increases stability, as does a large viscosity, because these factors tend to oppose fluid motion. The final factor is the value of the transport coefficient for the quantity causing the gradient. For a solute gradient, the diffusion coefficient is relevant, because it is a measure of how rapidly a perturbation in concentration will diffuse away. For a temperature gradient, the thermal diffusivity is the appropriate parameter.

The Rayleigh number Ra is a measure of the convectional stability of a fluid system and is defined for a vertical cylinder of radius r as (Cussler, 1984):

$$Ra = \frac{r^4 g}{\mu Tr}\frac{d\rho}{dz} \qquad (9.7)$$

Tr is the transport coefficient ($cm^2\ s^{-1}$) of the quantity causing the density gradient along the vertical (z) axis, and μ is the dynamic viscosity ($g\ s^{-1}\ cm^{-1}$). A thermally induced density gradient must be 100 times as large as a concentration-induced gradient to have the same effect on the stability because the diffusivity of heat is about 100 times as great as most diffusion coefficients. When both thermal expansion and change in chemical composition contribute to the density change, the total Rayleigh number is

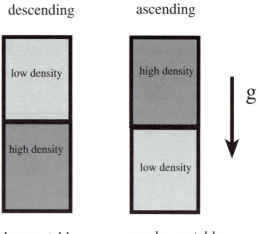

Figure 9.1 Density gradient induced by a descending wavefront is always stable if $\Delta\rho < 0$. Density gradient induced by an ascending wavefront may be unstable. The direction of gravitational force is indicated by g.

$$Ra = \frac{r^4 g}{\mu} \left(\frac{1}{\kappa} \frac{d\rho_T}{\partial z} + \frac{1}{D} \frac{d\rho_c}{dz} \right) \tag{9.8}$$

Taylor calculated that convection cannot occur for a cylinder with nonconducting sidewalls and free upper and lower boundaries as long as $Ra < 67.9$ (Taylor, 1954). This analysis indicates that there exists a maximum tube radius at which the wavefront propagates by pure reaction–diffusion processes. If the density gradients are very large, the tube will need to be very narrow to prevent convection. Any convective fluid flow will affect the wavefront velocity. Taylor predicted that the first convective mode that becomes unstable at the critical value $Ra = 67.9$ is the antisymmetric flow depicted in Figure 9.2. If the reaction is run in a wider tube, the Rayleigh number increases according to eq. (9.7) or eq. (9.8). Gershuni and Zhukhovitsky (1976) analyzed the modes of convection that occur at higher values of the Rayleigh number. When $Ra = 452.0$, the flat front is unstable to axisymmetric perturbations, with fluid rising in the center of the cylinder and descending along the sides. Such a flow would distort the wavefront into a parabolic shape as the less dense reacted solution wells up into the unreacted zone. Figure 9.2 presents both antisymmetric and axisymmetric flows.

One might expect that performing the experiment in a horizontal tube would eliminate convection. In fact, this is not true. The density distribution (shown in Figure 9.3) of such a system corresponds to that of a horizontal cylinder with the ends at different temperatures. The horizontal tube will have an antisymmetric fluid flow with the shape indicated in the diagram. The density gradient caused by the chemical reaction causes a fluid flow, which distorts the flat front from the shape that would result from reaction and diffusion alone. This configuration will always allow convection, no matter what the radius of the tube or how small the density gradient.

9.1.4 Experimental Example: The Iodate–Arsenous Acid System

The iodate–arsenous acid system, which we have encountered in Chapters 2 and 6, is an excellent system for study. The density changes have been measured under homogeneous conditions, and the two factors act in the same direction, that is, to decrease the density. Pojman et al. (1991b) studied simple convection in this reaction and found all the qualitative features described in the previous section,

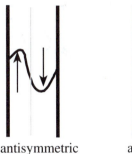

antisymmetric

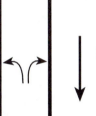

axisymmetric

g

Figure 9.2 (Left) Convective fluid flow and effect on the waveform, slightly above the critical Rayleigh number. (Right) Fluid flow at Rayleigh numbers greatly exceeding the critical value, where axisymmetric flow is expected. The waveform has a parabolic shape.

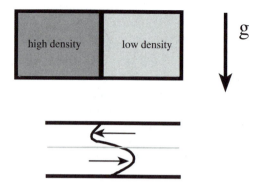

Figure 9.3 (Top) Wavefront propagating from left to right in a horizontal tube. (Bottom) An antisymmetric fluid flow results, causing a slanted wavefront.

except for the antisymmetric modes in ascending fronts. Masere et al. (1994) modeled the iodate–arsenous acid system and tested the prediction that in narrower tubes, antisymmetric convection should be observed. Their stability analysis assumes a sharp interface between the reacted and unreacted solutions and characterizes the stability in terms of a dimensionless number, S:

$$S = \frac{\delta g r^3}{\nu D} \tag{9.9}$$

where ν is the kinematic viscosity ($cm^2 s^{-1}$, μ/ρ) and δ is the fractional difference in density between reacted and unreacted solution. This approach has the advantage that it does not require a measurement of the front width, which is fraught with error and difficulty. Masere et al. predict that antisymmetric convection should occur when S exceeds 87.9.

In Figure 9.4, the front profiles in several tubes are shown. There is a clear transition from antisymmetric to axisymmetric convection for ascending fronts with increasing tube diameter. Descending fronts are flat, as predicted. From these experiments, the critical radius was determined to be 0.055 ± 0.005 cm, corresponding to a value of S of 77 ± 21, in agreement with the theoretical value. Pojman et al. (1991b) had predicted a critical radius of 0.06 cm.

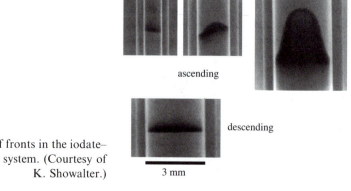

ascending

descending

Figure 9.4 Images of fronts in the iodate–arsenous acid system. (Courtesy of K. Showalter.)

3 mm

9.1.5 Multicomponent Convection

Bazsa and Epstein (1985) observed in the iron(II)–nitric acid system that in a vertical tube, descending fronts propagate more rapidly than ascending ones. The reaction is mildly exothermic. Surprisingly, not only do descending fronts propagate more rapidly than ascending ones and, necessarily, more rapidly than pure reaction–diffusion fronts, but also ascending fronts propagate more rapidly than pure reaction–diffusion fronts. Simple convection is clearly not the operative mechanism for this last observation. A mode of convection that increases the front velocity for both ascending and descending fronts, even when the density gradient is statically stable, must come into play. Such a form of convection, which can occur even with an apparently stable density gradient, is called *multicomponent* or *double-diffusive convection*.

Suppose that a wave propagates in a system for which $\Delta \rho_c$ and $\Delta \rho_T$ are of opposite signs, that is, the reaction is exothermic with an isothermal decrease in volume. Remarkably, in this configuration, even if the net density change is zero, or corresponds to what would appear to be a stable density gradient, convection may occur that significantly affects the mass transport.

In our previous case, the two components (heat and solutes) acted together in either destabilizing or stabilizing the fluid, depending upon the direction of wave propagation. Consider for a moment a simpler system that contains a single solute S. Instead of a complicated mixture of solutes, we treat $\Delta \rho_c$ (caused by the concentration gradients) as resulting from a concentration gradient of S, whose β [eq. (9.2)] is defined by

$$\frac{1}{\rho_0} \frac{d\rho_c}{dz} = \beta \frac{d[S]}{dz} \tag{9.10}$$

The case of simple convection with a wavefront propagating upward is analogous to the problem of hot, fresh water under cold, salty water in which

$$\beta \Delta[S] = \frac{\Delta \rho_c}{\rho_0} \tag{9.11}$$

Now consider hot, salty water above cold, fresh water, as depicted in Figure 9.5. The system may appear to be stable, if the density decreases with height. Yet, it may not be; kinetics matters, even when there is no chemical reaction taking place. Imagine that, at the interface between the layers, a small parcel of the upper solution were to deviate from its position by descending into the cold, fresh region. Because the temperature and concentration are higher than in the surrounding region, heat and salt will diffuse out. The heat will leave at a greater rate, because of the larger diffusivity of heat. Now the parcel is cool and dense; because of its higher concentration, it continues to sink. Similarly, if a parcel of cold, fresh water protrudes into the hot, salty layer, heat will diffuse in faster than the salt. This will leave the parcel less dense than the surrounding layer, and the parcel will rise further. What results are known as "salt fingers," which appear as long slender regions of alternately descending and ascending fluid.

If cold, fresh water overlies denser, hot, salty water, one would expect the mass flux to be equal to that caused by Fick's law. However, we again consider a

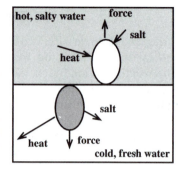

Figure 9.5 The mechanism of double-diffusive convection in the fingering regime. $\Delta\rho_c$ and $\Delta\rho_T$ are in the opposite direction, and the net density gradient appears to be stable. However, if a small parcel of the warm, salty solution enters the lower section, then because the heat diffuses faster than the salt, the parcel is left with a greater density than the surrounding medium, and it sinks.

perturbation at the interface, as depicted in Figure 9.6. The heat diffuses faster than the salt, leaving the parcel heavier than the surrounding region. It sinks and can continue through the interface into the hot, salty region, where it again heats up and rises. This oscillatory behavior will continue as long as the density gradients persist. This oscillatory process increases the effective area of the interface, thereby increasing the flux of salt. The heat released into the region above the interface can cause convection, which further increases the mass transport. The effective mass transport coefficient of salt under these circumstances may be ten times as great as the diffusion coefficient (Turner, 1965)!

Both of the above types of convection are categorized as "double-diffusive" convection ("thermohaline" when the two components are heat and salt), and, more generally, "multicomponent convection." It is not necessary that the two components be a solute and heat. Any two materials with different diffusivities can cause these phenomena, even if the differences are as small as with salt and sugar. This phenomenon has been extensively studied by oceanographers because of the role it plays in ocean current mixing. The two behaviors we have described correspond to the fingering and diffusive regimes of double-diffusive convection.

Nagypál et al. (1986) discovered a significant gravitational anisotropy in the chlorite–thiosulfate system. Plotting the front velocity vs. the angle of tube orientation, they prepared "pumpkin"-shaped diagrams (Figure 9.7). It was possible to reverse the anisotropy between ascending and descending fronts by changing the chlorite concentration, thus affecting the relative contributions of the chemical heat production and the physical heat to the density change.

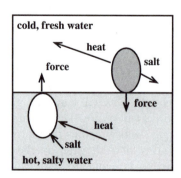

Figure 9.6 The mechanism of double-diffusive convection in the diffusive regime. The parcel of hot, salty solution that enters the cold, fresh region above loses heat, becoming more dense than the surrounding region. It sinks back to the hot region and regains heat, leading to an oscillatory motion.

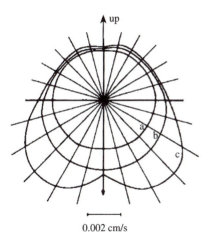

up

0.002 cm/s

Figure 9.7 Polar coordinate plot of front velocity vs. direction of propagation in the chlorite–thiosulfate system. $[ClO_2^-] = 0.009$ M, $[S_2O_3^{2-}] = 0.0045$, $[OH^-] = 0.002$ M. Distance from the origin is proportional to the speed of propagation. Tube diameter (i.d., mm): (a) 0.794, (b) 3.17, (c) 4.76. Reprinted with permission from Nagypál, I.; Bazsa, G.; Epstein, I. R. 1986. "Gravity Induced Anisotropies in Chemical Waves," *J. Am. Chem. Soc. 108*, 3635–3640. © 1986 American Chemical Society.)

Pojman et al. (1991c) studied the iron(II)–nitric acid system in detail. They measured an isothermal density change ($\Delta\rho_c$) of $5.4 \pm 0.6 \times 10^{-4}$ g cm^{-3} and a thermally induced density change ($\Delta\rho_T$) of -5.4×10^{-4} g cm^{-3} Thus, there was almost no net density difference across the front. A monotonic increase in velocity was observed with increasing nitric acid concentration (Figure 9.8). Convection-free fronts exhibit similar behavior in experiments (Póta et al., 1991) and simulations (Pojman et al., 1991a), but with much smaller velocities. The values of $\Delta\rho_c$ and $\Delta\rho_T$ are not affected by the nitric acid concentration. Why does the velocity increase? At higher nitric acid concentrations, the reaction occurs more rapidly. The amount of heat liberated is the same as at lower HNO_3 concentrations since the amount of reaction is determined by the concentration of the limiting reagent, iron (II), but the thermal gradient is larger because the heat has less time to diffuse. Also, the faster the heat is released, the smaller the fraction of the heat absorbed by the tube wall. The maximum temperature and the thermal gradient are therefore larger. The larger the thermal gradient, the more convection occurs above the front and the greater the mass transport. This increased convection-assisted mass transport is the origin of the increased front velocity.

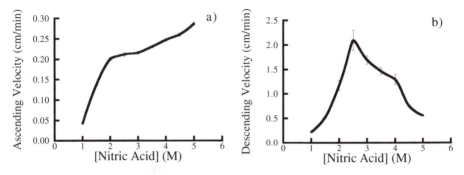

Figure 9.8 (a) Ascending velocity and (b) descending velocity of fronts in the iron(II)–nitric acid system as a function of initial nitric acid concentration.

The dependence of the velocity on the nitric acid concentration has a puzzling feature. A maximum occurs in the descending velocity at 2.5 M nitric acid, as shown in Figure 9.8. As noted above, the convection-free velocity increases monotonically with increasing nitric acid concentration, but with values an order of magnitude smaller. What is the origin of the maximum? Let us repeat our stability analysis for the fluid, with temperature and concentration gradients created by a descending wavefront. Suppose a warm parcel of the reacted solution enters the lower region by moving a distance δz. The time required for enough heat to diffuse out until the density of the parcel is greater than the unreacted solution is τ. If the reaction–diffusion wavefront velocity is v, then the time until the parcel is surrounded by solution of the same composition is $\delta z/v$. If $\tau > \delta z/v$, then no fingers can form. If the reaction is very exothermic, τ will be large and the descending fingers may be suppressed while the ascending ones still occur. That is, the descending fingers may not occur because the front "catches up" with the finger before it can completely form. However, the ascending finger can proceed and will be assisted by conversion of solution around it to hot, reacted solution.

Nagy and coworkers tested this interpretation in the chlorate–sulfite system (Nagy and Pojman, 1993). The isothermal density change was $+0.48\%$, and the temperature increase was a very large 40 °C. The "salt fingers" that are observed only ascend, in agreement with the above analysis (Figure 9.9).

9.1.6 Iodide–Nitric Acid System

Lest we leave the reader a false sense of security about predicting the type of convection that will occur in a front based solely on the relative signs of the isothermal and thermal density changes, we will now consider the oxidation of iodide by nitric acid. The thermal density change ($\Delta\rho_T < 0$) and the isothermal density change ($\Delta\rho_c < 0$) have the same sign. According to our previous discussion, simple convection should occur, as in the iodate–arsenous acid system (Pojman et al., 1991b). However, the shapes of the wavefronts suggest that something more is going on. In the iodate–arsenous acid system, the descending front is flat and the ascending one has a parabolic shape, while in the iodide–nitric acid system, the descending front is parabolic and the ascending one is flat. A signifi-

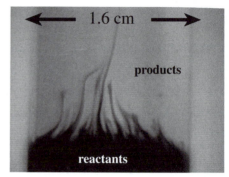

Figure 9.9 A descending front in the chlorate–sulfite system. Descending fingers are suppressed by the large thermal gradient.

cant anisotropy in front propagation is observed in the latter system. Despite the fact that only simple convection is expected, double-diffusive convection is observed with larger tube diameters. If simple convection were the only mechanism for fluid motion, then the descending front velocity would be the same as the pure reaction–diffusion velocity, but this is not the case.

The thermal gradient (with ΔT of 0.005–0.008 °C) in the iodide–nitric acid system is much smaller than in the iodate–arsenous acid system (where $\Delta T = 0.7$ °C). The thermal gradient plays no significant role in the former system as long as $\Delta \rho_c < 0$. The isothermal volume changes are comparable in the two systems.

These observations suggest that double-diffusive convection in the iodide–nitric acid system arises via another mechanism than the one we have sketched above. Such a route can be provided by two species having relatively large differences in their diffusion constants, as described by Turner (1985). In this case, we can use the same approach we took earlier to analyze an isothermal system consisting of two layers of isopycnic solutions containing two species with different diffusion coefficients, such as salt and sugar. Diffusion from one layer provides greater material flux to the layer with the less mobile component than diffusion in the reverse direction. A local density change thus occurs that can initiate fluid motion. This mechanism can create conditions for double-diffusive convection similar to the situation at the interface of the hot, salty water and the cold, fresh water. If the analysis above is valid, we should find fingering, even under unreactive conditions, if we have the same key species as in the iodide–nitric acid system. Nagy and Pojman demonstrated that the two species with different diffusion coefficients are the triiodide–starch complex and iodide (Nagy et al., 1994). The iodate–sulfite system also exhibits fingering because of the faster diffusion of sulfuric acid relative to the other components (Keresztessy et al., 1995; Pojman et al., 1996c).

9.1.7 Surface-Tension-Driven Convection

Even in the absence of buoyant forces, convection can occur, driven by gradients in surface tension at the interface of two fluids. In the experiments we have described in tubes, surface tension is not important. When some of the same reactions are performed in thin layers with a free aqueous–air interface, complex patterns can emerge. Evaporation from the surface can cause temperature gradients, so careful experiments need to be performed under an atmosphere saturated with the volatile component(s).

Surface tension is affected both by chemical concentration and by temperature. Figure 9.10 shows how a hot spot can cause convection by locally lowering the surface tension. The cooler fluid has a higher surface tension and draws the warm fluid towards itself. If the temperature gradient is perpendicular to the interface, both buoyancy-driven convection and Marangoni (surface-tension-driven) convection are possible (Antar and Nuotio-Antar, 1993).

There are two dimensionless numbers that characterize the stability of the configuration shown in Figure 9.10b. We have seen the Rayleigh number before but now formulate it in terms of the temperature difference, ΔT, and the layer thickness, d:

Figure 9.10 (a) A temperature gradient along an interface between two fluids causes a gradient in surface tension, which always results in fluid motion. (b) A difference in temperature between the interface and the bottom of the container may cause convection, depending on the temperature difference, the transport coefficients, and the depth of the container.

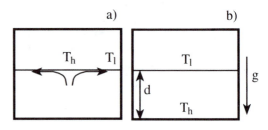

$$Ra = \frac{\partial \rho}{\partial T} \frac{\Delta T d^3}{\mu \kappa} \qquad (9.12)$$

We now introduce the Marangoni number,

$$Ma = \frac{\partial \sigma}{\partial T} \frac{\Delta T d}{\mu \kappa} \qquad (9.13)$$

where σ/T is the variation of the surface tension with temperature. Because the Marangoni number, which governs surface convection, depends on the first power of the layer depth, while the Rayleigh number, which describes bulk convection, depends on d^3, surface-tension-driven convection dominates in thin layers and in weightlessness. [This last fact was discovered in an experiment on the *Apollo 14* space mission (Legros et al., 1990).]

A chemical wave propagating in a thin layer causes thermal and concentration gradients. The surface tension is also a function of chemical concentration, so even an isothermal front will cause Marangoni convection. Simoyi and coworkers have elegantly demonstrated the relative effects of three modes of front propagation in the chlorite–thiourea–barium chloride reaction system. A front was initiated at the base of a barrier positioned at 45 °C to the vertical. The front tended to spread out horizontally via reaction and diffusion. It rose because of buoyancy. When it reached the surface of the fluid, it rapidly propagated horizontally because of Marangoni convection. The front caused a local temperature jump of 3 °C, which resulted in a 24×10^{-5} N cm^{-1} change in surface tension (Hauser and Simoyi, 1994a). By far the largest effect on front propagation was surface tension, as can be seen in Figure 9.11. The authors also confirmed this by lowering the surface tension dependence on temperature by covering the solution with hexane, which reduced the horizontal front velocity.

9.2 Electric and Magnetic Fields

We normally do not consider that electromagnetic fields of even moderate strength might affect chemical reactions or living organisms. However, the issue of whether fields from power lines and computer terminals can have negative consequences for human health is currently quite controversial (Pinsky, 1995),

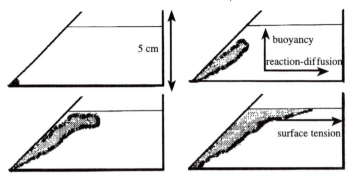

Figure 9.11 The relative effects of reaction–diffusion, buoyancy and Marangoni convection were demonstrated by Simoyi and colleagues in the chlorite–thiourea–barium chloride reaction system (Hauser and Simoyi, 1994b). The front rises along the tilted side of the container faster than it propagates horizontally. When the front reaches the interface, its horizontal propagation accelerates because of surface-tension-induced convection. (Courtesy of R. Simoyi.)

and, amazing as it seems, some animals are able to sense even the Earth's tiny magnetic field. Birds and sea turtles have been shown to use it for navigation (Lohmann, 1992). We consider here the effects of electric and magnetic fields upon chemical waves.

9.2.1 Electric Fields

When a charged species in solution experiences an electric field, it will move more rapidly than by simple diffusion in a direction determined by the polarity of the field and the charge on the ion. The magnitude of an ion's velocity in a constant electric field is proportional to the field strength; the proportionality constant is the *mobility* of the ion, which depends primarily on its charge and its size. Differences in mobility provide the basis for electrophoresis, an important technique used to separate proteins and DNA fragments (Gaal et al., 1980). Because most inorganic ions are about the same size, their diffusion coefficients are similar. In an electric field, the charge on a species is a more important determinant of its mobility than its size. Ions of different mobilities can move at different rates in a field and can be forced, by choosing the polarity of the field, to move in either direction, even against a concentration gradient.

Given that many of the important intermediates in the BZ system are ionic, we might expect an applied electric field to affect the propagation of waves in a BZ system. Schmidt and Ortoleva (1977, 1979, 1981) were the first to consider this problem theoretically. They also performed experiments in a quasi-two-dimensional configuration, but the results were only qualitative because of ohmic heating, evaporation and electrolytic side products (Feeney et al., 1981).

Ševčíková and Marek (1983) carried out experiments in capillary tubes under isothermal conditions. They observed qualitatively different phenomena depending on the polarity of the applied field. A negative field of -20 V cm^{-1} (meaning the front propagates toward the positive electrode) accelerates the waves from 2.3 mm min^{-1} to 5.7 mm min^{-1} (Figure 9.12). With a field of -40 V cm^{-1}, the waves propagate as much as five times faster than normal. The applied field causes the bromide ions to migrate ahead of the wavefront, decreasing their concentration faster than in the absence of a field. The lower bromide concentration allows the wave to propagate faster.

If the field is reversed, surprising phenomena result. Waves are slowed down, as we might expect, but if the field is high enough, waves are annihilated and/or new ones traveling in the opposite direction split off. For fields between about 10 and 20 V cm^{-1}, new waves appear. Figure 9.13 shows an example. Smaller numbers of new waves can be produced by applying the field for only a short time.

At higher positive field strengths, the original wave is annihilated, and a new one is formed that propagates in the opposite direction. Figure 9.14 shows a wave annihilated by a field of 20 V cm^{-1} and the generation of a new wave that propagates faster in the opposition direction.

The positive field causes bromide to migrate into the advancing wave, slowing down the autocatalytic oxidation of ferroin and, thus, wave propagation. At high positive fields, the positively charged Fe^{2+} and Fe^{3+} migrate ahead of the front, and the bromide ions build up in the refractory region behind it. The low bromide region is spread out until it is wide enough for a new wave to initiate and propagate. Figure 9.15 shows the predicted concentration profiles during the splitting process. Wave extinction occurs when sufficient bromide migrates into the wave-

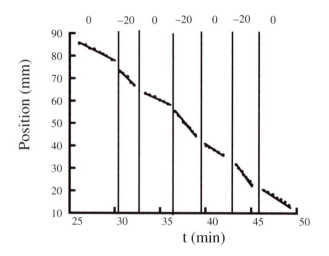

Figure 9.12 Dependence of wavefront position on time with and without an applied electric field (indicated on top in V). (Adapted from Ševčíková and Marek, 1983.)

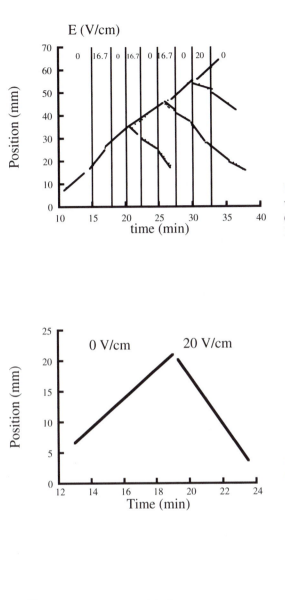

Figure 9.13 Generation of new waves by applying an electric field. (Adapted from Ševčíková and Marek, 1983.)

Figure 9.14 Annihilation and formation of a new wave by a $20\,\mathrm{V\,cm^{-1}}$ field. (Adapted from Ševčíková and Marek, 1983.)

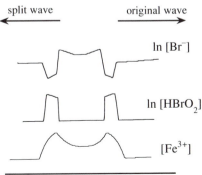

split wave ← → original wave

ln [Br⁻]

ln [HBrO₂]

[Fe³⁺]

Figure 9.15 Schematic concentration profiles during wave splitting. (Adapted from Ševčíková and Marek, 1983.)

front to increase the rate of the Br^-–$HBrO_2$ reaction enough to prevent the autocatalytic oxidation of ferroin.

Marek and coworkers simulated the experiments using an abstract two-variable propagator–controller model (Ševčíková and Marek, 1986). To include electric field effects, they added another term to the reaction–diffusion equation:

$$\frac{\partial C_i}{\partial t} = D_i \frac{\partial^2 C_i}{\partial x^2} + D_i z_i U \frac{\partial C_i}{\partial x} + R_i(\mathbf{C}) \tag{9.14}$$

where z_i is the charge on species i, U is the applied electric field, and R_i represents the dependence of the reaction terms on the concentration vector $\mathbf{C}$. By varying the charges assigned to the reacting species, they were able to obtain good qualitative agreement with experiment. In particular, they found acceleration and deceleration of wavefronts, wave splitting, and annihilation. An example of their calculated results is shown in Figure 9.16.

A more ambitious effort to examine the effects of electric fields was carried out by Münster et al. (1994). These authors modified the Brusselator model of eqs. (1.7)–(1.10) by assigning positive charges to the species X and Y:

$$A \rightarrow X^+ + C^- \qquad k_1 \tag{9.15}$$

$$B + X^+ \rightarrow Y^+ + D \qquad k_2 \tag{9.16}$$

$$2X^+ + Y^+ \rightarrow 3X^+ \qquad k_3 \tag{9.17}$$

$$X^+ \rightarrow X^{*+} \qquad k_4 \tag{9.18}$$

$$X^{*+} + C^- \rightarrow E \qquad k_5 \tag{9.19}$$

where the asterisk indicates an unstable intermediate. Comparison with the original Brusselator equations reveals that a new species C^- has been introduced in order to preserve the charge balance in each reaction. In order to maintain the kinetics, eq. (1.10) has been broken up into two steps, eqs. (9.18) and (9.19), with the last reaction taken to be very fast so that eq. (9.18) is rate-determining. The variable species concentrations are taken to be those of X^+, Y^+, and C^-. The form of the reaction–diffusion equations is similar to that of eq. (9.14).

In these calculations, the authors do not assume that the electric field strength U is constant. Rather, they examine this hypothesis, using the Nernst–Planck equation, which relates the fluxes of charged species to the electric field gradient. Even assuming local electric neutrality, that is, $[X^+] + [Y^+] = [C^-]$, they find nonnegligible spatial variations in the electric

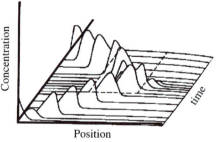

Figure 9.16 Simulated concentration profiles during wave reversal. (Adapted from Ševčíková and Marek, 1986.)

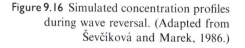

field intensity as a result of the nonconstant concentration profiles of the charged species. The behavior obtained by taking this effect into account differs significantly from simulations that assume a constant electric field. When an external electric field is applied, both the constant and the variable electric field models show the onset of traveling waves and transitions between striped and hexagonal Turing patterns like those seen experimentally in the chorite–iodide–malonic acid system (see Chapter 14).

Hasal et al. (1997) used similar techniques to model the effects of external electric fields on calcium waves like those shown in Figure 13.3. The phenomena observed resemble those found experimentally in the BZ system. Calcium waves play key roles in many biological phenomena and in neuronal–glial networks, for example, they are closely associated with local electrical field gradients arising from spatiotemporal patterns of ionic concentrations (Cooper, 1995).

9.2.2 Magnetic Fields

Nagypál and coworkers made the surprising discovery that some chemical waves are profoundly affected by moderate magnetic fields, such as the field produced by a large horseshoe magnet (Boga et al., 1990). They studied propagating fronts in the autoxidation of benzaldehyde by oxygen in glacial acetic acid catalyzed by cobalt(II). The initial solution is paramagnetic (meaning it is attracted into a magnetic field of increasing strength) because of the Co(II) and the O_2. The reacted solution is diamagnetic because of the Co(III). A capillary tube was placed beside the pole of an electromagnet. A front stopped on reaching the edge of the field, remaining there up to 8 h until the magnet was turned off; the front then resumed propagation.

Nagypál's group performed quasi-two-dimensional experiments with a magnetic field that they had quantitatively mapped. The field strength varied from 0.6 T to 0 over a distance of 8 cm. The front accelerated by as much as five times when going "downhill," that is, into regions of decreasing magnetic field strength. The front could not propagate up a gradient of increasing field strength. It is not the strength of the field that matters but the gradient. Because the unreacted solution is attracted up a gradient and the reacted solution is repelled down a field gradient, inhomogeneities in the magnetic field act against diffusion. In fact, a concentration gradient between Co(II) and Co(III) did not decay in an appropriate magnetic field gradient.

In the two-dimensional experiments, the inhomogeneity of the field causes a wave that would normally be circular to develop an elliptical shape. This field-induced distortion is shown in Figure 9.17, which also illustrates the variation in the field strength. He et al. (1994) demonstrated similar effects in a family of autocatalytic reactions in which hydrogen peroxide is reduced by complexes of Co(II) with polydentate organic ligands such as ethylenediaminetetra-acetic acid (EDTA). The mechanism for the magnetic field effect appears to be the same as in the benzaldehyde autoxidation reaction, but the hydrogen peroxide reduction, particularly with the EDTA complex, is an easier system to work with.

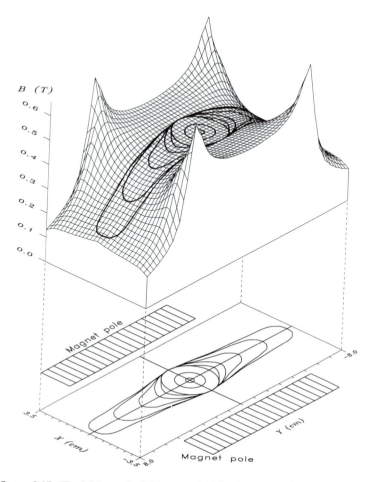

Figure 9.17 (Top) Magnetic field strength B in the central horizontal plane of two flat cylindrical magnetic poles (diameter 10 cm) when they are 8 cm apart. (Bottom) Time development of the wavefront in a solution containing benzaldehyde and cobalt(II) acetate in oxygen-saturated glacial acetic acid as it spreads in this quasi-two-dimensional system. (Reprinted with permission from Boga, E.; Kádár, S.; Peintler, G.; Nagypál, I. 1990. "Effect of Magnetic Fields on a Propagating Reaction Front," *Nature 347*, 749–751. © 1990 Macmillan Magazines Limited.)

9.3 Conclusions

In this chapter, we have considered three external forces that can affect chemical systems away from equilibrium. Gravity affects chemical reactions through natural convection, which changes the mass and heat transport through the fluid. Even without buoyancy, surface-tension gradients can significantly affect chemical waves. Electric fields, which alter the movement of ionic species, not only can change the speed of propagation, but also can even reverse the direction of a chemical wave. Magnetic fields from even simple bar magnets can change the velocity of, stop, or distort fronts if paramagnetic and diamagnetic species are involved. The practical effects of these phenomena are largely unknown, but they may be quite significant, particularly in living systems, where transport of ionic species is known to play a key role in such processes as morphogenesis (Créton et al., 1992) and signal propagation (Lechleiter and Clapham, 1992).

10

Delays and Differential Delay Equations

Mathematically speaking, the most important tools used by the chemical kineticist to study chemical reactions like the ones we have been considering are sets of coupled, first-order, ordinary differential equations that describe the changes in time of the concentrations of species in the system, that is, the rate laws derived from the Law of Mass Action. In order to obtain equations of this type, one must make a number of key assumptions, some of which are usually explicit, others more hidden. We have treated only isothermal systems, thereby obtaining polynomial rate laws instead of the transcendental expressions that would result if the temperature were taken as a variable, a step that would be necessary if we were to consider *thermochemical oscillators* (Gray and Scott, 1990), for example, combustion reactions at metal surfaces. What is perhaps less obvious is that our equations constitute an average over quantum mechanical microstates, allowing us to employ a relatively small number of bulk concentrations as our dependent variables, rather than having to keep track of the populations of different states that react at different rates. Our treatment ignores fluctuations, so that we may utilize deterministic equations rather than a stochastic or a master equation formulation (Gardiner, 1990). Whenever we employ ordinary differential equations, we are making the approximation that the medium is well mixed, with all species uniformly distributed; any spatial gradients (and we see in several other chapters that these can play a key role) require the inclusion of diffusion terms and the use of partial differential equations. All of these assumptions or approximations are well known, and in all cases chemists have more elaborate techniques at their disposal for treating these effects more exactly, should that be desirable.

Another, less widely appreciated idealization in chemical kinetics is that phenomena take place instantaneously—that a change in $[A]$ at time t generates a change in $[B]$ time t and not at some later time $t + \tau$. On a microscopic level, it is clear that this state of affairs cannot hold. At the very least, a molecular event taking place at point x and time t can affect a molecule at point x' only after a time of the order of $(x - x')^2/2D$, where D is the relevant diffusion constant. The consequences of this observation at the macroscopic level are not obvious, but, as we shall see in the examples below, it may sometimes be useful to introduce delays explicitly in modeling complex reaction networks, particularly if the mechanism is not known in detail.

Although mathematical biologists have been using delay models for some time (Sharpe and Lotka, 1923; May, 1974), relatively few studies have been undertaken of the effects of including delay in describing chemical kinetics, though interest in delay effects is growing rapidly. We believe that the time is ripe for chemists to exploit these approaches as well. In this chapter, we introduce some of the mathematical techniques available for treating systems with delay and look at some examples of chemical systems that show delay effects.

The equations of interest are *differential delay* or *differential difference equations* (DDEs), equations in which the time derivatives of a function depend not simply on the current value of the independent variable t, but on one or more earlier values $t - \tau_1, t - \tau_2, \ldots$ as well. In this chapter, we deal primarily with problems involving a single delay, since they are the most common and the most tractable mathematically. As our first example, we consider the prototype equation

$$dx(t)/dt = -kx(t - \tau) \tag{10.1}$$

Equation (10.1) is perhaps the simplest nontrivial differential delay equation. It is the analog of the rate equation that describes unimolecular decay. As we shall see, eq. (10.1) offers a far richer range of behavior.

10.1 Solving Differential Delay Equations

Since most readers will be unfamiliar with the mathematics of DDEs, we devote this section and the next to some fundamental notions of how to solve and assess the stability of solutions of such systems. Readers who are already conversant with this material may safely skip to section 10.3. Those who seek more detail should consult any of several excellent works on the subject. Macdonald (1989) is probably the most accessible for the reader without extensive mathematical background. The monograph by Bellman and Cooke (1963), while more mathematically oriented, contains a number of useful results. Hale's (1979) review article and several sections of Murray's (1993) excellent treatise on mathematical biology also contain readable treatments of differential delay equations.

Before we look at solution methods, we first rescale our prototype equation, (10.1), to a slightly simpler, dimensionless form. Let the dimensionless time be $s = t/\tau$ and the dimensionless rate constant be $q = k\tau$. We define a new function z

as $z(u) = x(\tau u)$ [or, equivalently, $z(s) = x(t)$]. If we substitute s, q, and z for t, k, and x, respectively, eq. (10.1) becomes

$$dz(s)/ds = -q\,z(s-1) \tag{10.2}$$

Thus, all equations of the form of eq. (10.1) can be put into a universal form with delay time 1. Our analysis reveals that a single dimensionless parameter q, the product of the rate constant and the time delay, determines the behavior of the system. We now explore several ways that one might seek to solve eq. (10.2) or its relatives.

10.1.1 Interval-by-Interval Integration

If eq. (10.2) were an ordinary differential equation (ODE), our first approach would be to separate the variables z and s and then to try to integrate both sides of the equation. It is by no means obvious how one might integrate $dz(s)/z(s-1)$. In fact, in this form, the task is impossible. There is, however, a key feature about DDEs that we have neglected to mention thus far, and this feature enables us to integrate the equation, at least in stages.

To solve a first-order ODE, we must be given an initial value—for example, the value of x at some some particular time t. The solution then depends upon this initial value, say, x_0. To solve a DDE like eq. (10.1) or eq. (10.2) also requires initial value information, but of a different type. We must have the value of the function x over an entire interval of length τ. In the case of eq. (10.2), we require $z(s)$ over some interval of length 1, for example, $[-1,0)$. In effect, the delay system, even if there is only a single dependent variable, has an infinite number of degrees of freedom because the solution must be specified at an infinite number of points on the initial interval. For this reason, the variety of dynamical behavior that can be exhibited by a differential delay equation is far wider than that accessible to the corresponding equation in the absence of delay. For example, even a single non-linear DDE can give rise to chaotic solutions, while at least three coupled first-order ODEs are required for chaos.

For simplicity in illustrating the various methods of solution, we take our initial condition as

$$z(s) = 1, \qquad -1 \le s < 0 \tag{10.3}$$

Our treatment will make it clear how to treat more complex initial conditions.

Now that we have eq. (10.3) to work with, we can write eq. (10.2) on the interval $[0,1)$, without explicitly including delay, as

$$dz/ds = -q, \qquad 0 \le s < 1 \tag{10.4}$$

which is easily integrated to yield

$$z(s) = -qs + \text{constant}, \qquad 0 \le s < 1 \tag{10.5}$$

How can we find the constant of integration? If we make the reasonable requirement that the solution be continuous at $s = 0$, the boundary between intervals, we easily find that the constant in eq. (10.5) must be 1.

We now have

$$z(s) = 1 - qs, \qquad 0 \le s < 1 \tag{10.6}$$

which implies that

$$dz/ds = -q + q^2(s - 1), \qquad 1 \le s < 2 \tag{10.7}$$

If we again integrate eq. (10.7) and obtain the constant of integration by requiring the solution to be continuous at $s = 1$, we obtain

$$z(s) = 1 - qs + q^2(s - 1)^2/2, \qquad 1 \le s < 2 \tag{10.8}$$

It is not difficult to continue this process, finding and solving the new ordinary differential equation in each interval from the solution in the previous interval and the condition that the solution be continuous everywhere. This, however, is a tedious process. What we need is either to see a pattern and deduce from it the general form of the solution for all s values or to use one of the other methods to be discussed below.

In the present case, comparing eqs. (10.3), (10.6), and (10.8) and, if necessary, extending the solution for another interval or two enables us to see that the general solution for any value of s may be written as

$$z(s) = \sum_{m=0}^{n} (-q)^m (s - m + 1)^m/m!, \qquad n - 1 \le s < n \tag{10.9}$$

or, using the heaviside step function, $\theta(u) = 0$ if $u < 0$, $\theta(u) = 1$ if $u \ge 0$,

$$z(s) = \sum_{m=0}^{\infty} (-q)^m (s - m + 1)^m \theta(s - m + 1)/m! \tag{10.10}$$

Obviously, the utility of the step-by-step integration approach will depend on the form of the right-hand side of the DDE. Only if we are rather fortunate (or very smart), will we be able both to do all the necessary integrations and to perceive the general pattern. Although step-by-step integration has considerable intuitive appeal, we seek a more powerful and general approach.

10.1.2 Laplace Transforms

A very valuable technique, useful in the solution of ordinary and partial differential equations as well as differential delay equations, is the use of *Laplace transforms* . Laplace transforms (Churchill, 1972), though less familiar and somewhat more difficult to invert than their cousins, Fourier transforms, are broadly applicable and often enable us to convert differential equations to algebraic equations. For rate equations based on mass action kinetics, taking the Laplace transform affords sets of polynomial algebraic equations. For DDEs, we obtain transcendental equations.

The Laplace transform $F(u)$ of a function $f(t)$ is defined as

$$F(u) = \int_0^{\infty} f(t) \exp(-ut)\, dt \tag{10.11}$$

What one does in solving an equation with Laplace transforms is to transform the equation by applying eq. (10.11), solve the resulting algebraic equation for the transform $F(u)$, and then invert the transform to obtain the solution $f(t)$. Calculating inverse Laplace transforms is not easy; it requires facility with contour integrals in the complex plane. On the other hand, many reference books, like the *Handbook of Chemistry and Physics*, contain extensive tables of Laplace transforms and their inverses. We recommend their use. It is easy to see from eq. (10.11) that the Laplace transform of $f(t - \tau)$ is just

$$\exp(-u\tau)[F(u) + \int_{-\tau}^{0} f(t) \exp(-ut) dt] \tag{10.12}$$

while the Laplace transform of $df(t)/dt$ is $uF(u) - f(0)$. Using these results, we can transform eq. (10.2) into an equation for the Laplace transform $Z(u)$:

$$uZ(u) - 1 = -q[Z(u) \exp(-u) + 1/u - \exp(-u)/u] \tag{10.13}$$

where the last two terms on the right-hand side result from the fact that $z(s) = 1$ for $-1 \leq s < 0$. We can solve for $Z(u)$ to obtain

$$Z(u) = (1/u)[1 - q/u + q \exp(-u)/u]/[1 - q \exp(-u)/u] \tag{10.14}$$

It is useful to expand the denominator on the right-hand side of eq. (10.14) as a power series in $q \exp(-u)/u$. If we then collect like powers of u, we obtain the following series expression for the transformed solution:

$$Z(u) = 1/u + \sum_{m=0}^{\infty} (-q)^{m+1} \exp(-mu)u^{m+2} \tag{10.15}$$

With a little help from a table of inverse transforms, we can invert eq. (10.15) term by term to obtain an expression for $z(s)$ identical to what we found by interval-by-interval integration in eq. (10.10).

As a result of eq. (10.12), the algebraic equations obtained by Laplace transforming DDEs are always transcendental. Sets of DDEs may be transformed into sets of simultaneous (transcendental) algebraic equations. Note that eq. (10.12) also implies that the value of the solution on the initial interval is required in order to perform the transforms.

10.1.3 Numerical Methods

Very few realistic sets of chemical rate equations can be solved analytically. In almost all cases, we must use numerical integration to obtain explicit solutions of the sets of ODEs that we encounter in chemical kinetics. Sets of chemically significant DDEs are a fortiori even less likely to afford analytic solutions. We discussed in Chapter 7 the powerful array of numerical methods that have been developed in recent years for treating even very large, stiff sets of ODEs. Unfortunately, numerical analysis of DDEs is still in a rather primitive state. No standard packages like GEAR or LSODE are, to our knowledge, available for DDEs, though these and similar packages provide convenient starting points for developing homemade software for these systems. Commercial packages should be coming shortly. It may prove more productive, especially in treating larger systems, for numerical analysts to design methods specifically aimed at DDEs rather than to adapt older techniques originally designed for treating ODEs.

To illustrate the ad hoc approach to numerical simulation of DDEs, we mention here two numerical techniques with which we have had some success (Epstein and Luo, 1991). The first approach is an adaptation of the popular GEAR program (Hindmarsh, 1974) for solving sets of ODEs. As additional input for DDEs, we require the time delay(s) and the value(s) of the delayed variable(s) over the initial interval. The standard ODE program saves time by stepping the integration through the longest time step compatible with the error specification. For DDEs, the program also computes and stores the variables at time intervals $\varepsilon\tau$, where ε is a number of the order of 10^{-2} or 10^{-3} and τ is the time delay. The delayed variables are stored in an array, and at each integration step the appropriate values of the delayed variables are chosen for use in computing the derivatives. The array of stored variables is updated at each step. The accuracy of the procedure can be increased either by interpolating the delayed variables from the previously computed stored variables, at the cost of increased computation time, or by decreasing ε, which requires additional time and storage.

Another approach is to use a Taylor series expansion. Each interval of length τ is divided into N steps. The nondelayed variables are stored in arrays of length N, while the delayed variables require arrays of length mN, where m is the order of the Taylor expansion. For example, a second-order expansion of eq. (10.1) would utilize the formula

$$\begin{aligned} x(T+a) &\approx x(T) + a\,dx/dt|_T + a^2/2\,d^2x/dt^2|_T \\ &= x(T) - ka\,x(T-\tau) + k^2a^2/2\,x(T-2\tau) \end{aligned} \tag{10.16}$$

We see that each additional term in the Taylor series requires storage of another set of N delayed variables. Our experience suggests that a fourth-order expansion generally gives satisfactory results and that increasing the number of steps per interval N is more efficacious with stiff DDE systems than is increasing the order of the Taylor series approximation.

While both of these methods work adequately with nonstiff systems, delay times that are long compared with characteristic times in the system, such as the period of oscillation, can lead to serious numerical difficulties in systems like the Oregonator that are very stiff in the absence of delay.

10.2 Linear Stability Analysis

In Chapter 2, we introduced linear stability analysis as a way of gaining important information about the dynamical behavior of a system of ODEs by studying how the system responds to small perturbations of its steady state(s). An analogous approach can be applied to systems of DDEs. The analysis is similar, but significantly more difficult.

Let us write our system in the general form

$$dx(t)/dt = \mathbf{f}[\mathbf{x}(t), \mathbf{x}(t - \tau)] \tag{10.17}$$

where the boldface signifies that $\mathbf{f}$ and $\mathbf{x}$ are vectors of length m, that is, there are m species in our system. A steady state of eq. (10.17) is one in which $\mathbf{x}(t) = \mathbf{x}(t - \tau)$ and all time derivatives vanish:

$$\mathbf{f}(\mathbf{x_s}, \mathbf{x_s}) = 0 \tag{10.18}$$

We are interested in the stability of the steady state $\mathbf{x_s}$ to small perturbations, so for some small α we write

$$\mathbf{x}(t) = \mathbf{x_s} + \alpha \exp(\omega t) \tag{10.19}$$

and substitute eq. (10.19) into eq. (10.17). We linearize by dropping terms of second and higher order in α. This procedure results in an equation that resembles eq. (2.56) for the eigenvalues ω obtained in linear stability analysis of a set of ODEs. There are, however, some crucial differences.

When we linearize eq. (10.17) we obtain

$$\det \left[\mathbf{J}(\mathbf{x_s}, \mathbf{x_s}) + \mathbf{J}_\tau(\mathbf{x_s}, \mathbf{x_s}) \exp(-\omega \tau) - \omega \mathbf{I} \right] = 0 \tag{10.20}$$

where $\mathbf{I}$ is the $m \times m$ identity matrix and we need to define two Jacobian matrices, $\mathbf{J}$, with respect to the instantaneous concentrations, and $\mathbf{J}_\tau$ with respect to the delayed concentrations:

$$J_{ij}(\mathbf{x_s}, \mathbf{x_s}) = \partial f_i[\mathbf{x}(t), \mathbf{x}(t - \tau)]/\partial x_j(t)|_{\mathbf{x}(t)=\mathbf{x}(t-\tau)=\mathbf{x_s}} \tag{10.21}$$

$$J_{\tau ij}(\mathbf{x_s}, \mathbf{x_s}) = \partial f_i[\mathbf{x}(t), \mathbf{x}(t - \tau)]/\partial x_j(t - \tau)|_{\mathbf{x}(t)=\mathbf{x}(t-\tau)=\mathbf{x_s}} \tag{10.22}$$

Systems without delay that obey mass action kinetics give rise to polynomial equations of order m for the stability eigenvalues $\{\omega_j\}_{j=1,...,m}$. As eq. (10.20) shows, whenever a derivative is taken with respect to a delayed variable in a DDE, the resulting term in the Jacobian $\mathbf{J}_\tau$ must be multiplied by a factor $\exp(-\omega \tau)$. The inclusion of delay leads to a transcendental equation which, in general, has an *infinite* number of roots ω_j.

Despite the infinite number of roots of eq. (10.20), it is often possible to determine analytically whether or not a given steady state $\mathbf{x_s}$ is stable. The problem is equivalent to deciding whether all the roots of an equation of the form [cf. eq. (10.20)]

$$g(\omega) + h(\omega) \exp(-\omega \tau) = 0 \tag{10.23}$$

where g and h are polynomials of degree m, have negative real parts (lie in the left half of the complex plane). Several general theorems (Bellman and Cooke, 1963; Macdonald, 1989) allow one to test algebraically the stability of a particular steady state. Very few results are available for systems of more than two variables, while those for two variables apply only to a limited subset of systems. When there is only a single concentration variable, powerful algebraic (Hayes, 1950) and geometric (Macdonald, 1989) techniques can be applied. In our prototype example, eq. (10.1), these methods easily yield the result (Epstein, 1990) that the unique steady state, $x_s = 0$, is stable if $k\tau < \pi/2$. For $k\tau > \pi/2$, the steady state is unstable and we have growing oscillations, as we shall see below.

10.3 Examples

In this section, we present a series of relatively simple examples chosen to illustrate how one might employ DDEs in chemical kinetics and how the introduction of delay can, in some sense, simplify the description of a system while increasing the complexity of the dynamics that the equations may display.

10.3.1 The Prototype Equation

We start by summarizing the behavior of our prototype DDE, eq. (10.1). The explicit solution is presented in eq. (10.10), but it is not immediately obvious what this function looks like. Numerical analysis combined with the algebraic approach to linear stability analysis described in the previous section yields the following results.

For $k\tau$ sufficiently small, $x(t)$ decays monotonically, though not exactly exponentially, to zero in a manner similar to the nondelayed system. As $k\tau$ is increased, we observe first damped and then growing oscillations. The transition from damped to growing oscillations occurs at $k\tau = \pi/2$, where the solution is periodic. In Table 10.1 and Figure 10.1, we summarize the behavior of $x(t)$ as a function of the parameter $k\tau$.

The periodic behavior at $q = \pi/2$ is structurally unstable: the smallest change in the parameter q leads to a qualitative change in the behavior. By adding a small nonlinear term, for example, $-ax(t)^3$, to the right-hand side of eq. (10.1), one obtains a system that gives structurally stable limit cycle oscillations, as illustrated in Figure 10.2, over a range of parameters q (Epstein, 1990).

10.3.2 A Sequence of First-Order Reactions

To illustrate how a complex mechanism described in terms of ODEs may be approximated by a simpler system in which delays are included, we consider a sequence of coupled irreversible first-order reactions. Epstein (1990) analyzed this system by introducing the concept of a *bottleneck intermediate*, a generalization of the notion of a rate-determining step.

Consider the set of consecutive reactions

$$A_i \xrightarrow{k_i} A_{i+1}, \qquad i = 1, 2, \ldots, n-1 \tag{10.24}$$

with initial conditions

Table 10.1 Qualitative Behavior of Solutions to Equation (10.1)

Range of Parameter $q = k\tau$	Behavior of Solution $x(t)$
$0 \le q < 1/e$	Monotonic decay to $x = 0$
$1/e \le q < \pi/2$	Damped oscillatory decay to $x = 0$
$q = \pi/2$	Periodic oscillation, period $= 4\pi$
$q > \pi/2$	Undamped, growing oscillation

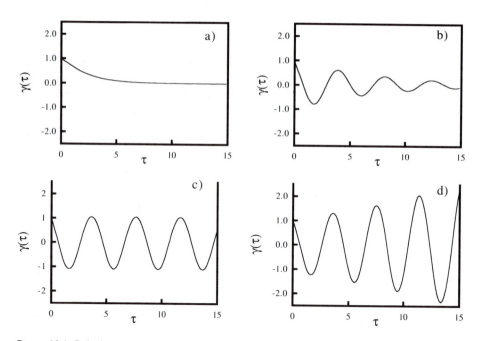

Figure 10.1 Solutions to eq. (10.1) for several values of the parameter $q = k\tau$ with $\tau = 1$. (a) $q = 0.3$, (b) $q = 1.3$, (c) $q = 1.5708$ ($\pi/2$), (d) $q = 1.7$. (Adapted from Epstein, 1990.)

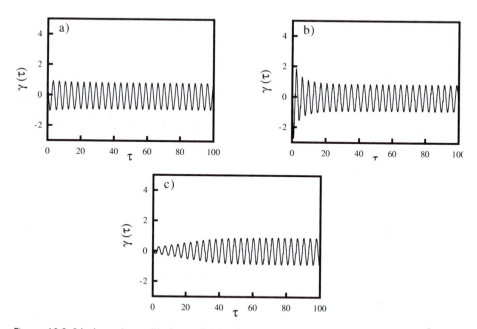

Figure 10.2 Limit cycle oscillations obtained by adding a nonlinear term $-ax(t)^3$, with $a \ll 1$, to the right-hand side of eq. (10.1). Here $q = 1.7$. Initial values $x_0 = 1$ (a), 5 (b), and 0.2 (c) all lead to a periodic oscillation with amplitude 1.87 and period 3.76. (Adapted from Epstein, 1990.)

$$A_1(0) = A_0 \tag{10.25a}$$
$$A_i(0) = 0, \qquad i = 2, 3, \ldots, n \tag{10.25b}$$

The resulting set of rate equations

$$dA_1(t)/dt = -k_1 A_1(t) \tag{10.26a}$$
$$dA_i(t)/dt = -k_i A_i(t) + k_{i-1} A_{i-1}(t), \qquad i = 2, 3, \ldots, n-1 \tag{10.26b}$$
$$dA_n(t)/dt = k_{n-1} A_{n-1}(t) \tag{10.26c}$$

constitute a set of linear differential equations with constant coefficients. Therefore, they can be solved analytically for each of the concentrations, yielding a sum of exponential terms for each A_i.

We now need to define the concept of a bottleneck intermediate. Qualitatively, what we mean is that we have a species whose rate of disappearance is rate-determining during some portion of the reaction. For example, if $k_j (j \neq 1)$ is the smallest of the rate constants in eq. (10.26), species A_j will be a bottleneck intermediate, because there will be a period of time during which its concentration will build up, and the progress of the system toward the final state in which only A_n remains will be determined by the rate at which A_j is converted to A_{j+1}. We can imagine having a series of pipes of different widths connected in series. The rate at which water flows through this arrangement depends upon the width of the narrowest pipe, the bottleneck. Is the species associated with the smallest rate constant the only bottleneck in the system? Not necessarily. If A_j lies fairly far downstream, and if there is a faster, but still slow, reaction further upstream, it will slow things up before A_j even has a chance to accumulate. Thus, this reaction, too, constitutes a bottleneck at a different, earlier, stage of the reaction.

The above analysis suggests an algorithm for defining the subset $\{A_{\alpha(p)}\}_{p=0,1,\ldots,P}$ of bottleneck intermediates. Let $\alpha(0) = 1$, and let $\alpha(P)$ be the index of that species i such that k_i is the smallest rate constant, that is,

$$k_{\alpha(P)} = \min(k_i), \qquad 1 \leq i \leq n-1 \tag{10.27}$$

If k_1 is the smallest rate constant, then $\alpha(P) = 1$, $P = 0$, and there are no bottleneck intermediates. If some other rate constant is the smallest, consider the remaining species for which $i < \alpha(P)$. We choose $\alpha(P-1)$ to be the index of that species whose rate constant is the smallest in this group:

$$k_{\alpha(P-1)} = \min(k_i), \qquad 1 \leq i \leq \alpha(P) - 1 \tag{10.28}$$

Again, if k_1 is the smallest of the remaining rate constants, we are done, $P = 1$, and there is a single bottleneck intermediate. If not, we continue the process until we are left with k_1 as the smallest remaining rate constant. The set of bottleneck intermediates is then complete.

Epstein (1990) shows that after a transient period, the exact solution of eq. (10.26) approaches arbitrarily close to the solution of the following, generally smaller set of DDEs:

$$dA_1(t)/dt = -k_1 A_1(t) \tag{10.29a}$$

$$
\begin{aligned}
dA_{\alpha(p-1)}(t)/dt = &-k_{\alpha(p-1)} A_{\alpha(p-1)}(t) \\
&+ k_{\alpha(p-1)-1} A_{\alpha(p-1)-1}(t - \tau_{\alpha(p-1)}), \quad p = 1, 2, \ldots, P
\end{aligned} \tag{10.29b}
$$

$$dA_{\alpha(P)}(t)/dt = k_{\alpha(P)} A_{\alpha(P)}(t - \tau_{\alpha(P)})$$

where the delay times α_p are given by

$$\tau_p = 1/k_{\alpha(p)} \ln \left[\prod_{m=\alpha(p-1)+1}^{\alpha(p)-1} \{k_m/(k_m - k_{\alpha(p)})\} \right], \quad p = 1, 2, \ldots, P \tag{10.30}$$

Except for a brief transient period, the DDE model, eqs. (10.29), reproduces the exponential decay of the reactant A_1; the rise and fall, after a delay, of each of the bottleneck intermediates $A_{\alpha(p)}$; and the delayed buildup of the product A_n; which constitute the essential aspects of the dynamics of the full system represented by eqs. (10.26). The model we have just treated is not particularly interesting in terms of its chemistry. It does, however, suggest that an approach based on treating systems with many intermediates in terms of DDE models with just a few key intermediates and delays that incorporate the effects of the nonessential intermediates may be worth exploring. The key problem to be solved in implementing such an approach lies in deriving relationships analogous to eq. (10.30) between the parameters of the system and the delay times.

Epstein (1990) also considers a model (Allnatt and Jacobs, 1968) for nucleation in solid-state reactions that is equivalent to a set of coupled first-order rate equations. He shows that by introducing delays to account for the fact that an n-particle nucleus cannot grow until an $(n-1)$-particle nucleus is formed, the model can be made more physically realistic with relatively little increase in computational effort.

10.3.3 Diffusion through a Membrane

At the beginning of this chapter, we suggested that delays associated with the motions of molecules from place to place may affect the dynamics of a system. That this can happen, *even in the absence of a chemical reaction*, is demonstrated by another simple model (Epstein, 1990). In Figure 10.3, we show two compartments that contain a species C and are connected by a membrane that permits the diffusion of C. Each compartment is assumed to be well stirred and homogeneous, so each is characterized by a single concentration variable, C_i, where $i = 1, 2$. The diffusion constant of C is D; the membrane has cross-sectional area A and length L; and the compartments have lengths l_1 and l_2, respectively.

If we assume that the time required for the molecules of C to cross the membrane is infinitesimal, we can apply Fick's First Law of Diffusion to obtain a pair of ODEs that describe the system:

$$dC_i/dt = (D/Ll_i)(C_j - C_i), \quad i, j = 1, 2, \quad j \neq i \tag{10.31}$$

In reality, of course, molecules require a finite time τ to traverse the membrane. The existence of such time lags is well known; it is possible to calculate the mean

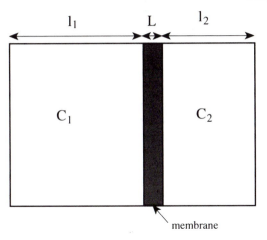

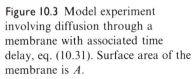

Figure 10.3 Model experiment involving diffusion through a membrane with associated time delay, eq. (10.31). Surface area of the membrane is A.

and even the distribution (since not all molecules take the same time to get across) of τ for various diffusion geometries (Crank, 1975). Instead of taking this fact into consideration in its full complexity, we examine instead a simpler model in which *all* molecules take the *same* time τ to cross the membrane. If we solve this problem for a general τ, we should be able to average the result over an appropriate distribution of time lags if we need to.

Since we are now taking the time lag explicitly into account, the rate of increase of C_j at time t as a result of diffusion now depends not upon $C_i(t)$ but upon $C_i(t - \tau)$. If we define $q_i = D/Ll_i$, we obtain the DDE version of our instantaneous diffusion model eqs. (10.31):

$$dC_i(t)/dt = q_i[C_j(t - \tau) - C_i(t)], \qquad i, j = 1, 2, \qquad j \neq i \qquad (10.32)$$

Specifying the values of the concentrations on the initial interval requires a bit of thought. Let us imagine the diffusion being turned on at $t = 0$ (e.g., by poking tiny holes in an impermeable membrane), and denote the concentrations before this time by a superscript zero. Since all molecules require a time τ to cross the membrane, the delay is irrelevant for $t < \tau$, and no molecules enter either compartment from the other during this initial interval. Therefore, the initial interval is characterized by a simple first-order decay of the concentrations, so that

$$C_i(t) = C_i^0 \exp(-q_i t), \qquad 0 \leq t < \tau, \quad i = 1, 2 \qquad (10.33)$$

We can solve eqs. (10.32) analytically (Epstein, 1990) in terms of "delayed exponentials," the solutions to our prototype DDE given in eqs. (10.9) or (10.10). Some surprising features of these solutions are illustrated in Figure 10.4. We expect the system to approach an equilibrium with equal concentrations in the two compartments. What is far from obvious, though, is that if the compartments have equal volumes, if τ is large enough, and if the initial concentrations are sufficiently different (Figure 10.4a), then the system approaches equilibrium in a damped oscillatory fashion. Even more remarkable (Figure

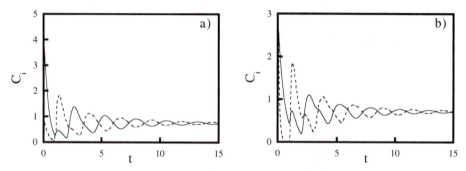

Figure 10.4 Oscillatory behavior in the model of diffusion through a membrane as shown in Figure 10.3. Solid line is $C_1(t)$, dashed line is $C_2(t)$. (a) Unequal initial concentrations: $q_1 = q_2 = 3$, $C_1(0) = 5$, $C_2(0) = 1$, $\tau = 1$; (b) unequal volumes: $q_1 = 2$, $q_2 = 8$, $C_1(0) = C_2(0) = 3$, $\tau = 1$. (Adapted from Epstein, 1990.)

10.4b) is the fact that, if the volumes of the compartments are sufficiently different, we can start with *equal* concentrations and the system will oscillate on the way to equilibrium.

This oscillatory behavior is counterintuitive and apparently nonphysical; it seems to violate the laws of thermodynamics. In fact, it does. One reason is that our assumption of equal transit times for all molecules across the membrane requires a Maxwell demon to help molecules avoid the collisions that produce a distribution of transit times.

Another clue to the source of the apparent paradox lies in the fact that the final equilibrium value lies significantly below the starting concentrations in the two compartments. A lot of material gets "stuck" in the membrane, and the oscillations are transients that occur while the initially empty membrane is filling up. Consider the case of equal initial concentrations with cell 1 much longer than cell 2. In the first interval $[0, \tau)$, the same number of molecules leave each cell, but because cell 2 is smaller, its concentration drops further. At time τ, when molecules begin to arrive at the other cell, the effect is reversed, and the concentration in cell 2 now rises more rapidly than in cell 1. Since molecules are now flowing in both directions, the effect will soon damp out. Calculations of the difference between the initial and final concentrations, using the value $\tau = L^2/6D$ appropriate for this geometry (Crank, 1975), show that the amount of material "missing," that is, remaining in the membrane at equilibrium, is proportional to $L/[3(l_1 + l_2) + L]$.

The reader may find the above result more convincing if he or she tries to picture how the initially equal densities of people in two rooms of very different sizes connected by identical doors through an anteroom will change in time if, when the doors are opened, people move from room to room with equal probability and equal velocity without collisions. It is also reassuring to observe that when the results are averaged over a realistic distribution of τ the oscillations disappear and the concentrations behave monotonically, as they should.

10.3.4 The Cross-Shaped Phase Diagram

The examples of DDEs that we have considered so far have all been linear. Linear systems allow a considerable amount of analysis and even exact solution, on occasion, but few real systems are linear. We now turn to some examples that involve nonlinear DDEs, starting with two familiar examples, the cross-shaped phase diagram and the Oregonator. In these two models we see how, much like in the case of the sequence of first-order reactions treated above, one can reduce the number of variables by introducing a delay. In the nonlinear case, however, the choice of the delay time is far more difficult than in linear models.

In Chapter 4, we discussed the role played by the cross-shaped phase diagram in the systematic design of chemical oscillators. A key element in that effort was the two-variable ODE model (Boissonade and De Kepper, 1980):

$$dx/dt = -(x^3 - \mu x + \lambda) - ky \qquad (10.34a)$$

$$dy/dt = (x - y)/T \qquad (10.34b)$$

We recall that, for a sufficiently long relaxation time T, the feedback variable y provides a delayed feedback that causes the system, which is bistable for appropriate values of λ and μ as a result of the cubic terms in the rate equation for x, to become oscillatory.

Epstein and Luo (1991) suggested that eqs. (10.34) might be replaced by a single DDE in which the delayed feedback of y, which tends to make the variables x and y approach each other, is mimicked by replacing $y(t)$ in eq. (10.34a) by $x(t - \tau)$. They performed a linear stability analysis of the resulting eq. (10.35),

$$dx(t)/dt = -[x(t)^3 - \mu x(t) + \lambda] - kx(t - \tau) \qquad (10.35)$$

using eq. (10.20), which leads to an equation for the stability eigenvalues ω of the form of eq. (10.23):

$$-3x_s^2 + \mu - k \exp(-\omega\tau) - \omega = 0 \qquad (10.36)$$

Analysis of eq. (10.36) leads to three conditions that must be satisfied for the stability of the steady states x_s:

$$3x_s^2 - \mu + 1/\tau > 0 \qquad (10.37)$$

$$3x_s^2 - \mu + k > 0 \qquad (10.38)$$

$$[(3x_s^2 - \mu)^2 + (a_1/\tau)^2]^{1/2} > 0 \qquad (10.39)$$

where a_1 the root of eq. (10.40) that lies between 0 and π,

$$a_1 = (-3x_s^2 + \mu)\tau \tan a_1 \qquad (10.40)$$

Equations (10.37) and (10.38) are identical to the conditions [eqs. (4.12) and (4.13)] for the stability of the steady states of the ODE model if we replace the feedback time T in eq. (10.34b) by the delay time τ in eq. (10.35). The third condition, eq. (10.39), is new and makes possible additional regions of instability of the steady states of eq. (10.35).

Examples of the results obtained are given in the phase diagrams of Figure 10.5. Figure 10.5a illustrates the situation for the case of the usual cross-shaped

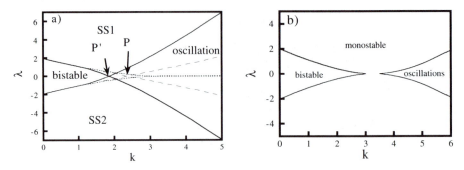

Figure 10.5 Phase diagrams in the k–λ plane obtained from linear stability analysis of eq. (10.35). (a) $\mu = 3$, $\tau = 1$. Dotted line separates regions in which there are one or three steady states. Dashed lines are stability boundaries given by eqs. (10.37) and (10.38), equivalent to ODE model, eqs. (10.34) with $T = 1$. Solid lines are stability boundary for full DDE model using condition (10.39) as well. (b) $\mu = 3$, $\tau = 0.3$. Only bistable region is found in ODE model. Oscillatory region results from stability condition, eq. (10.39). (Adapted from Epstein, 1990.)

phase diagram where $\tau > 1/\mu$. The behavior of the ODE and DDE models is qualitatively the same. We have regions of monostability, bistability, and oscillation. Now, however, the additional condition of eq. (10.39) has caused regions of the parameter space that gave stable steady states in the ODE model to become unstable. The oscillatory parameter range is increased, and the cross point has shifted from P to P'.

A more dramatic effect is seen in Figure 10.5b, where we examine the case $\tau < 1/\mu$. Here, the ODE model yields only a single wedge-shaped region of bistability, surrounded by a much larger region of monostability. In addition to these features, the DDE model possesses a region of oscillatory behavior at large k and small $|\lambda|$. Experiments confirm that by adding a delayed feedback to a bistable system, one can generate chemical oscillation.

10.3.5 The Oregonator

Just as the BZ reaction has become the experimental prototype for nonlinear chemical dynamics, the Oregonator model (Field and Noyes, 1974b), is easily the most familiar and thoroughly studied model system in nonlinear chemical dynamics. We recall from Chapter 5 that the model equations are

$$A + Y \rightarrow X \tag{10.41a}$$

$$X + Y \rightarrow P \tag{10.41b}$$

$$A + Y \rightarrow X \tag{10.41c}$$

$$B + X \rightarrow 2X + Z \tag{10.41d}$$

$$Z \rightarrow fY \tag{10.41e}$$

where the "reactant" (essentially bromate) concentrations A and B and the "product" concentrations P and Q (bromate and hypobromous acid) are fixed, f is a stoichiometric factor that specifies how many bromide ions are generated for each oxidized metal catalyst ion reduced, and the variable concentrations are $X = \text{HBrO}_2$, $Y = \text{Br}^-$, and $Z = \text{Ce(IV)}$ for the cerium-catalyzed version of the reaction. Chemically, the role of cerium (Z) is to provide a means of regenerating bromide ion (Y), whose level controls the periodic switching of the reaction between autocatalytic and nonautocatalytic pathways for the consumption of bromous acid (X). In dynamical terms, as we shall see, it is also essential that this feedback generate a delay.

Field and Noyes asked whether it is possible to simplify the model shown in eqs. (10.41) by combining eqs. (10.41d) and (10.41e) into a single step, eq. (10.42), thereby eliminating Z and reducing the model to two variables, X and Y:

$$B + X \rightarrow 2X + fY \tag{10.42}$$

Analysis showed that it was impossible to obtain oscillatory behavior in any such reduced model. However, might it be possible to combine the two steps and maintain oscillatory behavior if one also incorporated the effects of delay that species Z produces? It turns out that this can be done, and in the process one also gains valuable insight into the dynamics of the original Oregonator. Epstein and Luo (1991) examined two "delayed Oregonator" models consisting of eqs. (10.41a), (10.41b), (10.41c), and (10.42). In the first version, the rate of step, eq. (10.41a), was taken to be $k_a A Y(t - \tau)$, while all other rates were determined by the Law of Mass Action. In the second model, step (10.41a) had mass action kinetics, but step (10.41b) was assigned a rate $k_b X(t) Y(t - \tau)$. Thus, the first DDE model examines the effect of delay in the bromide feedback on the reaction of bromide with bromate, while the second looks at delay effects on the bromide–bromous acid reaction.

The results obtained in this study were unequivocal. A combination of linear stability analysis and numerical simulation demonstrates that the first model cannot give oscillatory solutions for *any* values of the delay time and initial concentrations, while the second model, with an appropriate choice of τ, yields oscillations very similar to those found in the three-variable ODE Oregonator model, as shown in Figure 10.6. Thus, the calculation not only shows that delay is essential in this system but also reveals just where in the mechanism that delay plays its role.

10.3.6 Bubble Growth and Gas Evolution Oscillators

The scheme of reducing the number of variables by introducing delays into a model can be used with systems that involve physical as well as chemical processes. When formic acid is added to concentrated sulfuric acid, the formic acid loses water, and carbon monoxide is produced. Under certain conditions, the CO gas bubbles are released from the solution in an oscillatory fashion (Morgan, 1916). The first successful efforts to model this phenomenon (Smith and Noyes, 1983) take explicit account of the process by which CO nucleates and forms

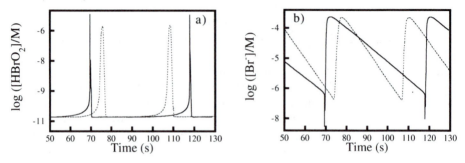

Figure 10.6 Oscillatory behavior in (solid lines) the ordinary Oregonator ODE model of the BZ reaction, and (dashed lines) the delayed Oregonator with $\tau = 2$ s as described in the text. (Adapted from Epstein and Luo, 1991.)

bubbles. This process is described by a series of rate equations that describe the concentrations of bubbles of different sizes.

While the results obtained in this "bubbelator" model are encouraging, it is necessary to include least twenty different classes of bubbles in order to get even qualitative agreement with experiment. Smith and Noyes suggest a more elegant DDE formulation that requires many fewer variables. In this approach, the change of concentration of dissolved gas, $C(t)$, is given by

$$dC(t)/dt = \Phi(t) - LFJ_n(t - \tau) \tag{10.43}$$

where L is the average number of moles of gas per bubble that escapes the solution, F is the fraction of nuclei that ultimately grow and escape,

$$\Phi(t) = k_c[\text{HCOOH}]_0 \exp(-k_c t) \tag{10.44}$$

is the rate of formic acid dehydration, and J_n is the rate of formation of new nuclei. The authors did not carry out any explicit simulations with this DDE formulation, but, despite its introduction of several phenomenological parameters (L, F, J_n, τ, k_c), the reduction in the number of variables and rate constants for growth of bubbles of different sizes makes this an attractive model for further exploration.

10.3.7 Delayed Feedback in an Illuminated Thermochemical Reaction

Our examples thus far have focused on models of experimentally characterized systems in which delays are introduced into the model in order to reduce the number of variables and/or to test our understanding of the dynamics. Another way in which delay effects have been explored is to carry out experiments in which delays are deliberately incorporated into the experiments in order to change the dynamics.

Photochemical reactions offer a convenient opportunity to introduce delays, because it is relatively easy to vary the intensity of illumination in response to the value of some measured property of the system at a specified time after the measurement is made. The decomposition of $S_2O_6F_2$ provides a very pretty example of such a system:

$$S_2O_6F_2 \leftrightarrow 2SO_3F \tag{10.45}$$

When gaseous $S_2O_6F_2$ is illuminated at 488 nm, only the product SO_3F absorbs the light, which is converted to heat. As a result of the heating, the equilibrium shifts to the right, causing more SO_3F to be produced, and increasing further the amount of light absorbed. The reaction thus exhibits a form of autocatalysis. If the incident light intensity is slowly varied, the system shows hysteresis between a high-monomer and a low-monomer steady state (Zimmerman and Ross, 1984).

Zimmerman, Schell, and Ross (1984) analyzed the consequences of modifying the intensity of the incident light according to the concentration of SO_3F with a delay τ:

$$\Phi_0(t) = C_1 + C_2[1 - A(t - \tau)]\Phi_0(t - \tau) \tag{10.46}$$

where Φ_0 is the incident light intensity, C_1 and C_2 are positive constants, and A is the light absorption of the solution, which can be calculated from (SO_3F) using Beer's law. The steady states of the modified system turn out to be the same as those of the system in the absence of the feedback. What is changed by the delay is the stability of these states. For relatively short delays, the analysis shows that the middle steady state which is unstable, and hence experimentally unobservable, in the absence of the feedback, can be stabilized. This prediction has been confirmed experimentally (Zimmerman et al., 1984), and a significant segment of the previously unstable branch of steady states has been observed. At longer delay times, the middle state again becomes unstable, as do the upper and lower states; the system becomes oscillatory.

Schell and Ross (1986) suggest a number of general features of illuminated thermochemical reactions with delayed feedback based on their study of several models similar to that of eq. (10.46). Although, as we have noted, delay systems possess an infinite number of degrees of freedom, like other chemical systems they tend to be strongly contracting, so that they have attractors of relatively low-dimension steady states and periodic orbits, or low dimensional chaos. Schell and Ross find the stabilization of an unstable state at short delays to occur in a wide variety of systems. Systems that oscillate periodically in the absence of delay exhibit chaos and hyperchaos (aperiodic behavior with more than one unstable direction in phase space) when subjected to delays that are long compared with the natural period of oscillation. These more exotic modes of behavior are attributed to memory effects caused by the delay, so that the system is repeatedly reinjected at different points into the neighborhood of a saddle focus.

10.3.8 The Minimal Bromate Oscillator

A second system that has been subjected to external perturbation with delay is the minimal bromate oscillator. In these experiments (Weiner et al., 1989), the quantity varied is the flow rate through a CSTR. In the oscillatory region of parameter space, varying the flow rate according to the expression

$$k_f = k_0\{1 + \beta([Ce^{4+}(t - D)]/C - 1)\} \tag{10.47}$$

where C is the average value of $[Ce^{4+}]$ in the free-runing oscillator at a constant flow rate equal to the reference rate k_0 and D is the time delay, leads to a sawtooth-shaped variation in the period of the oscillation as a function of the delay time.

A similar but more detailed study has been performed (Chevalier et al., 1991) in the bistable region with fixed flows of $MnSO_4$ and $NaBrO_3$ and a variable flow rate $f(t)$ of $NaBr$ governed by the delayed feedback

$$f(t) = \kappa\{1 + \varepsilon \sin [\omega x(t - \tau)]\} \tag{10.48}$$

where x is the voltage of the bromide-sensitive electrode, κ is the mean flow rate, ε sets the amplitude of the nonlinear feedback, and ω determines the frequency of the feedback. In these experiments, the flow rate was updated for the change in $[Br^-]$, according to eq. (10.48), every 0.22 s.

The variety of dynamical behavior observed in this apparently simple system is truly remarkable. As τ is increased, there are Hopf bifurcations to simple periodic behavior, then period-doubling sequences to chaos, as well as several types of multistability. One of the most interesting findings, shown in Figure 10.7, is the first experimental observation (in a chemical system) of *crisis* (Grebogi et al., 1982), the sudden expansion of the size of a chaotic attractor as a parameter (in this case, τ) is varied. Simulations with the delayed feedback of eq. (10.48) taken into account give qualitative agreement with the bifurcation sequences observed in the experiments.

10.3.9 Biological Models

One might argue that, while thinking in terms of delays may provide useful insights into the workings of a handful of chemical systems, descriptions that

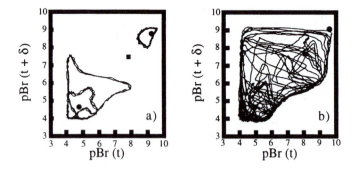

Figure 10.7 Experimental phase portraits showing crisis. Axes represent potential of bromide-sensitive electrode at times t and $t + 32.9$ s. $\tau = $ (a) 54.925 s, (b) 109.85 s. Filled circles represent initially ($\tau = 0$) stable fixed points, filled squares are initially unstable fixed points. As τ is increased, the lower attractor collides with an unstable fixed point and undergoes sizable expansion (crisis). (Adapted from Chevalier et al., 1991.)

take explicit account of delay seem artificial for the vast majority of chemical reactions. Living systems, on the other hand, afford many instances in which inclusion of delays provides a natural and revealing approach to describing phenomena. The literature of mathematical biology (e.g., Macdonald, 1989; Murray, 1993) offers many examples of biologically significant delay models. To give some of the flavor, we mention two here.

The idea of reducing the number of variables in a model by introducing delays is extremely attractive in biological systems, where there are often many intermediates whose identity is uncertain or whose concentrations cannot be monitored. In attempting to describe the transcriptional and translational processes involved in DNA replication during infection by T3 and T7 bacteriophages, Buchholtz and Schneider (1987) find that even a seventeen-step ODE model cannot account for the observed dynamics. By introducing delay times that are associated with transport processes involving linear diffusion of proteins along macromolecules and that can be estimated from experimental data, they are able to model successfully the abrupt rise in the concentrations of certain proteins and DNA species. In addition, they find that the delay times may be correlated with physical data about the position of genes.

Physicians have identified a class of diseases caused by abnormalities in the temporal organization of physiological processes. One such *dynamical disease* (Glass and Mackey, 1979) is Cheyne–Stokes syndrome, a pathological respiratory condition characterized by dangerously large variations in ventilation, the volume of air taken in during breathing.

A simple DDE model (Mackey and Glass, 1977) that takes into account the known physiological features of the respiratory system gives remarkably good agreement with clinical data on Cheyne–Stokes patients. The delay time in the model is identified with the time required for information from the CO_2 sensors in the brainstem to be transmitted to the lungs. The higher the concentration of carbon dioxide in arterial blood, the more air is taken in. The relationship between this volume of air, $V(t)$, and the CO_2 concentration, $c(t)$, is given in the model by

$$V(t) = V_{max} c^m(t - \tau)/[a^m + c^m(t - \tau)] \tag{10.49}$$

The parameters a and m, which describe the saturation of the ventilation as $[CO_2]$ increases, and V_{max}, the maximum ventilation rate, can be determined experimentally. Carbon dioxide is generated by metabolic processes in the body at a nearly constant rate, p. Taking into account the production of CO_2 and its loss through respiration, Mackey and Glass construct the following DDE to describe the dynamics of respiration:

$$dc(t)/dt = p - bV(t)c(t) = p - bV_{max}c(t)c^m(t - \tau)/[a^m + c^m(t - \tau)] \tag{10.50}$$

The CO_2 removal coefficient b can also be obtained from experimental data.

Equation (10.50) has a unique steady state, corresponding to a fixed (and presumably healthy) level of ventilation. Linear stability analysis shows that this state is stable if the delay time is short enough and if the dependence of V on c at the steady state is not too steep, that is, if m is not too large. If these conditions are violated, the steady state becomes unstable, and pathological oscillations like those seen in Cheyne–Stokes syndrome can result. The parameters that determine the stability of the steady-state solution are governed by both genetic and environmental factors. The model and its analysis may suggest promising avenues for treatment of the disease by pharmacological modification of these parameters so as to stabilize the state of constant ventilation.

10.4 Future Directions

Chemists, and indeed scientists in general, have only begun to explore the uses of delays and delay models in probing the systems they wish to understand. As the mathematical and computational aspects of DDEs come to seem less formidable and as advances in microcomputer technology make introduction of delayed feedback into experiments a straightforward task, one can expect many advances beyond the examples presented here.

We anticipate further use of delays both to analyze and to simplify complex reactions with many intermediates, particularly in biological systems. The most difficult part of reducing complex systems by the introduction of delays is to understand how the delay is related to the kinetic parameters (rate constants, steady-state concentrations, etc.) of the intermediates being "hidden" in the delay. Results on the BZ reaction (Epstein and Luo, 1991) suggest that the time between concentration maxima of the intermediate being eliminated and the species being delayed may be a useful estimate of the delay. However, this quantity may not be experimentally accessible. What is needed is an *experimental* approach to assessing the number and magnitude of the time lags in a system. As we have seen in our biological examples, it is often easier to identify the delay with an experimentally measurable quantity in a biological system.

One very attractive idea (Chevalier et al., 1991) is that the imposition of a delayed feedback can be used to (1) determine the elements of the steady-state Jacobian and, thus, (2) differentiate among competing mechanisms. The Oregonator example presented above represents a similar approach in that the use of a delay model made it possible to demonstrate at which stage of the BZ reaction the delay in the feedback becomes essential for oscillation.

Experimentally, delayed feedback is a potentially powerful tool for the analysis of dynamical systems. It can be used either to stabilize otherwise unstable steady states (Zimmerman et al., 1984) or to cause a steady state to become unstable, leading to periodic or chaotic behavior. By extension, it should be possible to use delayed feedback to stabilize unstable periodic orbits, providing an alternative to other methods for controlling chaos (Grebogi et al., 1982; Peng et al., 1991).

11

Polymer Systems

In the classic 1967 film "*The Graduate*," the protagonist, Benjamin (Dustin Hoffman), is attempting to plan his postcollege path. His neighbor provides one word of advice, "Plastics." This counsel has become part of American culture and is often parodied. But, it is good advice, because not since the transformations from stone to bronze and then to iron have new materials so completely transformed a society. Plastics made from synthetic polymers are ubiquitous, from Tupperware to artificial hearts. About half the world's chemists work in polymer-related industries.

In this chapter, we will survey some of the work that has been done in applying nonlinear dynamics to polymerization processes. These systems differ from those we have considered so far because they do not involve redox reactions. We will consider polymerization reactions in a CSTR that exhibit oscillations through the coupling of temperature-dependent viscosity and viscosity-dependent rate constants. Emulsion polymerization, which produces small polymer particles dispersed in water, can also oscillate in a CSTR. Both types of systems are important industrially, and their stabilities have been studied by engineers with the goal of eliminating their time-dependent behavior. Our favorite oscillating system, the Belousov–Zhabotinsky reaction, can be used to create an isothermal periodic polymerization reaction in either a batch or continuous system. This, however, is not a practical system because of the cost of the reagents.

In most industrial processes, nonlinear behavior is seen not as an advantage but as something to be avoided. However, we will look at several reaction–diffusion systems that have desirable properties precisely because of their nonlinear behavior. Replication of RNA is autocatalytic and can occur as a traveling front.

Since not all RNA molecules replicate equally well, faster mutants gradually take over. At each mutation, the front propagates faster. Evolution can be directly observed in a test tube. Propagating polymerization fronts of synthetic polymers may be useful for making new materials, and they are interesting because of the rich array of nonlinear phenomena they show, with pulsations, convection, and spinning fronts. Finally, we will consider photopolymerization systems that exhibit spatial pattern formation on the micron scale, which can be used to control the macroscopic properties.

II.I Sources of Feedback

Nonlinear phenomena in any system require some type of feedback. The most obvious source of feedback in polymerization reactions is thermal autocatalysis, often called "thermal runaway" in the engineering literature. The heat released by the reaction increases the rate of reaction, which increases the rate of heat release, and so on. This phenomenon can occur in almost any reaction and will be important when we consider thermal frontal polymerization.

Free-radical polymerizations often exhibit autoacceleration at high conversion via the isothermal "gel effect" or "Trommsdorff effect" (Norrish and Smith, 1942; Trommsdorff et al., 1948). These reactions occur by the creation of a radical that attacks an unsaturated monomer, converting it to a radical, which can add to another monomer, propagating the chain. The chain growth terminates when two radical chains ends encounter each other and terminate. Each chain grows only briefly and then becomes unreactive. As the degree of polymerization becomes high, the viscosity increases. The diffusion-limited termination reactions are slowed, increasing the overall rate of reaction, even though the rate constant of the propagation step is unaffected. The propagation step involves a monomer diffusing to a polymer chain. The addition of a radical to a double bond is a relatively slow process, having a rate constant of 10^4 (M s)$^{-1}$. The termination process requires two polymer chains to diffuse together. As the concentration of polymer molecules increases, the rate of diffusion drops dramatically for both chains. The rate of termination decreases because the polymer chains entangle, but monomers can diffuse through the chains to maintain propagation.

The gel effect represents a different type of feedback than we have encountered so far, because it is not true autocatalysis. No species is produced that increases its own rate of production. In fact, the rate at which a polymer chain grows (recall that each chain has a very short lifetime) actually goes down slightly. The rate at which the chains are terminated decreases even more, so the overall rate of monomer consumption increases.

Both DNA and RNA are biological polymers that exhibit classic isothermal autocatalysis, just as we saw in Chapter 1 in the rabbit–lynx model. One DNA or RNA molecule will produce one copy of itself during each step of replication.

II.2 Industrial Reactors

11.2.1 Thermal Oscillations

Industrial reactors are prone to instability because of the slow rate of heat loss for large systems caused by the low surface-to-volume ratio. Because the consequences of an unstable reactor can be disastrous, industrial plants are often operated under far from optimal conditions to minimize the chance of unstable behavior. Teymour and Ray (1992a) studied vinyl acetate polymerization in a CSTR. The monomer and initiator were flowed into the reactor, which was maintained at a sufficiently elevated temperature for the initiation of polymerization. As the degree of conversion increased, the rate of polymerization increased. The higher rate of reaction meant that the heat produced had less time to dissipate, so the temperature rose. The reaction might have reached a new steady state with higher conversion had the higher temperature not lowered the viscosity. The decrease in viscosity increased the rate of termination. Because these competing processes occurred on different time scales, the system did not reach a steady state, but exhibited temporal oscillations in temperature and conversion. Figure 11.1 shows a time series and the experimental phase plot for vinyl acetate. The period of oscillation is long, about 200 min, which is typical for polymerization in a CSTR.

The same type of oscillations seen in laboratory-scale reactors have been reported for industrial copolymerization reactors (Keane, 1972). In a model of vinyl acetate polymerization in an industrial-scale reactor, Teymour and Ray (1992b) discovered a wide range of dynamical behavior, including a period-doubling route to chaotic oscillations. Oscillations in temperature ranged in amplitude from 70 to 140 °C. The extent of conversion oscillated from about 0.5 to almost 1. Obviously, behavior of this type would be detrimental to the operation of a plant.

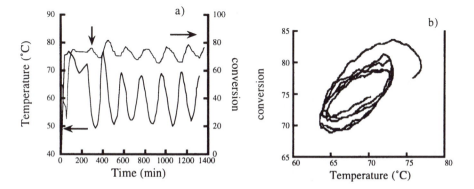

Figure 11.1 (a) Time series for conversion and temperature for vinyl acetate polymerization in a CSTR. Vertical arrow indicates change in residence time from 60 to 90 min. (b) Experimental phase plot for data in part a. (Adapted from Teymour and Ray, 1992a.)

11.2.2 Emulsion Polymerization

Another class of oscillating polymerization reactions in a CSTR was observed in emulsion polymerization (Rawlings and Ray, 1987; Schork and Ray, 1987). In this process, a water-insoluble monomer–polymer is dispersed throughout the aqueous phase with the aid of a surfactant. The surfactant is an aliphatic chain connected to a charged head, represented by the circles in Figure 11.2. A commonly used surfactant is sodium dodecyl sulfate.

Potassium persulfate is dissolved in the aqueous phase and thermally dissociates into free radicals that can initiate polymerization when they diffuse into monomer micelles. As the polymer particle grows by absorbing monomer micelles, surfactant leaves the aqueous phase to cover the increased surface area of the particle. Schork and Ray (1987) demonstrated slow oscillatory behavior in the conversion and in the surface tension of the aqueous phase. (The surface tension increases with decreasing surfactant concentration.)

Periodic behavior is shown in Figure 11.3. Schork and Ray propose the following explanation. At 300 min, conversion increases rapidly because new polymer particles are formed and old ones grow. Additional surfactant adsorbs on the increased surface area of the particles. Micelles dissociate to contribute surfactant.

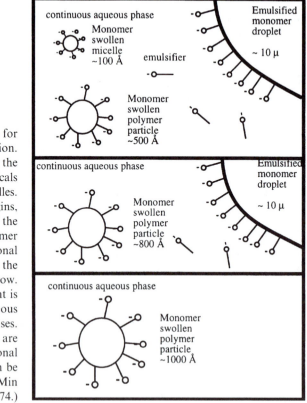

Figure 11.2 The mechanism for emulsion polymerization. Initiator decomposes in the aqueous phase, and the radicals diffuse into monomer micelles. Polymerization begins, increasing the size of the particles as new monomer diffuses in. Additional surfactant diffuses to the surface as the particles grow. When all the surfactant is consumed from the aqueous phase, the surface tension rises. When all the micelles are consumed, no additional polymer particles can be formed. (Adapted from Min and Ray, 1974.)

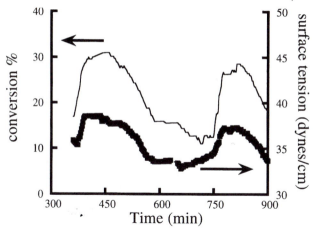

Figure 11.3 Oscillations in the polymerization of methyl methacrylate with a concentration of surfactant $[S]$ = 0.02 M, [initiator] = 0.01 M. Residence time = 47 min. (Adapted from Schork and Ray, 1987.)

When all the micelles are gone, no additional polymer particles are formed, so the rate of polymerization slows. The surface tension rises as the aqueous phase is no longer saturated with surfactant. As the particles are washed out of the reactor, the extent of conversion decreases. With the total surface area decreasing and new surfactant flowing in, the surface tension decreases. When the aqueous phase becomes saturated with surfactant, micelles form, they adsorb free radicals, and polymerization can commence in them. The process begins anew.

This cycle occurs because there is always enough monomer, but not enough surfactant, to form micelles that can adsorb initiating radicals. The nonlinear surfactant consumption vs. the linear rate of surfactant input results in oscillations. Surfactant consumption increases rapidly, as many particles are initiated and grow, but the process overshoots the capacity of the surfactant in the aqueous phase to support polymerization. Once the aqueous phase is no longer saturated, no new particles form, but the old ones continue to grow and consume surfactant. When they are washed out of the reactor and the surfactant saturates the aqueous phase again, new particles can form.

11.3 Biopolymers: Propagating Fronts of RNA Replication

Direct autocatalysis occurs in biological polymerizations, such as DNA and RNA replication. In normal biological processes, RNA is produced from DNA. The RNA acts as the carrier of genetic information in peptide synthesis. However, RNA has been found to be able to replicate itself, for example, with the assistance of the $Q\beta$ replicase enzyme. Bauer and McCaskill have created traveling fronts in populations of short self-replicating RNA variants (Bauer et al., 1989; McCaskill and Bauer, 1993). If a solution of monomers of triphosphorylated adenine, guanine, cytosine, and uracil is placed in a tube and some RNA is added at one end with the necessary enzymes, more RNA will be produced. This RNA can diffuse into adjacent regions and act as a template for further RNA synthesis. The presence of double-stranded RNA was detected using ethidium bromide, which intercalates in the double strand of RNA and fluoresces. The progress of the RNA replication wave was monitored by tracking the position of the fluorescence. Figure 11.4 shows the progress of two waves. They propagate slowly, with velocities on the order of 1 μm s^{-1} (0.36 cm h^{-1}).

The authors gained two useful pieces of information from these fronts. First, they found that the velocity could be described by the Fisher equation (Fisher, 1937):

$$v = 2\sqrt{\kappa D} \tag{11.1}$$

where κ is the rate constant for the autocatalytic homogeneous kinetics and D is the diffusion constant of the autocatalyst. From the homogeneous kinetics and the front velocity, they were able to determine the effective diffusion coefficient for the autocatalyst, and they concluded that this species was the RNA–enzyme complex. Second, they observed that changes in front velocity occurred, indicating that the RNA had mutated into a variant with different reproduction kinetics. In Figure 11.4, at 170 min there is a change in the slope of the position vs. time curve that indicates such a mutation.

McCaskill and Bauer (1993) then used this system to study the spontaneous evolution of systems without an initiating RNA source. Using capillary tubes filled only with the bases, they found that about 5% of the samples would spontaneously organize into RNA and initiate propagating fronts.

11.4 Synthetic Polymers: Frontal Polymerization

Synthetic polymers range from styrofoam cups to high-performance composites in aircraft. Free-radical polymerizations are the most common because of the

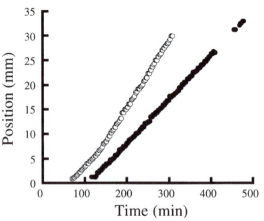

Figure 11.4 RNA replication wavefront position vs. time. Closed circles represent MNV$_{11}$ RNA concentration. In the other experiment (open circles) with de novo RNA, a change in velocity occurred at 170 min. (Adapted from Bauer et al., 1989.)

relative ease with which they can be carried out. You have performed one yourself if you have ever used BONDO to repair your car. The "cream hardener" you mixed with the resin was benzoyl peroxide, a source of free radicals to initiate the reaction. For more detailed information on polymer chemistry, we refer the reader to several texts on the subject (Allcock and Lampe, 1981; Odian, 1991).

If a solution of monomer and initiator, such as benzoyl peroxide, is placed in a test tube and the top is heated, the peroxide will produce free radicals. These radicals will initiate polymerization, which, in turn, will produce heat. The heat will diffuse into the underlying regions and decompose more peroxide, initiating further polymerization. A constant velocity front of polymerization can occur.

Frontal polymerization (propagating fronts of polymerization) was first developed by Davtyan and colleagues, who published an excellent review of the pioneering work done in the former Soviet Union up to 1984 (Davtyan et al., 1984). Since 1990, much work in this field has been performed in Hattiesburg, Mississippi (Khan and Pojman, 1996; Pojman et al., 1996b).

11.4.1 Basic Phenomena

Let us consider the mechanism of propagation and some of the necessary conditions for the existence of the frontal mode. To start with, there must be no significant homogeneous reaction at ambient temperatures. Further, a localized input of heat must be able to initiate a wave of reaction such that there is an autocatalytic regeneration of the stimulus to maintain the front. One can find a number of such chemical systems having free-radical or cationic mechanisms. The input of heat causes formation of reactive intermediates that add to the monomer–resin to produce growing chains. The addition of monomer to those chains releases heat, which diffuses into neighboring unreacted regions and initiates further reaction, which releases heat, and so on. (Information on performing frontal polymerization in an undergraduate lab is presented in Appendix 2).

An experiment can be performed in a glass tube filled with reactants. An external heat source, when applied at the top of the tube, starts a descending front that appears as a slowly moving ($\sim$1 cm min^{-1}) region of polymer formation (Figure 11.5). In the absence of significant homogeneous reaction, the front moves with a constant velocity (Figure 11.6). As we discuss in a later section, various instabilities can occur, especially with liquid systems.

Frontal polymerization occurs in a wide variety of systems and has been demonstrated with neat liquid monomers, such as methacrylic acid (Pojman, 1991; Pojman et al., 1992a, 1996b, d), n-butyl acrylate (Pojman et al., 1996d), styrene, methyl methacrylate, and triethylene glycol dimethacrylate (Pojman et al., 1995b). Fronts can also be obtained with solid monomers, whose melting points are below their adiabatic reaction temperature, such as acrylamide (with initiator) (Pojman et al., 1993) or transition-metal nitrate acrylamide complexes (without initiator) (Savostyanov et al., 1994). Pojman et al. (1996a) demonstrated frontal polymerization of acrylamide, methacrylic acid, and acrylic acid, each in dimethyl sulfoxide or dimethyl formamide, which are excellent systems for studying convection induced by fronts.

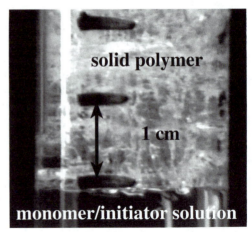

Figure 11.5 A front of triethylene glycol dimethacrylate propagating downward in a 2.2-cm (i.d.) test tube with 1% m/v benzoyl peroxide initiator.

Unfortunately, we do not know the necessary and sufficient conditions for self-sustaining fronts. However, we can identify several factors that favor the frontal mode. The monomer must have a boiling point above the front temperature to prevent heat loss from vaporization and bubbles that can obstruct the front. (The boiling point can be raised by applying pressure.) The front temperature should be high. Thus, highly exothermic reactions are the most likely candidates for frontal polymerization because the heat production must exceed the heat losses, especially if a reinforced or filled composite is sought. The reaction rate must be

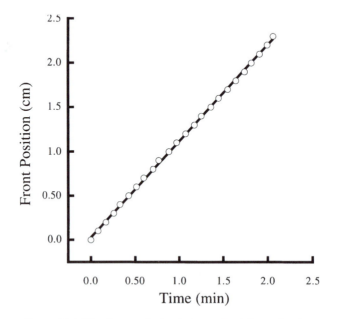

Figure 11.6 The front velocity is determined from the slope of a plot of front position vs. time. $(AIBN)_0 = 0.00457$ molal, initial temperature = 18.9 °C.

vanishingly small at the initial temperature but rapid at the front temperature. In this respect, free-radical polymerization is ideal because for most peroxide and nitrile initiators, the rate of polymerization is low at ambient temperature but high at elevated temperatures.

Figure 11.7 shows the temperature history at a fixed point in the reaction tube as a front passes. The temperature at this point is ambient when the front is far away and rises rapidly as the front approaches. Hence, a polymerization front has a very sharp temperature profile (Pojman et al., 1995b). Figure 11.7 shows five temperature profiles measured during frontal free-radical polymerization of methacrylic acid with various concentrations of BPO initiator. Temperature maxima increase with increasing initiator concentration. For an adiabatic system, the conversion is directly proportional to the difference between the initial temperature of the unreacted medium and the maximum temperature attained by the front. The conversion depends not only on the type of initiator and its concentration but also on the thermodynamic characteristics of the polymer (Pojman et al., 1996b).

A problem with propagating fronts is that the high front temperature causes rapid initiator decomposition or "burnout," which can lead to very low conversion (< 75 %) (Davtyan et al., 1984). One method to improve conversion is to use two initiators with sufficiently different energies of activation that one initiator will start decomposing only after the other initiator has been consumed (Pojman et al., 1995b). The advantage of this dual system is that the front velocity is determined by the least stable initiator (providing a fast front), but the conversion is determined by the more stable one.

11.4.2 Mechanism of Polymerization

A number of radical polymerization reactions are highly exothermic and are able to support frontal polymerization. Free-radical polymerization with a thermal initiator can be approximately represented by a three-step mechanism. First, an

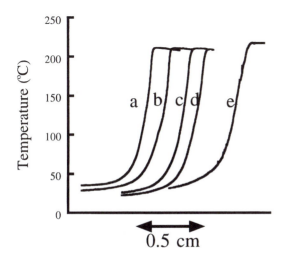

Figure 11.7 Spatial temperature profiles for methacrylic acid fronts at different initial temperatures: (a–d) 2% w/v of BPO, (e) 1.5% v/v of *tert*-butyl peroxide.

unstable compound I, usually a peroxide or a nitrile, decomposes to produce radicals:

$$I \rightarrow 2fR^{\bullet} \tag{11.2}$$

where f is the efficiency, which depends on the initiator type and the solvent. A radical can then add to a monomer to initiate a growing polymer chain:

$$R^{\bullet} + M \rightarrow P_1^{\bullet} \tag{11.3}$$

$$P_n^{\bullet} + M = P_{n+1}^{\bullet} \tag{11.4}$$

The propagation step (11.4) continues until a chain terminates by reacting with another chain (or with an initiator radical):

$$P_n^{\bullet} + P_m^{\bullet} \rightarrow P_n + P_m \text{ (or } P_{n+m}) \tag{11.5}$$

The major heat release in the polymerization reaction occurs in the propagation step. However, the propagation reaction does not have a sufficiently high activation energy, that is, a large enough reaction rate difference between the reaction and cold zones, to support a frontal regime. Frontal polymerization autocatalysis takes place through the initiator decomposition step because the initiator radical concentration is the main control for the total polymerization rate, rather than the gel effect or direct thermal polymerization that may also be present in the frontal polymerization process. The steady-state approximation can be applied to the polymerization model, eqs. (11.2)–(11.5), to give an approximate relationship between the effective activation energy of the entire polymerization process and the activation energy of the initiator decomposition reaction:

$$E_{\text{eff}} = E_p + (E_i/2) - (E_t/2) \tag{11.6}$$

where E_p, E_i and E_t are the activation energies of the propagation, initiator decomposition, and termination steps, respectively.

The second term on the right-hand side of eq. (11.6) depends on the initiator. Because it has the largest magnitude, this value is the major determinant of the effective activation energy. As a result, the initiator plays a significant role in determining if a front will occur, in addition to the temperature profile and the velocity of the front.

11.4.3 Problems with Frontal Polymerization

A significant problem in frontal polymerization is the formation of bubbles at the front. These bubbles affect the front velocity (Pojman et al., 1996b) and can cause voids in the final product. The high front temperature can cause boiling of some monomers at ambient pressures. The main source of bubbles in free-radical systems is the initiator, because all nitriles and peroxides produce volatile by-products.[1] The bubbles make a porous polymer, which may have less than optimal properties.

---

[1]Persulfate initiators used with frontal acrylamide polymerization in DMSO do not produce bubbles (Pojman et al., 1996a).

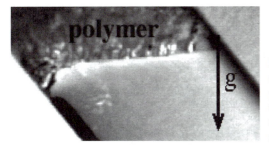

Figure 11.8 A propagating front of triethylene glycol dimethacrylate polymerization in a 2.2-cm (i.d.) test tube. The tube has been tilted, and simple convection causes the front to reorient.

Another problem is one we have encountered in Chapter 9—that is, convection. For a descending front with a solid product, the front will always orient itself perpendicular to gravity. The front shown in Figure 11.8 is a crosslinked material, so that the product is rigid. Many polymers are not crosslinked and can be affected by instabilities. Methacrylic acid forms a polymer that is insoluble in the monomer and is a solid at the front temperature. Nonetheless, the reaction zone can be subject to a double-diffusive instability that appears as descending slender drops ("fingers") of polymer (Figure 11.9) (Pojman et al., 1992a). These fingers not only contaminate the unreacted medium, but also decrease the front velocity significantly by removing heat from the reaction zone. Nagy and Pojman (1996) developed a technique to suppress fingering in methacrylic acid fronts by rotating the tube around the axis of front propagation.

A Rayleigh–Taylor instability (Taylor, 1950; Rayleigh, 1899), which also appears as "fingers," can ensue if the resulting polymer is molten at the front temperature, such as with n-butyl acrylate. The more dense molten polymer streams down from the reaction zone and destroys the front (Figure 11.9). The only current method to study fronts with polymers that are molten at the front temperature is to add a filler, such as silica gel. However, Pojman et al. (1997a) suspected that silica gel could affect the molecular weight distribution, so there is no way on earth (literally!) to study a pure poly(n-butyl acrylate) front. Instead, they carried out a frontal

1 cm

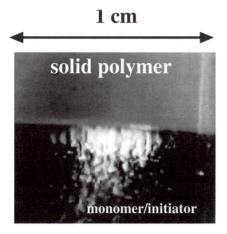

Figure 11.9 A descending front of methacrylic acid polymerization (with benzoyl peroxide initiator). Descending fingers of polymerizing material occur under some conditions.

polymerization on a sounding rocket flight with 6 min of weightlessness (Pojman et al., 1997a).

What about ascending fronts? If a front were to propagate upward, then the hot polymer–monomer solution in the reaction zone could rise because of buoyancy, removing enough heat at the polymer–monomer interface to quench the front. With a front that produces a solid product, the onset of convection is more complicated than the cases that we considered in Chapter 9, because the critical Rayleigh number is a function of the velocity (Volpert et al., 1996). Bowden et al. (1997) studied ascending fronts of acrylamide polymerization in dimethyl sulfoxide. As in the iodate–arsenous acid fronts, the first unstable mode is an antisymmetric one followed by an axisymmetric one. Unlike that system, in the polymerization front the stability of the front depends on both the solution viscosity and the front velocity. The faster the front, the lower the viscosity necessary to sustain a stable front.

11.4.4 Periodic Modes

In stirred reactions, we saw that a steady state could lose its stability as a bifurcation parameter was varied, leading to oscillations. Propagating thermal fronts can show analogous behavior. The bifurcation parameter for a thermal front is the Zeldovich number (Zeldovich et al., 1985). Zeldovich assumed that the reaction occurs in an infinitely narrow region in a single step with activation energy E_{eff}, initial temperature T_0, and maximum temperature T_m:

$$Z = \frac{T_m - T_0}{T_m} \frac{E_{\text{eff}}}{RT_m} \qquad (11.7)$$

A great deal of theoretical work has been devoted to determining the modes of propagation that occur (Shkadinsky et al., 1971; Matkowsky and Sivashinsky, 1978; Sivashinsky, 1981; Margolis, 1983; Shcherbak, 1983; Bayliss and Matkowsky, 1987, 1990). In a one-dimensional system, the constant velocity front becomes unstable as Z is increased. A period-doubling route to chaos has been shown numerically (Bayliss and Matkowsky, 1987). A wide array of modes has been observed in self-propagating high-temperature synthesis (thermite) reactions (Strunina et al., 1983).

Decreasing the initial temperature of the reactants increases the Zeldovich number (if the conversion does not change) and drives the system away from the planar front propagation mode into nonstationary regimes. The stability analysis of the reaction–diffusion equations describing the propagation of reaction waves along the cylindrical sample predicts the existence of spin modes with different numbers of heads (Sivashinsky, 1981). Some of the nonplanar frontal polymerization regimes described above, such as single and multipoint spin modes, have very similar topologies to those predicted theoretically; namely, the perturbed solutions show the front structure with temperature maxima rotating as a whole around the cylinder axis with constant angular velocity. Depending on the Zeldovich number and the tube diameter (provided the kinetic parameters are kept constant), the number of spinning heads can vary from one to infinity as the tube diameter increases.

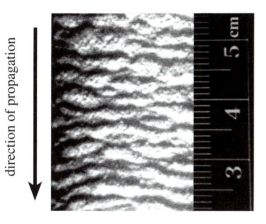

direction of propagation

Figure 11.10 Surface tracks of the multipoint spin modes. The experiment was performed in a 2.2-cm-diameter test tube for methacrylic acid with 2% w/v of benzoyl peroxide (BPO).

11.4.4.1 Experimental Observations

Begishev et al. (1985) studied anionic polymerization fronts with ε-caprolactam.[2] There were two interesting aspects to this system. First, the polymer crystallizes as it cools, which releases heat. Thus, a front of crystallization follows behind the main front. Volpert et al. (1986) investigated this two-wave system. Second, a "hot spot" moved around the front as it propagated down the tube, leaving a spiral pattern in the product. The entire front propagated with a velocity on the order of 0.5 cm min^{-1}, which was a function of the concentrations of activator and catalyst. The hot spot circulated around the outside of the 6-cm (i.d.) front sixteen times as rapidly as the front propagated.

Experimental study of frontal polymerization of methacrylic acid has shown the existence of a rich variety of periodic regimes (Pojman et al., 1995a, 1996b). At ambient initial temperature with heat exchange to room-temperature air, only a stable planar-front mode exists. Decreasing the initial temperature of the reactants, as well as increasing the rate of heat loss, leads to the occurrence of periodic modes. The magnitude of heat losses is a convenient control parameter for studying periodic regimes.

We now describe the sequence of modes observed during frontal polymerization for a 2% w/v benzoyl peroxide solution in methacrylic acid as a function of the initial temperature and heat loss intensity. Lowering the water bath temperature to 21 °C causes a multipoint spin mode (Figure 11.10). This regime manifests itself as many spinning heads adjacent to the tube wall. Multiple heads appear and disappear on the sample surface, moving along spiral trajectories (2–5 mm long) and leaving a characteristic reticulated pattern on the surface.

The initial monomer temperature cannot be lowered beyond 21 °C by immersion in a water bath because heat loss quenches the front. Further decrease of the initial solution temperature is possible by using the experimental setup shown in Figure 11.11. The clearance, Δ, between the front and the coolant interface, along with the coolant temperature, determines the radial and longitudinal temperature gradients. At an initial solution temperature of 0 °C, with a 1.5-mm clearance, it is

---

[2]The product is a form of nylon.

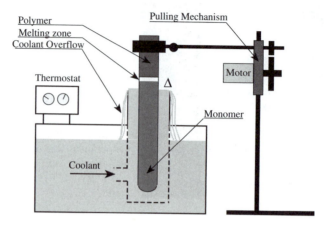

Figure 11.11 Experimental setup for studying periodic regimes at different initial temperatures. As the front propagates downward, the test tube is removed from the cooling bath to maintain constant clearance, Δ, between the front and coolant surface.

possible to observe a few-head spin mode, which has the pattern shown in Figure 11.12. It is difficult, however, to separate heat loss and initial temperature effects in this case; they probably act together.

The most impressive pattern (a large translucent spiral as shown in Figure 11.13) has a single-head spin mode that can be realized by decreasing the clearance to 0.5 cm at 0 °C initial temperature. The single "hot spot" propagates around the front (itself moving with a velocity of 0.66 cm min^{-1}) with a period of 1.3 min. The direction of the single-head spinning is arbitrary and can change during front propagation. The spin mode is best observed with an infrared (IR) camera. Figure 11.14 shows montages of a single head spiraling around the tube,

Figure 11.12 Pattern on the surface of a polymer sample left by the several-head spin mode. The experiment was performed in a 2.2-cm-diameter test tube containing methacrylic acid with 2% w/v of BPO.

direction of propagation

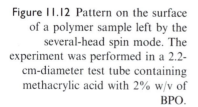

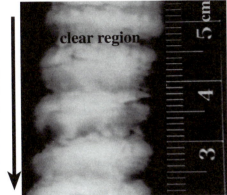

Figure 11.13 Track of the single-head spin mode. The experiment was performed in a 2.2-cm-diameter test tube containing methacrylic acid with 2% w/v of BPO.

a)

b)

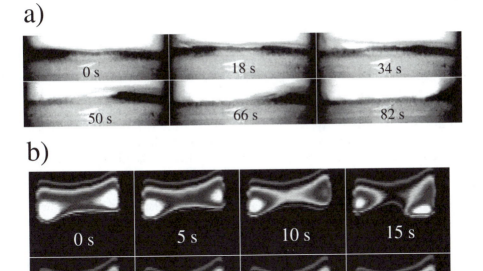

Figure 11.14 Montages of (a) visible images and (b) IR images of a single-head spin mode (same conditions as in Figure 11.13).

in visible light and IR. In the visible range, spin modes can be inferred from a perturbation in the front shape.

In general, the behavior of the polymerizing system is determined by the initial temperature of the reactants and the magnitude of heat losses. The experimental propagation mode depends on the coolant temperature and clearance Δ for a given initiator concentration and tube diameter. The phase diagram of the experimental stability boundaries is shown in Figure 11.15.

11.4.4.2 Qualitative Model

What is the physical mechanism for these instabilities? First, we consider the one-dimensional case of uniform pulsations. For a given activation energy, if the initial temperature is "high enough," then heat diffuses ahead of the front and initiates polymerization at a temperature near the initial temperature T_i. A constant front velocity is observed with a front temperature equal to $T_i + |\Delta H/C_p|$ (assuming 100% conversion). If the initial temperature is low, the heat diffuses and accumulates until a critical temperature is reached (the ignition temperature for the homogeneous polymerization); the polymerization then occurs very rapidly, consuming all the monomer. Again, the front temperature is $T_i + |\Delta H/C_p|$. Then, because all the monomer in the reaction zone has been consumed, the rate of reaction plummets, and the velocity drops. Heat then diffuses into the unreacted region ahead of the front, repeating the cycle.

Another way of looking at the problem is to compare the rates of heat production and heat transport. When the initial temperature is low, the rate of heat production can exceed the rate of heat transport along the direction of propagation, causing an overshoot in the temperature. When all the monomer has been consumed, the rate of heat transport exceeds the rate of heat production, causing a drop in the temperature and the propagation velocity. This lack of a balance is reflected in the changing temperature gradient. When the initial temperature is high, the heat production is balanced by diffusion, and a constant velocity and temperature profile can exist.

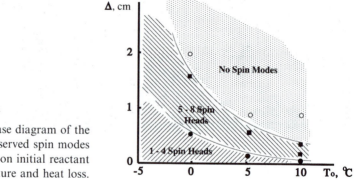

Figure 11.15 Phase diagram of the experimentally observed spin modes depending on initial reactant temperature and heat loss.

The spinning mode is more complicated. Let us consider the plane front and a localized temperature perturbation on it. As we discussed earlier, at the spot of the perturbation there are two competing processes: heat production and heat transfer. Again, if the energy of activation is sufficiently large or the initial temperature is sufficiently low, the temperature can begin to increase. As it increases, the reactants are consumed more rapidly. From where are the reactants consumed? There are two possibilities: (1) to take the reactants from ahead of the front (i.e., to move in the direction perpendicular to the front), and (2) to take the reactants from the left or from the right of the perturbed spot (i.e., to move along the front to the left or right). In some cases, it is more effective to move along the front because the reactants there are preheated, while in the perpendicular direction they are cold. This motion of the temperature perturbation along the front gives a two-dimensional instability that can be a spinning mode or some other mode, depending on the geometry and the values of the parameters.

11.4.5 Applications of Propagating Fronts

Polymerization fronts may be useful in three practical respects:

1. Polymerization fronts generate product rapidly with essentially no energy input.
2. No solvent is required, so less waste is produced. Because no solvent needs to be evaporated from the polymer, less energy is required.
3. Novel materials can be produced. Multicomponent systems often undergo phase separation because upon polymerization the components become immiscible. The very rapid reaction that occurs in the front can "freeze in" the components via crosslinking reactions. An example is the preparation of a thermochromic plastic that consists of cobalt, glycerin, and TGDMA–acrylamide (Nagy et al., 1995b). If the reaction is performed homogeneously, the glycerin phase separates before significant reaction can occur (Figure 11.16). A poor-quality product is formed. However, the product generated in the frontal reaction is hard and machinable. A similar result was obtained when preparing a conductive composite (Szalay et al., 1996).

0.9 cm

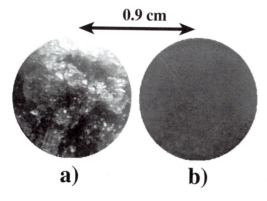

a)   b)

Figure 11.16 Morphology of copolymer composite (a) by homogeneous polymerization performed at 60 °C, (b) by frontal polymerization in a 9.0-mm (i.d.) tube. Initial composition: 54.5 w/w % acrylamide, 26.7 w/w % TGDMA, 16.3 w/w % glycerol, 1.4 w/w % Co(II)-chloride, 1.1 w/w % AIBN.

11.4.6 Manufacture of Thick Composites

The manufacture of thick composites is difficult because the internal temperature of the thick section lags behind the surface temperature. If the surface cures faster than the interior regions, resin can be trapped in the interior, which leads to nonuniform composition and voids. Another problem is thermal spiking, during which the heat produced in the interior cannot escape and therefore builds up. The autocatalytic nature of the curing process can lead to thermal runaway and nonuniform composition. Very slow heating rates can solve this problem, but at the expense of increased processing time.

The continuous production of composites may be the most promising technological application of frontal polymerization. This approach reduces imperfections and production costs for a very high value-added product. White and colleagues developed two continuous curing processes for producing graphite fiber composites (White and Kim, 1993; Kim et al., 1995). The first is used for large rectangular objects (Figure 11.17). Layers of prepreg (graphite fiber material impregnated with the epoxy thermoset resin) are placed into a mold that is heated from below. Pressure (25 psi) is applied until the front approaches the top of the mold. The pressure is then released, and another layer of prepreg is inserted. The sequence is repeated until the desired thickness is attained. White has shown that this process yields superior product in less time.

The second approach allows the production of cylindrical objects, such as flywheels for energy storage. A filament winding procedure has been developed in which resin-impregnated fibers are wound onto a heated mandrel at a rate that matches that of the expanding cure front. Korotkov et al. (1993) proposed a model for such a process.

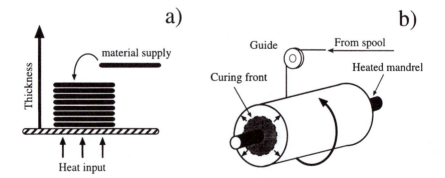

Figure 11.17 Schematic diagram of White's method for studying propagating fronts of epoxidation in thick materials. (a) Layers of epoxy-impregnated fiber are added as the front propagates. (b) Illustration of a continuous filament winding process.

II.5 Polymerization Coupled to Oscillating Reactions

Given the ubiquity of the BZ reaction in nonlinear chemical dynamics, it would seem that this reaction might find a way to participate in nonlinear polymerization reactions. Indeed, it does! Váradi and Beck (1973) observed that acrylonitrile, a very reactive water-soluble monomer, inhibits oscillations in the ferroin-catalyzed BZ reaction while producing a white precipitate, thereby indicating the formation of free radicals. Pojman et al. (1992c) studied the cerium-catalyzed BZ reaction [ferroin is a poor catalyst because it can complex with poly(acrylonitrile) in a batch reactor]. Because poly(acrylonitrile) is insoluble in water, the qualitative progress of the polymerization was monitored by measuring the relative decrease in transmitted light due to scattering of an incandescent light beam passed through the solution. The ESR data (Figure 11.18) show the oscillations in the malonyl radical concentration (Venkataraman and Sørensen, 1991). Oscillations in the bromine dioxide concentration have a greater amplitude (10^{-6} M) and are out of phase with those of the malonyl radical (10^{-8} M).

Acrylonitrile halts oscillations for a time proportional to the amount added. However, no polymer precipitates until oscillations return in both the platinum electrode potential and the bromide concentration. Then, a white precipitate forms periodically during the oscillations (Figure 11.19). Even if acrylonitrile is added in excess of its solubility limit, the oscillations continue.

Oscillations and polymerization occur in both batch and flow reactors into which acrylonitrile is continuously flowed along with the other BZ reactants. Polymerization occurs periodically, in phase with the oscillations (Pojman et al., 1992c).

To eliminate the possibility that the precipitation process is itself nonlinear, acrylonitrile polymerization was initiated by cerium(IV)/H_2SO_4 and malonic acid;

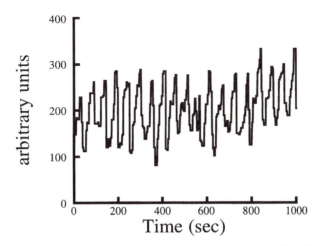

Figure 11.18 Oscillations in the concentration of malonyl radicals. The magnitude of oscillations is less than 10^{-8} M. (Adapted from Venkataraman and Sørensen, 1991.)

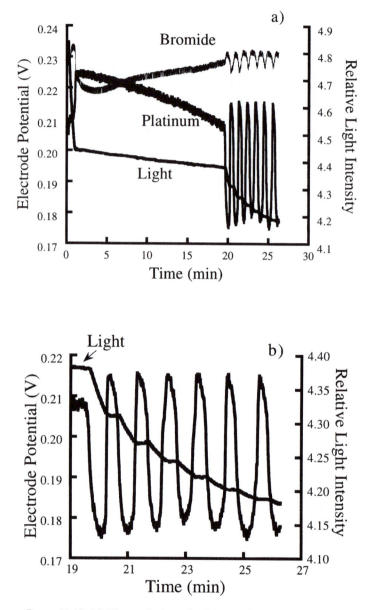

Figure 11.19 (a) The evolution of a BZ reaction in which 1.0 mL acrylonitrile was present before the Ce(IV)/H$_2$SO$_4$ solution was added. [NaBrO$_3$]$_0$ = 0.077 M, [Malonic Acid]$_0$ = 0.10 M; [Ce(IV)]$_0$ = 0.0063 M, [H$_2$SO$_4$]$_0$ = 0.90 M. (b) An enlargement of the oscillatory region. The light intensity is inversely related to the turbidity of the solution.

polymer appears monotonically as the cerium is reduced. This experiment also indicates that malonyl radicals formed by the oxidation of malonic acid (Ganapathisubramanian and Noyes, 1982b; Brusa et al., 1985; Försterling and Noszticzius, 1989; Venkataraman and Sørensen, 1991; Försterling and Stuk, 1992) can initiate polymerization. Initiation is not from $BrO_2^{\bullet}$ radicals formed during the oxidation of cerium. Using cerium(III)/H_2SO_4 and bromate solution of the same concentrations as with the Ce(IV) experiments, acrylonitrile did not polymerize. Also, cerium(IV) does not react at a significant rate with acrylonitrile.

While these experiments are interesting, it remains to be seen if using an oscillating reaction to initiate polymerization can be more useful than current approaches.

11.6 Ordering Phenomena in Phase Separation of Polymer Mixtures

Most polymers are incompatible with one another in the melt and can undergo phase separation, depending on temperature, mole fractions, and molecular weights of each component. As in the case of metallic alloys (Gunton et al., 1983), there exist two distinct mechanisms for the phase separation of polymer mixtures. One is the nucleation-and-growth process and the other is the spinodal decomposition process. The former is initiated by large fluctuations in composition (nuclei), whereas the latter proceeds with infinitesimal fluctuations. Though completely different at the beginning, both processes lead to the same phase equilibrium, resulting in two-phase random morphology (Strobl, 1996). By controlling the kinetics of the spinodal decomposition process, multiphase polymer materials with various co-continuous structures can be designed.

11.6.1 Photo-crosslinking of Polymer Blends *in situ*

Tran-Cong and coworkers have demonstrated that a wide variety of ordered structures in the micrometer range can be prepared and controlled by photo-crosslinking one polymer component in a binary miscible blend (Tran-Cong and Harada, 1996). These materials belong to a class of molecular composites called *semi*-interpenetrating polymer networks (*semi*-IPNs), where the networks of one polymer component are formed by appropriate chemical reactions and trap the other polymer component inside. In the experiments, polystyrene/poly(vinyl methyl ether) (PS/PVME) mixtures were used as polymer blends where PS was labeled with anthracenes by reacting with the chloromethyl styrene moieties incorporated into the PS chains with the content 15 mol % via copolymerization. Upon irradiation of the blend in the one-phase region, anthracenes undergo photodimerization, forming a semi-IPN with PVME trapped inside the PS networks. The morphology resulting from this photo-crosslinking reaction is mainly controlled by the competition between two antagonistic processes, i.e., phase separation and photo-crosslinking reactions. By adjusting the strengths of these two interactions via the reaction temperatures and the blend compositions, polymers with co-continuous, labyrinthine, nucleation-assisted spinodal structures and concentric (target) patterns were obtained (Tran-Cong and Harada, 1996). Figure 11.20 shows an example of the target patterns observed in a PS/PVME (50/50) blend.

The mechanism of pattern formation in photo-crosslinked polymer blends is not completely understood. Besides the competition between phase separation and photo-crosslinking reactions, elastic stress in the sample due to network formation, as well as an autocatalytic feedback arising from concentration fluctuations amplified by the crosslinking reaction, may play an important role in these ordering processes.

11.6.2 Phase Separation of Polymer Blends in a Temperature Gradient

Phase separation of polymer blends under thermally nonuniform conditions is not only an interesting subject related to pattern formation far from equilibrium, but is also crucially important for morphology and/or quality controls of many industrial products. Though not popular in polymer materials science, studies of these phenomena have been extensively carried out in the field of metallurgy known as directional solidification (Mullins and Sekerka, 1964; Caroli et al., 1992). Furukawa (1992) performed two-dimensional computer simulations for phase-separating binary mixtures where the boundary between the stable and the unstable phases was shifted at a constant velocity. Depending on the magnitude of this velocity, three types of morphology were found. As the shift velocity of the phase boundary decreases, these patterns change from randomly co-continuous structures to undulatory lamellae perpendicular to the shifting direction and

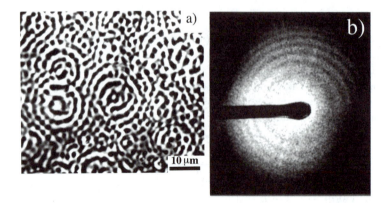

$$2 \times 10^4 cm^{-1}$$

Figure 11.20 (a) Optical micrograph and (b) the light-scattering pattern of the target patterns emerging from a PS/PVME (50/50) blend irradiated in 600 min at 90 °C. (Courtesy of Q. Tran-Cong.)

eventually to columnar morphology parallel to the moving direction of the boundary. Experimentally, Tran-Cong and coworkers studied the phase separation of a poly(2-chlorostyrene)/poly(vinyl methyl ether) (P2CS/PVME) blend of critical composition in the presence of a temperature gradient (Okinaka and Tran-Cong, 1995). The critical point of the blend was set between the two ends of a temperature gradient. The blend was first allowed to reach phase equilibrium in an initial temperature gradient in order to develop a stationary boundary dividing the stable region from the two-phase region. Subsequently, this phase boundary was shifted toward the low-temperature side by increasing the temperature gradient. As the front of the boundary propagated, phase separation occurred in the region of the sample with lower temperatures. The phase separation along the temperature gradient was anisotropic (Figure 11.21). The length scales of the morphology in the direction of the gradient are smaller than those in the perpendicular direction. Furthermore, this structural anisotropy of the blend morphology becomes less remarkable toward the high-temperature side. Thus, in the presence of a temperature gradient, the length scale and the anisotropy of the

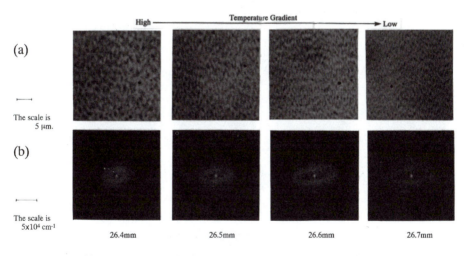

(a) Polymer morphology under a temperature gradient.
(b) Corresponding power spectra obtained by FFT.

High heat source : $60°C \longrightarrow 210°C$ (Heating rate = 0.5°C/min.)
Low heat source : $40°C \longrightarrow 40°C$

Figure 11.21 (a) Morphology and (b) the corresponding power spectra obtained by two-dimensional fast Fourier transform (lower) for a P2CS/PVME blend undergoing phase separation under a temperature gradient. The larger number indicated below the power spectra corresponds to the smaller distance from the macroscopic interface between the miscible and the phase-separated regions. (Courtesy of Q. Tran-Cong.)

phase morphology become functions of the propagation velocity of the phase boundary.

II.7 Summary

Polymer systems exhibit all the dynamic instabilities that we have studied: oscillations, propagating fronts, and pattern formation. Some of the instabilities, such as those in a CSTR, have been studied with the goal of eliminating them from industrial processes. A new trend is developing to harness the instabilities to create new materials or to create old materials in new ways.

12

Coupled Oscillators

We have thus far learned a great deal about chemical oscillators, but, except in Chapter 9, where we looked at the effects of external fields, our oscillatory systems have been treated as isolated. In fact, mathematicians, physicists, and biologists are much more likely than are chemists to have encountered and thought about oscillators that interact with one another and with their environment. Forced and coupled oscillators, both linear and nonlinear, are classic problems in mathematics and physics. The key notions of resonance and damping that arise from studies of these systems have found their way into several areas of chemistry as well. Although biologists rarely consider oscillators in a formal sense, the vast variety of interdependent oscillatory processes in living systems makes the representation of an organism by a system of coupled oscillators a less absurd caricature than one might at first think.

In this chapter, we will examine some of the rich variety of behaviors that coupled chemical oscillators can display. We will consider two approaches to coupling oscillatory chemical reactions, and then we will look at the phenomenology of coupled systems. We begin with some general considerations about forced oscillators, which constitute a limiting case of asymmetric coupling, in which the linkage between two oscillators is infinitely stronger in one direction than in the other. As an aid to intuition, picture a child on a swing or a pendulum moving periodically. The forcing consists of an impulse that is applied, either once or periodically, generally with a frequency different from that of the unforced oscillator. In a chemical oscillator, the forcing might occur through pulsed addition of a reactive species or variation of the flow rate in a CSTR. Mathematically, we can write the equations describing such a system as

$$dx/dt = \mathbf{f}(\mathbf{x}) + \varepsilon\, \mathbf{g}(\mathbf{x}, t) \qquad (12.1)$$

where the vector $\mathbf{x}$ contains the concentrations, the vector function $\mathbf{f}(\mathbf{x})$ contains all the rate and flow terms in the absence of forcing, $\mathbf{g}(\mathbf{x})$ represents the appropriately scaled temporal dependence of the forcing, and the scalar parameter ε specifies the strength of the forcing. When $\varepsilon = 0$, the unforced system has a periodic solution with period τ_0.

It is often useful to specify the instantaneous state of an oscillator by its *phase*, that is, the point along its closed orbit in phase space at which it currently finds itself. Since this orbit is topologically equivalent to a circle, we may choose the starting point $\phi = 0$ anywhere we like on the orbit—for example, at the maximum of a particular variable. If we then measure time from this point and rescale the time variable by dividing by the natural period τ_0, the phase of our oscillator is now periodic with period 1. Obviously, the choice of phase is not unique. We could have picked a different origin or a different time-scaling—for example, to make the period of the phase variable 2π. We could even have chosen a nonlinear scaling of the time. Despite these ambiguities, the phase proves to be an extremely valuable concept.

Experiments in which a nonlinear, limit cycle oscillator is subjected to a single pulse are often analyzed by constructing a *phase-resetting curve*. These experiments, which can yield a great deal of information about the nature of an oscillator, consist of applying a single pulse of a fixed strength to the oscillator at different initial phases. In general, the oscillator will be knocked off its limit cycle by the pulse. If the pulse is not too strong, or if the system has no other stable attractors, the oscillator will return to the cycle after some period of time. When it does return, the phase of the oscillator will almost certainly differ from what it would have been had the system not been perturbed. We can write

$$\phi = \phi_0 + \delta \qquad (12.2)$$

A plot of the new phase ϕ as a function of the initial phase ϕ_0 at fixed pulse strength ε constitutes the phase-resetting curve. The system will be more sensitive to perturbation at some points on its cycle than at others, and this is the information that is contained in the resetting curve. A set of curves at different values of ε can be a valuable guide to the nature of an oscillator. Biologists have often exploited this technique in their studies of both neurons and circadian rhythms.

Since single-pulse experiments are of limited relevance to the study of coupled oscillators, we shall say no more about them here, but refer the interested reader to the work of Marek and collaborators (Dolnik et al., 1986; Marek and Schreiber, 1991). Of greater bearing on the subject of this chapter are experiments in which the forcing function $\mathbf{g}(\mathbf{x}, t)$ in eq. (12.1) is itself periodic with a frequency ω that differs from the unforced or natural frequency $\omega_0 = 2\pi/\tau_0$. A number of possible behaviors can arise, depending upon the coupling strength ε and the frequency ratio $r = \omega/\omega_0$. If r is a rational number, that is, if whole numbers of oscillator and forcing periods can fit exactly into the same time interval, then we have *frequency locking* or *entrainment*, and the system behaves periodically with a period equal to the least common multiple of τ and τ_0. If r is irrational, so that the periods or frequencies are incommensurate, and if the coupling is weak, then we have *quasiperiodicity*. The system oscillates aperiodically, but with two well-defined frequencies, ω and ω_0 (as well as their harmonics, sums, and differences), in its Fourier spectrum. If the coupling is somewhat stronger, then frequency-locking can occur to a rational ratio near the true r value. For example, if $r = 2^{1/2} = 1.414\ldots$, and ε is large enough, the system may lock at a ratio of $7/5$, meaning that the oscillator will undergo exactly seven oscillations for every five cycles of the forcing. Whether or not the system will frequency-lock depends not only on r and ε but also on the detailed dynamics of the oscillator. In general, frequency-locking is more likely to occur as the frequency ratio approaches a rational number with a small integer as denominator (1:1 is the most favorable ratio) and as the coupling strength increases.

The behavior of a typical forced nonlinear oscillator is shown in Figure 12.1. The shaded regions, in which frequency-locking occurs, are known as *Arnol'd tongues* in honor of the Russian mathematician who did much to characterize the behavior of coupled and forced oscillators. Note that the tongues start as points corresponding to the rational numbers on the r axis, where there is zero forcing, and broaden as ε increases. The light areas between the tongues contain an infinite number of additional tongues that are too narrow to be seen, corresponding to the infinite number of other rationals that lie between any pair of rational numbers, as well as regions of quasiperiodicity, where the forcing is too weak to bring on frequency-locking. Above a certain critical value of ε, the tongues broaden so far that they begin to overlap, filling the entire range of r values and creating the possibility not only of multiple stable modes of frequency-locked oscillation, but also of chaos.

In some cases, one can learn a great deal about the behavior of a system of coupled oscillators, even one that contains a large number of oscillatory subunits, by studying a set of equations in which each oscillator is described by only a single variable—its phase. Strogatz's study (Mirollo and Strogatz, 1990) of the synchronization of flashing fireflies is a beautiful example of this approach. Unfortunately, chemical oscillators are almost never this simple, and it is generally necessary to keep at least two variables (which may be thought of as con-

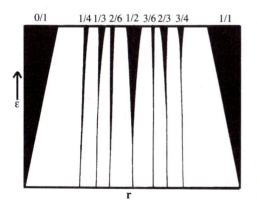

Figure 12.1 Arnol'd tongue structure for a typical forced nonlinear oscillator.

centrations or as amplitude and phase) for each oscillator in order to describe successfully the dynamics of a system of coupled chemical oscillators.

12.1 The Nature of the Coupling

How might we couple together a pair of chemical oscillators? If we were talking about springs or swings, it would be easy. A rigid rod or another spring provides a nice connection between two springs. Two children swinging can hold hands, or we could tie their swings together with a length of string. Hooking two oscillating reactions together seems like a more challenging problem, but there are several ways in which it can be done.

We might take a lesson from the way in which nature couples together pairs of neurons in a living organism. Nerve cells, many of which are oscillators, communicate with each other via two fundamentally different mechanisms: *electrical coupling* and *synaptic coupling.* In electrical coupling, sometimes referred to as coupling via gap junctions, the two cells share a cell membrane, and any difference between the potentials of the cells tends to decay to zero as ions move across the membrane to equalize their concentrations. It is as if two electrical oscillators are coupled together by a wire that has the resistance and capacitance of the membrane or junction. Thus, the cells tend to follow one another's behavior, though there may be a delay owing to the membrane properties. Synaptic coupling is somewhat more complex and comes in many forms. Whereas electrical coupling is symmetric, synaptic coupling is decidedly asymmetric, with information from the *presynaptic* cell being transmitted to the *postsynaptic* cell, but not vice versa, though it is possible to have mutual coupling in which there are independent routes for transmission in both directions. Electrical coupling is always *excitatory* in that an increase in the potential of one cell will always excite an increase in its partner's potential. Synaptic coupling can be either excitatory or *inhibitory*, meaning that a potential increase in the presynaptic cell can yield either an increase or a *decrease* in the potential of the postsynaptic neuron. In synaptic coupling, a change in the potential of the presynaptic cell causes the release of a chemical substance called a *neurotransmitter*—for example, acetylcholine. The molecules of transmitter diffuse across the narrow space, the *synapse*, between the pre- and postsynaptic cells. When they arrive at the postsynaptic cell, they become attached to receptor molecules in the membrane of that neuron, setting in motion a chain of events that results in conductivity and, hence, potential changes in the target cell.

In considering the coupling of chemical oscillators in flow reactors, we have found it useful to think in terms of two modes of coupling that are analogous to electrical and synaptic coupling in neurons. We call these approaches *physical* and *chemical coupling*, respectively. The division of coupling methods into physical and chemical is by no means the only possible classification. One can imagine experimental designs that fit into neither category or that have aspects of both. Nevertheless, as we hope to show in this chapter, this scheme provides a fruitful way of thinking about experiments and simulations involving two or more coupled chemical oscillators.

12.1.1 Physical Coupling

In physical coupling of chemical reactions (the analog of electrical coupling of neurons), one has two or more subsystems and links them together by some sort of transport mechanism, typically diffusion or facilitated diffusion. One possible experimental arrangement (Crowley and Epstein, 1989) is shown in Figure 12.2, where two CSTRs are linked through a common wall containing an opening whose size can be adjusted by the experimenter, thus providing a tunable coupling strength.

In such a system, the rate equations may be written

$$dx_1/dt = f(x_1) - c(x_1 - x_2) \tag{12.3a}$$

$$dx_2/dt = g(x_2) - c(x_2 - x_1) \tag{12.3b}$$

where x_1 and x_2 represent the concentrations in the two subsystems; the vector functions f and g describe the kinetics of the reactions (which may be identical)

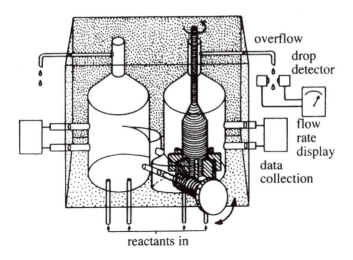

Figure 12.2 Schematic diagram of an apparatus consisting of two CSTRs for studying physically coupled oscillating reactions. A needle valve controls the flow between the reactors. Inputs to the reactors are independently controlled. Drop detectors ensure that liquid flows out of the two reactors at the same rate so that there is no *net* mass transfer from one to the other. Reprinted, in part, with permission from Crowley, M. F.; Epstein, I. R. 1989. "Experimental and Theoretical Studies of a Coupled Chemical Oscillator: Phase Death, Multistability, and In-Phase and Out-Of-Phase Entrainment," *J. Phys. Chem.* **93**, 2496–2502. © 1989 American Chemical Society.)

and the flow to and from external reservoirs in the two reactors; and c is the coupling strength, which in the configuration of Figure 12.2 is proportional to the mass transfer rate between reactors.

Other experimental arrangements have also been utilized in physical coupling experiments. Perhaps the most common consists of using an additional peristaltic pump to create a flow from one CSTR to another. Two experimental difficulties must be taken into consideration here. First, the flows must be carefully balanced to avoid net transfer from one reactor to another. Second, the liquid from one reactor requires a finite amount of time to arrive at the other reactor. This transit time introduces a delay in the system. In eq. (12.3a), for example, $x_2(t)$ should really be written as $x_2(t - \tau)$, where τ is the time required for material to travel from reactor 2 to reactor 1. If τ is very small compared with the time scale of the reaction (e.g., the period of oscillation), then this delay effect can safely be neglected. If not, the coupling may introduce some of the complications discussed in Chapter 10. The multiple reactor – multiple pump configuration is an attractive one, because it can be utilized to couple arbitrarily many CSTRs, though the expense and experimental headaches increase at least in proportion to the number of reactors. As early as 1982, Stuchl and Marek coupled up to seven CSTRs containing the BZ reagent, and examined the sets of final states obtained when the input flows corresponded to concentrations in the bistable range. More recently, Laplante and Erneux (1992) studied a system of sixteen physically coupled chlorite–iodide reactions.

In a series of experiments that correspond remarkably well with the electrically coupled neuron analogy, Crowley and Field (1986) studied the BZ reaction in two CSTRs, using the difference in redox potential between the two reactors to determine the amount of current that flows from one to the other. They obtained a wealth of dynamical behavior, including entrainment, quasiperiodicity, and chaos. In this study, as in all other physical coupling experiments to date, the same reaction is run in all of the subsystems.

12.1.2 Chemical Coupling

In a chemically coupled system (the counterpart of synaptic coupling in neurons), the subsystems are coupled through chemical reactions between their reactants, products, or intermediates. The simplest way to carry out a chemical coupling experiment is to employ the organic chemist's "one-pot" approach and to feed the reactants of two independently oscillatory reactions into a single CSTR. For example, the chlorite–iodide and bromate–iodide reactions both show oscillation as well as bistability in a flow reactor. The bromate–chlorite–iodide reaction in a CSTR constitutes the most thoroughly studied example of a chemically coupled oscillator (Alamgir and Epstein, 1983). It exhibits an astonishing variety of dynamical behavior, including three stationary and two oscillatory states and various combinations of bistability among them, as summarized in the phase diagram of Figure 12.3. Several of the more complex behaviors indicated in this figure will be discussed in detail later in this chapter.

The bromate–chlorite–iodide system represents a form of parallel coupling in that, schematically, we have subsystems $(A + B)$ and $(C + B)$ coupled through the

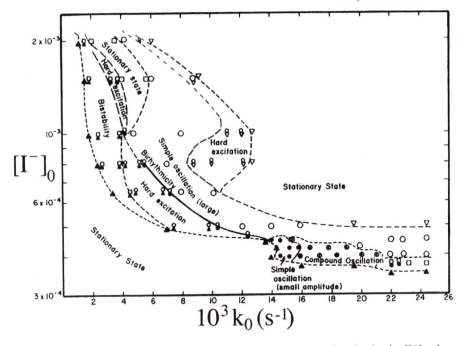

Figure 12.3 Phase diagram of the bromate–chlorite–iodide reaction in the k_0–$[I^-]_0$ plane. Fixed constraints: $[BrO_3^-]_0 = 2.5 \times 10^{-3}$ M, $[ClO_2^-]_0 = 1.0 \times 10^{-4}$ M, $[H_2SO_4]_0 = 0.75$ M. Symbols: open circles, low-frequency oscillatory state; filled circles, high-frequency oscillatory state; open triangles, low-potential stationary state; filled triangles, high-potential stationary state; open squares, intermediate-potential stationary state. Combinations of two symbols imply bistability between the corresponding states. (Reprinted with permission from Alamgir, M.; Epstein, I. R. 1983. "Birhythmicity and Compound Oscillation in Coupled Chemical Oscillators: Chlorite–Bromate–Iodide System," *J. Am. Chem. Soc.* 105, 2500–2502. © 1983 American Chemical Society.)

common reactant *B* (iodide). One can also envisage a series scheme in which the two subsystems are (*A* + *B*) and (*C* + *D*), but *C* is a product of the *A* + *B* reaction, and the input to the reactor consists of *A*, *B*, and *D*, but not *C*. Decroly and Goldbeter (1982) studied a model enzymatic system of this type in which enzyme 1 served to convert substrate 1 into substrate 2 in an oscillatory fashion. Substrate 2 was then converted into product in a second oscillatory reaction catalyzed by enzyme 2. The input flow for the system consisted of substrate 1 plus both enzymes. Other forms of coupling can be imagined in which the "cross talk" between subsystems takes place via reactions between intermediates. There is clearly a wide variety of possibilities, whose range is limited only by the chemistry of the systems involved. Chemical coupling is always asymmetric and nearly always nonlinear. Unlike physical coupling, chemical coupling can be extremely difficult to characterize, since it requires a knowledge of all possible reactions between components of the two subsystems. The strength of chemical coupling is essentially inaccessible to the experimenter; one cannot tune the strengths of chemical reactions at will, though one may be able to exert some influence by varying the concentrations used.

Although it is impossible to quantify the strength of chemical coupling, one can obtain some qualitative insights. Alamgir and Epstein (1984) conducted a comparative study of two chemically coupled systems: the bromate–chlorite–iodide reaction and the bromate–iodide–manganous reaction. The latter consists of the bromate–iodide and minimal bromate (bromate–bromide–manganous) oscillators. The authors concluded that the bromate–chlorite–iodide system represents a case of strong coupling, as evidenced by the existence of several dynamical states (birhythmicity, compound oscillation, chaos; see Figure 12.3 and discussion below) in the coupled system that cannot be identified as originating in the uncoupled subsystems. On the other hand, the phase diagram of the bromate–iodide–manganous system, shown in Figure 12.4, is essentially a superposition of the nearly cross-shaped phase diagrams of the component subsystems, with a region of tristability at intermediate bromate concentrations where the bistable

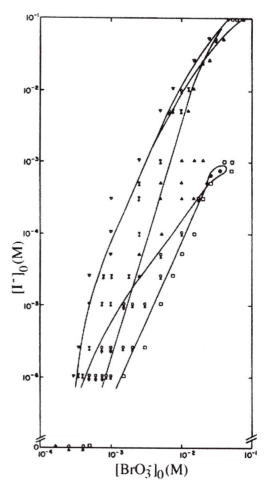

Figure 12.4 Phase diagram of the bromate–iodide–manganous reaction in the $[BrO_3^-]_0$–$[I^-]_0$ plane. Fixed constraints: $[Mn^{2+}]_0 = 1.0 \times 10^{-3}$ M, $[H_2SO_4]_0 = 1.5$ M. Symbols: open squares, high-potential–low-iodine (minimal bromate) steady state; filled triangles, intermediate-potential–intermediate-iodine (hybrid) steady state; open triangles, low-potential–high-iodine (bromate–iodide) steady state; filled circles, minimal-bromate-type oscillation; open circles, bromate–iodide-type oscillation. Combinations of symbols signify multistability among the corresponding states. (Reprinted with permission from Alamgir, M.; Epstein, I. R. 1984. "Experimental Study of Complex Dynamical Behavior in Coupled Chemical Oscillators," *J. Phys. Chem.* **88**, 2848–2851. © 1984 American Chemical Society.)

regions of the subsystems overlap. At high iodide concentrations we have bromate–iodide-type behavior, while at low $[I^-]_0$ we have dynamics resembling that of the minimal bromate reaction.

To what extent are we justified in thinking of a chemically coupled oscillator system as consisting of the two independent subsystems plus a set of cross-reactions that provide the coupling? A partial answer can be found in a mechanistic investigation of the bromate–chlorite–iodide reaction (Citri and Epstein, 1988). The mechanisms that had been derived for the bromate–iodide and chlorite–iodide systems in independent studies of these reactions are shown, respectively, in Tables 12.1 and 12.2.

In Table 12.3, we show the mechanism constructed for the coupled bromate–chlorite–iodide system. The reactions designated BC are cross-reactions between bromine and chlorine species. These provide the coupling. We see that only minimal modifications have been made in the component mechanisms to build the description of the full system. Reaction B13 was found to be unnecessary and was therefore deleted. Small reductions were made in some of the B rate constants, consistent with work done on the BZ reaction (Field and Försterling, 1986) after the original bromate–iodide study, which indicated that several reactions of bromine-containing species were slower than previously thought. However, the equilibrium constants of reactions B3, B7, and B9 were held fixed. In the chlorite–iodide mechanism, the rate of step C3 was increased, and the new reaction C9 was added, which allows the bromate–iodide and chlorite–iodide oscillations to merge, giving rise to the compound oscillations discussed later in this chapter. Reactions C6, C7, and C8 were found to be unnecessary for describing the behavior of the coupled system. Finally, the three new reactions BC1, BC2, and BC3 provide the "glue" to fit the coupled system together. Thus, the twenty reactions of the mechanism for the coupled system consist of twelve of the thirteen reactions from the bromate–iodide mechanism, five of the eight (one is common to both

Table 12.1 The Bromate–Iodide Mechanism[a]

Number	Reaction
B1	$2H^+ + BrO_3^- + I^- \rightarrow HBrO_2 + HOI$
B2	$HBrO_2 + HOI \rightarrow HIO_2 + HOBr$
B3	$I^- + HOI + H^+ \leftrightarrow I_2 + H_2O$
B4	$BrO_3^- + HOI + H^+ \rightarrow HBrO_2 + HIO_2$
B5	$BrO_3^- + HIO_2 \rightarrow HBrO_2 + IO_3^-$
B6	$HOBr + I_2 \leftrightarrow HOI + IBr$
B7	$IBr + H_2O \leftrightarrow HOI + Br^- + H^+$
B8	$HBrO_2 + Br^- + H^+ \rightarrow 2HOBr$
B9	$HOBr + Br^- + H^+ \leftrightarrow Br_2 + H_2O$
B10	$2H^+ + BrO_3^- + Br^- \leftrightarrow HBrO_2 + HOBr$
B11	$HIO_2 + Br^- + H^+ \leftrightarrow HOBr + HOI$
B12	$HOBr + HIO_2 \rightarrow Br^- + IO_3^- + 2H^+$
B13	$BrO_3^- + IBr + H_2O \rightarrow Br^- + IO_3^- + HOBr + H^+$

[a] Source: Citri and Epstein, 1986.

Table 12.2 The Chlorite–Iodide Mechanism[a]

Number	Reaction
C1	$H^+ + Cl(III) + I^- \rightarrow HOCl + HOI$
C2	$I^- + HOI + H^+ \leftrightarrow I_2 + H_2O$
C3	$HClO_2 + HOI \rightarrow HIO_2 + HOCl$
C4	$HOCl + I^- \rightarrow Cl^- + HOI$
C5	$HIO_2 + I^- + H^+ \leftrightarrow 2HOI$
C6	$2HIO_2 \rightarrow IO_3^- + HOI + H^+$
C7	$HOI + HIO_2 \rightarrow I^- + IO_3^- + 2H^+$
C8	$HOCl + HIO_2 \rightarrow Cl^- + IO_3^- + 2H^+$

[a] Source: Citri and Epstein, 1987.

mechanisms) from the chlorite–iodide reaction, one new chlorite–iodide reaction and three new coupling reactions.

Figure 12.5 shows the phase diagram calculated with the mechanism of Table 12.3 under the same experimental conditions as in Figure 12.3. The agreement is far less than perfect, as might be expected for such a complex system and with no adjustment of the experimental parameters. Nevertheless, the qualitative features are, on the whole, reproduced. In the next section, when we discuss some of the exotic phenomenology of the bromate–chlorite–iodide system, we shall see that the mechanism stands up remarkably well.

Table 12.3 The Bromate–Chlorite–Iodide Mechanism[a]

Number	Reaction
B1	$2H^+ + BrO_3^- + I^- \rightarrow HBrO_2 + HOI$
B2	$HBrO_2 + HOI \rightarrow HIO_2 + HOBr$
B3, C2	$I^- + HOI + H^+ \leftrightarrow I_2 + H_2O$
B4	$BrO_3^- + HOI + H^+ \rightarrow HBrO_2 + HIO_2$
B5	$BrO_3^- + HIO_2 \rightarrow HBrO_2 + IO_3^-$
B6	$HOBr + I_2 \leftrightarrow HOI + IBr$
B7	$IBr + H_2O \leftrightarrow HOI + Br^- + H^+$
B8	$HBrO_2 + Br^- + H^+ \rightarrow 2HOBr$
B9	$HOBr + Br^- + H^+ \leftrightarrow Br_2 + H_2O$
B10	$2H^+ + BrO_3^- + Br^- \leftrightarrow HBrO_2 + HOBr$
B11	$HIO_2 + Br^- + H^+ \leftrightarrow HOBr + HOI$
B12	$HOBr + HIO_2 \rightarrow Br^- + IO_3^- + 2H^+$
C1	$H^+ + Cl(III) + I^- \rightarrow HOCl + HOI$
C3	$HClO_2 + HOI \rightarrow HIO_2 + HOCl$
C4	$HOCl + I^- \rightarrow Cl^- + HOI$
C5	$HIO_2 + I^- + H^+ \leftrightarrow 2HOI$
C9	$HOI + HOCl \rightarrow HIO_2 + Cl^- + H^+$
BC1	$HOCl + Br^- \rightarrow Cl^- + HOBr$
BC2	$HBrO_2 + HOCl \rightarrow Cl^- + BrO_3^- + 2H^+$
BC3	$HOBr + HClO_2 \rightarrow HOCl + HBrO_2$

[a] Source: Citri and Epstein, 1988.

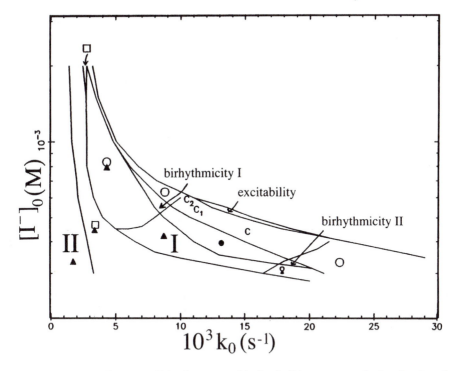

Figure 12.5 Phase diagram of the bromate–chlorite–iodide system calculated using the mechanism in Table 12.3. Fixed constraints and symbols as in Figure 12.3. C signifies compound oscillation, C_j signifies one compound oscillation plus j chlorite–iodide oscillations per cycle. (Reprinted with permission from Citri, O.; Epstein, I. R. 1988. "Mechanistic Study of a Coupled Chemical Oscillator: The Bromate–Chlorite–Iodide Reaction," *J. Phys. Chem.* **92**, 1865–1871. © 1988 American Chemical Society.)

12.2 Phenomenology of Coupled Oscillators

We have suggested that it is not totally absurd to think of living systems as arrays of coupled oscillators. Although it is not difficult to believe that chemistry is the driving force behind most of the key life processes, it seems a perilous leap from the sorts of crude coupled systems we have described in the previous section to even the simplest of living organisms. Perhaps the best way to get a sense of the amazing range of behavior that coupled oscillators can generate is to look at a series of examples taken from experimental and theoretical studies of relatively simple coupled oscillator systems.

12.2.1 Entrainment

Entrainment is probably the most transparent behavior of coupled oscillators. If two oscillators have nearly the same frequencies and we couple them together, it makes sense intuitively that as we increase the coupling strength we will eventually reach a point where they oscillate with the same frequency. In fact, if the coupling is physical, by flow between two CSTRs, we must reach such a state eventually no matter what the uncoupled frequencies are, because in the limit of infinite coupling we have a single reactor. At intermediate coupling strengths, we find entrainment at unequal but rationally related frequencies, as indicated in the Arnol'd tongue diagram of Figure 12.1. Figure 12.6a shows a closed, periodic trajectory in which one oscillator goes through three cycles while the other carries out seven cycles of oscillation. The model used in the calculations (Hocker and Epstein, 1989) consists of two Boissonade–De Kepper two-variable models coupled diffusively by a parameter D. Figures 12.6b and 12.6c, respectively, illustrate how weaker coupling leads to quasi periodicity while stronger coupling can result in chaos. The model is given by

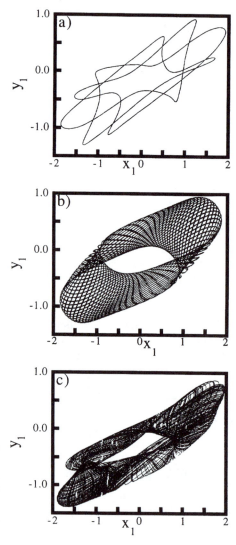

Figure 12.6 Trajectories for the coupled oscillator system of eqs. (12.4). Parameters: $\mu_1 = \mu_2 = 3$, $\lambda_1 = 0.02$, $\lambda_2 = 2$, $k_1 = 3.3$, $k_2 = 10$. (a): 3:7 Entrainment, $D = 0.22$; (b) quasi periodicity, $D = 0.1$; (c) chaos, $D = 0.4$. (Adapted from Hocker and Epstein, 1989.)

$$dx_i/dt = -(x_i^3 - \lambda_i x_i - \mu_i) - k_i y_i - D(x_i - x_j) \qquad (12.4a)$$
$$dy_i/dt = (x_i - y_i)/\tau_i - D(y_i - y_j) \qquad (12.4b)$$

with $i, j = 1, 2$ and $i \neq j$.

Even the simplest case of one-to-one entrainment at a common frequency offers a surprisingly rich array of possibilities. Crowley and Epstein (1989) studied the behavior of two coupled BZ oscillators in the experimental configuration shown in Figure 12.2. They used acetylacetone instead of malonic acid to prevent the formation of carbon dioxide bubbles. The compositions of the feedstreams (sodium bromate, cerous nitrate, sulfuric acid, acetylacetone) were the same for the two CSTRs, except that the acetylacetone concentration was 0.015 M for one reactor and 0.016 M for the other. This difference gave rise to a difference in uncoupled frequencies of about 15%, with $\tau_0 = 99.6$ s for one reactor and 112 s for the other. In Figure 12.7, we see what happens as the coupling parameter ρ, the ratio of the mass flow between reactors to the mass flow through each CSTR, is increased.

In the first segment, with zero coupling, the subsystems oscillate independently, and the difference signal in the top row is large and apparently quasi-periodic. When the coupling is raised to an intermediate value, the oscillators lock together. The difference signal is now periodic, but it still has large amplitude. The oscillators have entrained at the same frequency, but they are approximately 180° out of phase with each other. The period of this behavior is, surprisingly, 139 s—longer

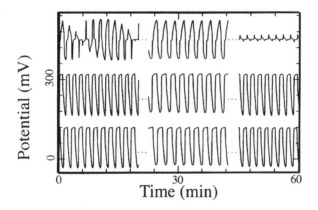

Figure 12.7 Experimental traces of platinum (Pt) electrode potential in a physically coupled BZ oscillator experiment. Bottom trace is potential V1 from reactor 1; middle trace is potential V2 from reactor 2, shifted up by 200 mV. Top trace is (V1 − V2)/2, and is shifted so that 0 mV corresponds to 450 mV on the millivolt axis. 0–20 min: uncoupled oscillations, $\rho = 0$; 23–45 min: out-of-phase entrainment, $\rho = 0.5$; 45–60 min: in-phase entrainment, $\rho = 0.75$. (Adapted from Crowley and Epstein, 1989.)

than the periods of either of the uncoupled oscillators. Finally, if the coupling is increased further, the out-of-phase oscillation gives way to an in-phase oscillation seen in the third segment. The difference signal is now quite small, and shrinks as the coupling increases. The period is 105 s—roughly the average of the uncoupled periods.

The existence of two different entrained modes, in-phase and out-of-phase, is perhaps unexpected, but there is still more to be learned about this "simple" system. One indication of the complexity that awaits us is that the two entrained modes can exist at the same value of ρ; there is hysteresis as one varies the coupling parameter, resulting in bistability between two oscillatory modes. We shall discuss this phenomenon of *birhythmicity* a bit further on, but first we consider something even more puzzling that happens in these experiments as the coupling is varied.

12.2.2 Oscillator Death

Figure 12.8 shows an experiment on the coupled BZ system that we have been discussing, in which the coupling is varied in a different way than in Figure 12.7. Notice especially what happens at 5 min, when ρ jumps from 0.5 to 0.75. The system suddenly ceases its out-of-phase oscillation and reaches an asymmetric steady state with oscillator 2 at a high potential and oscillator 1 at a low potential. We have coupled together two oscillating subsystems and ended up with a system that does not oscillate at all!

But wait! The system is nearly symmetric; the feedstreams are almost identical for the two reactors. If the system can stop with one oscillator "up" and the other "down," why can't it stop in the opposite configuration? In fact, it can. We only need to figure out how to get it from one asymmetric steady state to the other. We see in Figure 12.8 how to do this. At 11 min, we turn off the coupling and allow

Figure 12.8 Experimental traces of Pt electrode potential in a physically coupled BZ oscillator experiment. Bottom trace is potential V1 from reactor 1; upper trace is potential V2 from reactor 2, shifted up by 200 mV. 0–5 min: out-of-phase entrainment, $\rho = 0.5$; 5–11 min: steady state I, $\rho = 0.75$. Coupling is switched off from 11 min to 11 min 40 s. 11.7–19.5 min: steady state II, $\rho = 0.75$; 19.5–22.5 min: steady state II, $\rho = 0.65$; 22.5–36 min: out-of-phase entrainment, $\rho = 0.56$. (Adapted from Crowley and Epstein, 1989.)

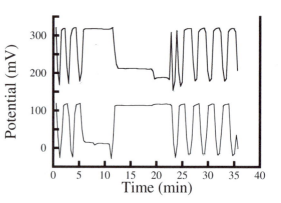

each oscillator to run freely for approximately half its natural cycle. We then switch the coupling back on to its previous value. The system stops again, but now in a steady state that is nearly the mirror image of the previous one. If we decrease the coupling gradually, the steady state persists for a while, but then the out-of-phase oscillations are reborn.

Alternatively, if we increase the coupling from the steady-state value, the system eventually begins to oscillate, but now the oscillations are in-phase. On decreasing the coupling from this state, we observe that in-phase oscillation persists through the entire range of ρ that gives steady-state behavior and through much of the out-of-phase entrainment region. The behavior of the system is illustrated in Figure 12.9.

Is there an intuitive way of understanding this behavior? Computer simulation combined with a somewhat oversimplified mechanical analogy offers some insight. Figure 12.10 shows a simulated trajectory of one of the two oscillators, obtained from model calculations using two diffusively coupled Oregonators. The limit cycle is essentially unchanged as we vary the coupling parameter, and because of the near symmetry of the system, the limit cycles of the two oscillators are almost identical in shape. Note that the actual models are three- rather than two-dimensional, so that the trajectory shown is a projection of the real one, which "bends over" into the third dimension.

Now imagine our two oscillators moving around on the limit cycle in Figure 12.10. Each oscillator experiences two "forces." The first results from its chemistry and tends to push it around the limit cycle at its natural (uncoupled) frequency. The second is caused by the physical coupling and tends to pull the oscillators toward each other. When the coupling if turned off or is small, the chemical force dominates the coupling force, and the oscillators go around the cycle nearly independently; the behavior is quasi-periodic. As the coupling increases, its force becomes more important relative to the chemical force. In particular, at certain points on the cycle in the neighborhood of A and B, the coupling force points almost directly opposite to the chemical force. (This would be easier to see if we had a three- rather than a two-dimensional picture of the limit cycle.) What happens is that the system tends to slow down when one oscillator is near A and the other is near B, and, when the coupling gets strong enough, the two points march in lockstep so that one reaches A just as the other reaches B. This is the out-of-phase oscillatory mode.

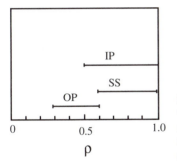

Figure 12.9 Experimental stability diagram for physically coupled BZ system showing ranges of ρ for which different behaviors are stable. IP = in-phase entrained oscillations, SS = complementary pair of steady states, OP = out-of-phase entrained oscillations. (Adapted from Crowley and Epstein, 1989.)

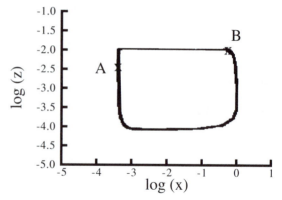

Figure 12.10 Phase plane showing the limit cycle for one of two coupled Oregonator oscillators. Points *A* and *B* are the steady states at which the oscillator can come to rest when oscillator death occurs. (Adapted from Crowley and Epstein, 1989.)

As the coupling increases, the forces balance better and better at *A* and *B*. The system spends more and more time getting past these points. The period continues to lengthen. This phenomenon is the origin of the fact, which we remarked on earlier, that the out-of-phase mode has a period longer than either of the uncoupled periods. At a certain value of the coupling, the match between the chemical and coupling forces is so good that the system "hangs up," with one oscillator at *A* and the other at *B*. *Oscillator death* (Ermentrout and Kopell, 1990) has thus occurred. If we increase the coupling still further, it overwhelms the chemistry. The oscillators "snap together." The steady state loses stability and the oscillators trot around the cycle together, in the in-phase oscillatory mode.

If we now decrease the coupling, because the oscillators start off together we cannot reach the steady state of oscillator death. The coupling is sufficient to hold them together until we get to quite low values of ρ, where the oscillators can drift apart and oscillate in the out-of-phase mode again. Thus, our picture also gives us insight into the origin of the observed hysteresis and multistability.

12.2.3 Rhythmogenesis

We have just seen that coupling two oscillating systems can cause them to stop oscillating. Can coupling work in the opposite direction? Can we couple together two steady-state systems in such a way as to make them oscillate? This question can be answered in the affirmative, first in a trivial way and then in a highly nontrivial way. To see the distinction, consider the schematic cross-shaped phase diagram in Figure 12.11.

Suppose that we physically couple two subsystems whose constraint parameters place them at points *A* and *B* in Figure 12.11, so that neither is oscillating and each is in one of the two steady states of the system. If the subsystems have equal volume and if we make the coupling strong enough, the coupled system should start to behave like a single system whose parameters are the average of the parameters of the uncoupled system. That is, the coupled system should have the characteristics associated with point *C*; it should oscillate! We have thus created an oscillating system by coupling two subsystems in different stable steady

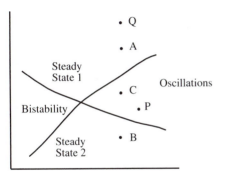

Figure 12.11 Schematic cross-shaped phase diagram (see text for discussion).

states. While, at first, this result may have seemed surprising, our analysis of Figure 12.11 now makes it almost self-evident that this feat can be accomplished.

A more challenging (but potentially more significant) trick is to produce *rhythmogenesis* by relatively weak coupling between two subsystems in *the same* steady state, for example, at points *A* and *Q* of Figure 12.11. This certainly looks like a harder job, but why do we claim that this sort of rhythmogenesis would have special significance? If weak diffusive coupling of two systems in slightly different steady states can produce oscillations, then oscillation might also arise in an unstirred system in which there were concentration gradients, but where each individual point in the system had concentrations that would give rise to a steady state in the absence of diffusion. This assertion follows from looking at two physically coupled subsystems as a first approximation to the discretization of a continuous medium. We can think of representing each half of the (inhomogeneous) medium by a homogeneous subsystem having the average concentrations of that half. The two halves are then coupled by a diffusion-like term. Next, we divide the system further into four, then eight, then sixteen, etc. In the limit of an infinite number of coupled subsystems, we would be modeling the system exactly. Just as reaction and diffusion can combine to turn homogeneous steady-state behavior into *spatial* periodicity in the Turing phenomenon to be discussed in Chapter 14, here we would have a diffusion-induced *temporal* periodicity.

Boukalouch et al. (1987) analyzed the model of two physically coupled Boissonade–De Kepper oscillators in eq. (12.4) using singular perturbation theory. Their calculations predicted that there should be a set of parameters and a range of the coupling strength *D* in which the sort of rhythmogenesis that we have been describing should occur. The theory suggests, moreover, that the amplitude of the oscillations should be quite asymmetric—large in one subsystem and small in the other. Figure 12.12 shows the results of an experiment on a system consisting of two chlorite–iodide reactions in physically coupled CSTRs.

12.2.4 Birhythmicity

We have discussed many examples of bistability between steady states or between a steady and an oscillatory state. Bistability can also occur between a pair of oscillatory states, a phenomenon referred to as *birhythmicity*. The physically

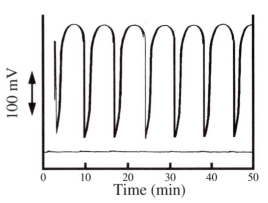

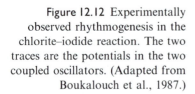

Figure 12.12 Experimentally observed rhythmogenesis in the chlorite–iodide reaction. The two traces are the potentials in the two coupled oscillators. (Adapted from Boukalouch et al., 1987.)

coupled BZ oscillator experiments summarized in Figure 12.9 represent one example of birhythmicity, since there is a range of parameters in which the in-phase and the out-of-phase oscillatory modes are both stable.

Chemical coupling experiments are another source of birhythmicity. In fact, the phenomenon was first suggested to occur in a realistic chemical system, and the term birhythmicity was coined in a model calculation (Decroly and Goldbeter, 1982) on two oscillatory enzyme–substrate subsystems, in which the product of the first serves as the substrate for the second.

Probably the best characterized example of birhythmicity is found in the parallel-coupled bromate–chlorite–iodide system. Figure 12.13 shows the behavior observed experimentally as the flow rate is first increased and then decreased. We see clearly that two very different modes of oscillations occur at $k_0 = 7.14 \times 10^{-3}$ s^{-1}, and we observe a sharp transition between these states at $k_0 = 7.26 \times 10^{-3}$ s^{-1}.

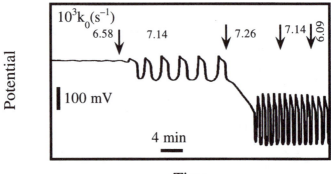

Time

Figure 12.13 Experimentally observed birhythmicity in the bromate–chlorite–iodide reaction. At the times indicated by arrows, the flow rate is changed. The flow rate in each time segment is shown at the top. At $k_0 = 7.14 \times 10^{-3}$ s^{-1}, both types of oscillation are stable. $[I^-]_0 = 6.5 \times 10^{-4}$ M. Other constraints as in Figure 12.3. (Adapted from Alamgir and Epstein, 1983.)

The bromate–chlorite–iodide mechanism detailed in Table 12.3 does a remarkably good job of describing the birhythmic behavior of this system. Figure 12.14 shows the results of calculations carried out using this model and the same parameter values as in Figure 12.13.

Systems of coupled neural oscillators frequently display bi- or even multirhythmicity. These complex entities must be able to generate and maintain different modes of oscillation, which result in different behaviors of the organism, in response to variations in their environment.

12.2.5 Compound Oscillation

It is not difficult to understand why a chemically coupled oscillator system can display birhythmicity for some range of parameters. If the coupling is not too strong, it may distort the shape and location of the limit cycle and the range of

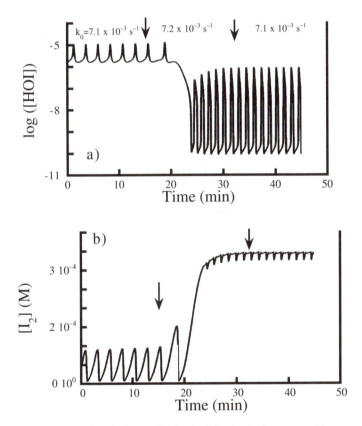

Figure 12.14 Simulations of birhythmicity in the bromate–chlorite–iodide reaction for (a) [HOI] and (b) [I$_2$]. [I$^-$]$_0$ = 7 × 10^{-4} M. Other constraints as in Figure 12.13. At k_0 = 7.1 × 10^{-3} s^{-1}, both types of oscillation are stable. (Adapted from Citri and Epstein, 1988.)

oscillation a little, but the cycles associated with the initial subsystems maintain their basic integrity. If the two (distorted) regions of oscillation overlap in parameter space, their intersection becomes a region of birhythmicity.

We now consider something else that can happen (though it is much less likely) as we move around in the parameter space of a chemically coupled system. As we vary the parameters, the limit cycles will change their shapes and will move around in concentration space. It is possible that for some parameter values, the limit cycles will come very close to each other, or even collide. What happens then? This turns out to be an extremely difficult question, with many possible answers depending on the details of the system. One possible outcome is shown in Figure 12.15.

If things fit together just right, the two limit cycles can merge to form a single cycle. Mathematicians have studied this kind of behavior in abstract models and refer to it as a *gluing bifurcation* (Meron and Procaccia, 1987); the two limit cycles become stuck together. Experimentally, what one observes, as shown in Figure 12.16, is a *compound oscillation* that appears to be formed by forcing the two modes of oscillation to have the same frequency and then sticking one type-B oscillation on top of each type-A cycle.

Again, the model of Table 12.3, "glued together" from the independent bromate–iodide and chlorite–iodide mechanisms, does an excellent job of reproducing the observations, as we see in Figure 12.17.

12.2.6 Complex Oscillation

In addition to the compound oscillation depicted in Figures 12.16 and 12.17, the collision of two limit cycles may lead to other scenarios. One possibility is complex periodic oscillation in which one cycle of one type is followed by several of another type. This type of behavior is analogous to the *bursting* mode of oscillation of neural oscillators, in which a period of relative quiescence is followed by a series of action potentials. In Figure 12.18, we compare a membrane potential trace from a crab neuron with potential oscillations in a pair of physically coupled chlorine dioxide–iodide oscillators.

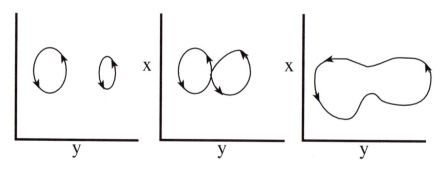

Figure 12.15 A gluing bifurcation leading to compound oscillation. Successive frames represent phase portraits in concentration space for different sets of constraint parameters.

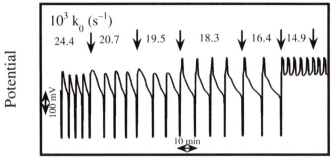

Figure 12.16 Experimentally observed compound oscillation in the bromate–chlorite–iodide reaction. $[I^-]_0 = 4 \times 10^{-4}$ M, other constraints as in Figure 12.3. (Adapted from Alamgir and Epstein, 1983.)

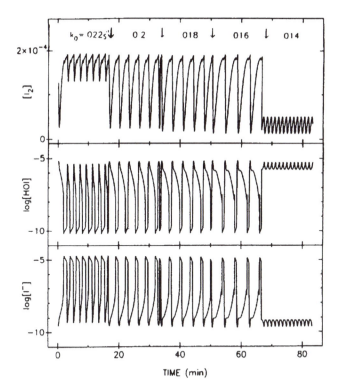

Figure 12.17 Simulated compound oscillation in the bromate–chlorite–iodide reaction. All constraints as in Figure 12.16. (Reprinted with permission from Citri, O.; Epstein, I. R. 1988. "Mechanistic Study of a Coupled Chemical Oscillator: The Bromate–Chlorite–Iodide Reaction," *J. Phys. Chem. 92*, 1865–1871. © 1988 American Chemical Society.)

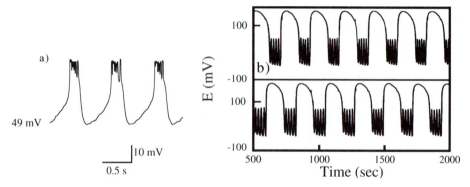

Figure 12.18 Bursting behavior in (a) the transmembrane potential of a neuron in the crustacean stomatogastric ganglion (Adapted from Sharp, 1994.); (b) two coupled CSTRs, each of whose flow rates is modified according to the iodide concentration in the other reactor. Input concentrations in each reactor: $[ClO_2]_0 = 1 \times 10^{-4}$ M, $[I^-]_0 = 4.2 \times 10^{-4}$ M. Note the similarity between the neural and chemical traces if one reverses the sign of the potential in one of the recordings. (Adapted from Dolnik and Epstein, 1993.)

A fascinating example of complex oscillation occurs in the bromate–chlorite–iodide chemically coupled system, where we observe periodic behavior consisting of one compound oscillation followed by n small amplitude oscillations. Trajectories for two of these C_n "animals" are shown in Figure 12.19.

12.2.7 Chaos

Not surprisingly, systems that can generate the intricate variety of periodic oscillations we have been describing are also capable of producing chaotic behavior. We have described above how physical coupling of two simple nonlinear oscillators will give rise to chaos when the coupling becomes strong enough that the entrainment bands begin to overlap. If the oscillators have interesting dynamics of

Figure 12.19 Trajectories in the bromate–chlorite–iodide system of complex oscillations in a space whose axes are absorbance at 460 nm (primarily due to I_2) and potential of a Pt redox electrode. (a) C_2: $[BrO_3^-]_0 = 2.5 \times 10^{-3}$ M, $[I^-]_0 = 4.0 \times 10^{-4}$ M, $[ClO_2^-]_0 = 1.0 \times 10^{-4}$ M, $[H_2SO_4]_0 = 0.1$ M, $k_0 = 3.0 \times 10^{-3}$ s^{-1}; (b) C_6: $[BrO_3^-]_0 = 3.0 \times 10^{-3}$ M, $[I^-]_0 = 4.0 \times 10^{-4}$ M, $[ClO_2^-]_0 = 1.0 \times 10^{-4}$ M, $[H_2SO_4]_0 = 0.04$ M, $k_0 = 5.8 \times 10^{-4}$ s^{-1}. (Adapted from Alamgir and Epstein, 1984.)

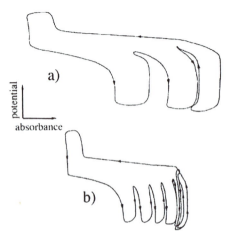

their own, extremely complex behavior can result. For example, the apparently simple model of eqs. (12.5) was developed by Degn and Harrison (1969) to describe the respiratory behavior of a *Klebsiella aerogenes* bacterial culture.

$$A \rightarrow Y \tag{12.5a}$$

$$B \leftrightarrow X \tag{12.5b}$$

$$X + Y \rightarrow P \tag{12.5c}$$

With the concentrations of A and B held constant, the differential equations for X and Y are

$$dx/dt = b - x - xy/(1 + qx^2) \tag{12.6a}$$

$$dy/dt = a - xy/(1 + qx^2) \tag{12.6b}$$

When two Degn–Harrison systems are physically coupled with the same diffusive coupling parameter, c, for both species, we obtain the phase diagram shown in Figure 12.20, where we plot the values of x on the attractor against the coupling strength. Note the multiple regions of chaos in the middle of the diagram and the sequences of forward and reverse period-doubling.

Chemically coupled oscillators can also give rise to chaotic behavior. Again, we choose an example from the bromate–chlorite–iodide system. Although this system may well show isolated regions of chaos, the best characterized chaotic behavior in this reaction occurs as part of a period-adding sequence in which, as the

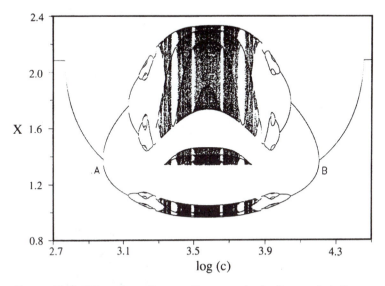

Figure 12.20 Bifurcation diagram for two physically coupled Degn–Harrison oscillators. (Reprinted with permission from Lengyel, I.; Epstein, I. R. 1991. "Diffusion-Induced Instability in Chemically Reacting Systems: Steady-State Multiplicity, Oscillation, and Chaos," *Chaos 1*, 69–76. © 1991 American Institute of Physics.)

flow rate is decreased, there is an increase in the number of small-amplitude oscillations in a series of C_n oscillations like those shown in Fig 12.19. The chaotic behavior is found in the narrow regions separating C_n from C_{n+1} oscillation. Figure 12.21 illustrates the chaos found between the C_2 and C_3 regions.

12.3 Multiple Oscillators

In this chapter, we have focused our attention on experiments and simulations involving a single pair of oscillators, both because such systems are simpler to understand than those involving more subsystems and because the vast majority of investigations have been carried out on two-oscillator systems. However, there does exist a significant literature on the behavior of many-oscillator systems. The two most commonly studied problems involve (1) trying to determine conditions under which a large group of oscillators with a given set of properties can lock together, and (2) attempting to describe the dynamics of a set of physically coupled subsystems under conditions that would yield bistability in the absence of coupling.

The issue of entrainment of large numbers of oscillators has been investigated almost exclusively in nonchemical contexts. We have mentioned Strogatz's efforts to explain the synchronous flashing of fireflies. Potential applications to neural systems, particularly the brain, are obvious. Physicists have been interested in the properties of large numbers of coupled Josephson junctions (Hamilton et al., 1987) or laser modes (Wiesenfeld et al., 1990). The problem of how to treat the dynamical behavior of a large number of coupled oscillators is a difficult one, requiring very different techniques from the ones used to elucidate single oscillator behavior. Perhaps what is needed is a conceptual breakthrough that enables us to utilize statistical methods to predict bulk properties of the coupled system from knowledge about the subsystems and the couplings, much as statistical mechanics makes it possible to carry out a similar program for molecules. Wiesenfeld and Hadley (1989), for example, have shown that, in general, the number of stable limit cycles for a coupled system tends to grow exponentially with the number of nonlinear oscillator subsystems, a daunting result that reinforces the need for a qualitatively different, statistical approach to such problems.

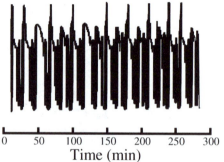

Figure 12.21 Chaotic oscillations in the bromate–chlorite–iodide reaction. $[BrO_3^-]_0$ = 2.5×10^{-3} M, $[I^-]_0$ = 4.0×10^{-4} M, $[ClO_2^-]_0$ = 1.0×10^{-4} M, $[H_2SO_4]_0$ = 0.1 M, k_0 = 2.5×10^{-3} s^{-1}. (Adapted from Maselko et al., 1986.)

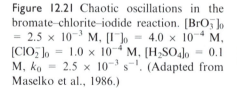

Time (min)

Studies of coupled bistable systems are somewhat easier. They have tended to focus on the sorts of attractors that are available to the coupled system and on the transitions from one such attractor to another. The early experiments of Stuchl and Marek (1982) on as many as seven coupled BZ oscillators were aimed at determining which of the 2^7 possible combinations of oxidized and reduced states were stable. Epstein and Golubitsky (1993) recently gave an analytical proof that a linear array of identical, diffusively coupled subsystems will always have steady states whose concentration patterns reflect the symmetry of the system. In what is to date certainly the most impressive experiment involving physically coupled oscillators, Laplante and Erneux (1992) used sixteen linearly coupled CSTRs containing a bistable chlorite–iodide composition to investigate the issue of *propagation failure*. This phenomenon, which is of considerable importance for the transmission of neural impulses, occurs when all subsystems start off in the same steady state and one is switched by an external perturbation to the other bistable state. The question is: What are the conditions that determine how far down the line the transition between states will propagate? Finally, we recall from our discussion of birhythmicity that a series of diffusively coupled subsystems is a useful approximation (which can be made as accurate as desired) to a continuous medium.

12.4 Coupled Patterns

When we think of coupling two systems together, we tend to imagine two homogeneous systems that change in time but not in space. Real systems, particularly those in living systems, tend not to be homogeneous, but to have spatial structure as well. The problem of studying coupled spatiotemporal systems is obviously much more difficult than that of looking at coupled temporal oscillators. Nonetheless, some interesting preliminary results on this problem have been obtained.

The first experiments of this sort were done by Showalter and collaborators. They impregnated a Nafion cation exchange membrane with ferroin and immersed it in a solution containing the remaining ingredients of the BZ reaction (Winston et al., 1991). As might be expected, target patterns and spiral waves appeared on the surfaces of the membrane where the catalyst came into contact with the other reagents. What was most interesting was that the patterns interacted. They could be either entrained in an in-phase mode, so that the reduced regions in one pattern were directly across from the corresponding regions of the pattern on the other surface, or they could be transverse, so that corresponding regions on the two surfaces crossed each other at about a 90° angle. A simulation (Gáspár et al., 1991) of two weakly coupled two-dimensional model systems gave similar crossing patterns. An example of an experimentally observed tranverse coupling pattern is shown in Figure 12.22.

Zhabotinsky et al. (1990, 1991) observed coupled patterns in the ferroin-catalyzed BZ reaction in a different configuration. In a layer of silica gel saturated with BZ solution, waves first begin to propagate along the bottom of the layer.

a)

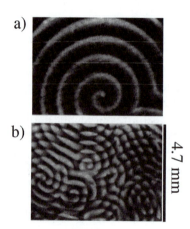

Figure 12.22 Crossing wave patterns in the BZ reaction on the two surfaces of a Nafion membrane. (a) Nafion membrane loaded with ferroin to 16.7% capacity gives strong coupling. (b) Nafion loaded to 38.7% capacity gives weak coupling. (Reprinted with permission from Winston, D.; Arora, M.; Maselko, J.; Gáspar, V.; Showalter, K. 1991. "Cross-Membrane Coupling of Chemical Spatiotemporal Patterns," *Nature 351*, 132–135. © 1991 Macmillan Magazines Limited.)

b)

4.7 mm

After 10–15 min, waves emerge in the top part of the layer and also propagate parallel to the gel surface. The middle portion of the gel remains refractory to wave propagation throughout the experiment. In many cases, the top and bottom waves are transverse to one another, giving rise to a crossing pattern similar to that observed in Showalter's membrane experiments. Recent studies (Zhabotinsky et al., 1994) suggest that this phenomenon results from the effect of oxygen on the BZ system. In an excitable BZ layer open on the top to the air, two gradients arise. The concentration of O_2 decreases from the top of the layer, where it enters the system, to the bottom. Bromine, on the other hand, can escape at the top, causing the concentration of Br_2, as well as the concentrations of bromo derivatives of malonic acid, to increase from top to bottom. Since the threshold of excitability increases, (i.e., the ability of the system to support waves decreases) with the concentrations of both oxygen and the bromo derivatives, a nonmonotonic profile of the excitability threshold can arise, with a maximum in the middle of the layer. Thus, waves can propagate at the top and at the bottom, but not in the middle where there is a poorly excitable sublayer acting as a barrier between the two excitable regions. Computer simulations with a modified Oregonator model of the BZ reaction and a nonmonotonic profile of the stoichiometric parameter q, which depends on the concentrations of both O_2 and BrMA, support this interpretation.

12.5 Geometrical Phases

We conclude this chapter by considering a phenomenon that has attracted a great deal of attention in the physics community, but remains almost unknown to chemists. It is an effect that arises when the constraint parameters of a nonlinear oscillator are slowly varied in a cyclic fashion. We may think of it as occurring in a situation where an oscillator is periodically forced with a frequency much less than its natural frequency, or when two oscillators with very different periods are coupled together. First pointed out by physicist Michael Berry in the context of quantum mechanical systems such as magnetic resonance and the neutron interferometer, the phenomenon is known at the *geometric phase* or the *Berry phase* (Berry, 1984; Shapere and Wilczek, 1989).

To get a physical feel for the geometric phase, try the following experiment with a pencil and a basketball or similar spherical object. Place the pencil at the "north pole" of the sphere and observe the direction in which the pencil is pointing. Now, without rotating the pencil, slide it down to the "equator." Next, move the pencil along the equator by about one quarter of the circumference. Finally, move the pencil back up to the pole. When you arrive back at the top, the pencil should be pointing in a different direction from the one in which you started. The change in direction (which is related to the famous Foucault pendulum experiment) is the physical analog of the geometric phase shift.

What Berry showed was that if one slowly varies two (or more) parameters that affect a nonlinear oscillator and eventually return to the original set of parameter values, there will be an extra phase shift. Since the period of the oscillator may depend on the parameters being varied, one expects, and can calculate, a dynamical phase shift associated with the fact that the period of oscillation changes with the parameters. The surprising result is that there is also an additional phase shift, the geometric phase shift, associated with the cyclic motion of the system in parameter space, like the change in direction of the pencil. Berry showed that the geometric phase shift is independent of the rate at which the parameter change is carried out (so long as it is slow enough to be considered adiabatic) and that the magnitude of the shift is proportional to the area traversed by the system during its circuit in phase space. The proportionality constant depends on the detailed dynamics of the oscillator. Berry's result has been extended from quantum systems to classic conservative systems (Hannay, 1985) and to dissipative systems (Kepler and Kagan, 1991).

Kagan et al. (1991) used the Lengyel–Epstein model of the chlorite–iodide–malonic acid (CIMA) reaction to demonstrate that geometric phases should, in principle, be observable in a real oscillating chemical reaction. They varied two of the constraint parameters (equivalent to feed concentration ratios) around a rectangular path in parameter space. By taking the difference between results obtained traversing the parameter space path in the clockwise and counterclockwise directions, and by extrapolating their results at several rates of traversal to infinitely slow parameter variation, they were able to eliminate nonadiabatic effects and obtain reliable geometric phase shifts. The results suggest that improvements in the stability of current CSTR technology will be required to observe geometric phase shifts experimentally, since the conditions must be maintained constant for many hundreds of cycles. Nonetheless, it seems likely that it will soon prove possible to measure this phenomenon in chemical, as well as in physical, oscillators.

13

Biological Oscillators

Including a chapter on biological oscillators was not an easy decision. In one sense, no book on nonlinear chemical dynamics would be complete without such a chapter. Not only are the most important and most numerous examples of chemical oscillators to be found in living systems, but the lure of gaining some insight into the workings of biological oscillators and into the remarkable parallels between chemical and biological oscillators attracts many, perhaps most, new initiates to the study of "exotic" chemical systems. On the other hand, it is impossible for us to do even a minimal job of covering the ground that ought to be covered, either in breadth or in depth. To say that the subject demands a whole book is to understate the case badly. There are indeed whole books, many of them excellent, devoted to various aspects of biological oscillators. We mention here only four of our favorites, the volumes by Winfree (1980), Glass and Mackey (1988), Murray (1993) and Goldbeter (1996).

Having abandoned the unreachable goal of surveying the field, even superficially, we have opted to present brief looks at a handful of oscillatory phenomena in biology. Even here, our treatment will only scratch the surface. We suspect that, for the expert, this chapter will be the least satisfying in the book. Nonetheless, we have included it because it may also prove to be the most inspiring chapter for the novice.

The range of periods of biological oscillators is considerable, as shown in Table 13.1. In this chapter, we focus on three examples of biological oscillation: the activity of neurons; polymerization of microtubules; and certain pathological conditions, known as dynamical diseases, that arise from changes in natural biological rhythms. With the possible exception of the first topic, these are not

Table 13.1 Some Biological Oscillators

Type	Period
Neural	0.01–10 s
Cardiac	1 s
Calcium	1 s to several minutes
Glycolytic	several minutes
Microtubules	2–10 min
cAMP in *Dictyostelium*	5–10 min
Cell cycle (vertebrates)	24 h
Circadian	24 h
Human ovarian	28 days
Annual (e.g., plants)	1 year
Animal populations	years

among the best-known nor the most thoroughly studied biological oscillators; they have been chosen because we feel that they can be presented, in a few pages, at a level that will give the reader a sense of the fascinating range of problems offered by biological systems. Our discussion of these examples is preceded by a short overview of some of the major areas in this field—areas that are treated in depth in the books mentioned above. First, we offer an important caveat. Although the similarities between chemical and biological (or cosmological, or geological, . . .) phenomena can be striking, one should be cautious about attributing phenomena that look alike to related mechanisms. Similar behavior can certainly arise from similar causes, but it does not necessarily do so; further evidence is required before mechanistic conclusions, even very general ones, can be drawn. It is, however, indisputable that biological oscillation has its origins in the chemical reactions that occur in living systems.

One of the most carefully studied biological oscillators is the process of *glycolysis*, in which the sugar glucose is converted via a series of enzyme-catalyzed reactions to pyruvate, with the concomitant production of ATP. Glycolysis serves as a major source of metabolic energy in nearly all organisms. Most of the details of the glycolytic pathway had been elucidated by 1940 (Stryer, 1995), but it was not until much later that oscillations in such quantities as the pH and the fluorescence of NADH were discovered when glycolysis takes place in cell-free yeast extracts (Duysens and Amesz, 1957). The function of glycolytic oscillations in vivo is not well understood, but it is clear that they arise from the complex regulatory properties of the enzymes involved in the pathway, particularly phosphofructokinase (Ghosh and Chance, 1964). Goldbeter (1996) gives an excellent discussion of efforts to understand the mechanism of glycolytic oscillation.

Another well-characterized system is the periodic generation of pulses of cyclic adenosine monophosphate (cAMP) during the aggregation of the cellular slime mold *Dictyostelium discoideum* (Gerisch, 1968). This humble organism has a life cycle that might prove instructive to our own species. During times of plenty, it exists as independent, single cells. When conditions are unfavorable, the cells act cooperatively to ride out the hard times. The pulses of cAMP serve as a signal that causes the cells to come together to form a fruiting body carrying spores that require no nutrient supply and are relatively impervious to harsh environments. When good times return, the spores rupture, and individual cells are free to pursue their normal activities again. During the aggregation process, the cells organize themselves into spiral patterns, like the one shown in Figure 13.1, which are strikingly similar to those seen in the BZ reaction (cf. Figure 6.7). Like glycolysis, the *Dictyostelium* system has been thoroughly studied and modeled. Many of these efforts have been summarized by Winfree (1980) and Goldbeter (1996).

Typically, the cycle of cell division in eukaryotic cells (Murray and Hunt, 1993) is a periodic process. The cell cycle is one of the most important processes in living systems. When it functions properly, it governs the growth and reproduction of the organism; when it goes awry, cancer is one result. There are three essential events in the cell cycle, shown schematically in Figure 13.2: chromosome replication, chromosome segregation, and cell division. Biologists have learned a great deal about the biochemical and structural changes that occur in each of these events, and new information is constantly being discovered. The cycle can be thought of as proceeding through four phases: synthesis (S), during which DNA is synthesized and the chromosomes replicate; mitosis (M), during which the replicated chromosomes segregate into two sets and the cell divides into two daughters, each containing a set of chromosomes; and two "gaps," G1 following mitosis and G2 following synthesis. For typical vertebrates, a full cycle takes about 24 h, with the S and G2 phases lasting about 6 h each, the G1 phase lasting about 12 h, and mitosis lasting only about 30 min. Later in this chapter, we examine one aspect of mitosis, the formation of microtubules.

There have been many models proposed of the dynamics of the cell cycle. Here we mention only a recent approach by Tyson et al. (1996), who suggest (p. 81), in contrast to the view taken by more biologically oriented modelers, that "The emerging picture is too complex to comprehend by box-and-arrow diagrams and casual verbal reasoning. Standard chemical-kinetic theory provides a disciplined method for expressing the molecular biologists' diagrams and intuitions in

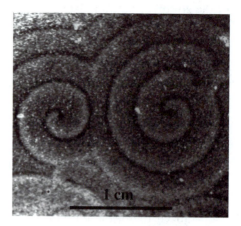

Figure 13.1 Spiral patterns in aggregating *Dictyostelium*. (Courtesy of S. Müller.)

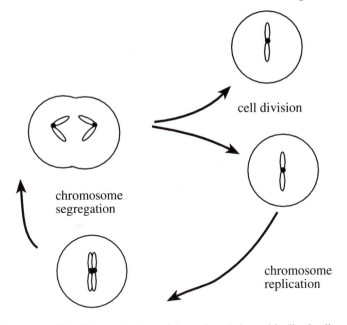

Figure 13.2 Schematic view of the cell cycle in an idealized cell with a single chromosome. The dark circle on each chromosome is the kinetochore, the protein–DNA complex that connects the chromosomes to the machinery responsible for segregating the chromosomes to the two daughter cells. (Adapted from Murray and Hunt, 1993.)

precise mathematical form, so that qualitative and quantitative implications of our 'working models' can be derived and compared with experiments." They then go on to use the techniques developed in this book to analyze a biochemical model of M-phase control.

One organ whose oscillatory function is absolutely essential is the heart. Pacemaker cells (Jalife and Antzelevitch, 1979) have been identified in the sino-atrial node, a region of specialized tissue in the right atrium. So long as the remainder of the heart oscillates in synchrony with the pacemaker, normal activity is maintained. A wide variety of arrhythmias exist, some caused by ectopic pacemakers, which are abnormal sites of oscillatory activity that interfere with the primary pacemaker rhythm. Fibrillation, the irregular contraction of the heart muscle, which can result in sudden cardiac death, has been suggested to occur as the result of circulating waves of electrical activity, similar to the waves of chemical activity in the BZ system (Winfree, 1983). The mechanism of oscillation of individual cardiac cells is thought to be quite similar to that of neurons, a topic discussed in the next section.

One of the most widespread forms of biological oscillation involves the behavior of calcium ions. Oscillations in (Ca^{2+}) can arise either spontaneously or as the result of stimulation by hormones or neurotransmitters. Although their function has not been completely elucidated, it appears that calcium oscillations and waves play an important role in intracellular and intercellular signaling (Berridge, 1989). The mechanism of calcium oscillations is gradually being unraveled; given the number and variety of systems in which they occur, there may be multiple mechanisms. One promising set of models involves the messenger molecule inositol 1,4,5-triphosphate (IP_3), which causes the release of Ca^{2+} from intracellular stores (Meyer and Stryer, 1988). The rate of synthesis of IP_3 is enhanced by higher levels of Ca^{2+} in the cytosol, thereby providing a positive feedback loop that leads to oscillations in both (Ca^{2+}) and (IP_3). Goldbeter (1996) presents a detailed treatment of modeling approaches to calcium oscillations. Recent advances in microscopy and in the use of fluorescent dyes have made it possible to obtain spatially resolved data on calcium waves in cells (Lechleiter et al., 1991). An example is shown in Figure 13.3

13.1 Neural Oscillators

Nerve cells, or *neurons*, are the primary means by which signaling and information transfer occur in living organisms (Levitan and Kaczmarek, 1997). The signals that neurons generate and transmit are electrical in nature; moreover, they are often oscillatory. In this section, we take a brief look at how neurons work and how some of the ideas that we have developed for studying chemical oscillators can usefully be applied to understanding neural oscillations.

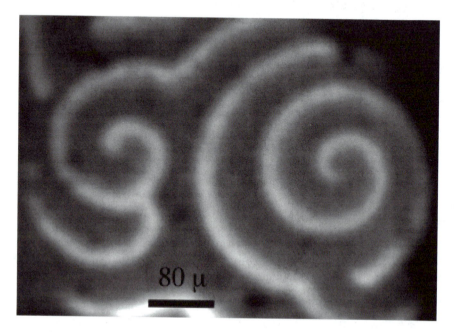

Figure 13.3 Calcium waves on the surface of a frog oocyte. (Courtesy of P. Camacho and J. Lechleiter.)

Very crudely speaking, we may consider a neuron as an aqueous solution of ions—primarily sodium, potassium, calcium, and chloride—enclosed in an insulating container (the plasma membrane) and surrounded by another aqueous solution of these ions, but at different concentrations. The situation is shown schematically in Figure 13.4. We have, of course, neglected a great deal of neuroanatomy, but the essential dynamics of neurons arise from the transfer of ions across the membrane, so our diagram captures the most important features.

The concentration of K^+ inside the neuron is significantly greater than the concentration of this species in the surrounding medium, while Na^+, Ca^{2+}, and Cl^- are present in higher concentrations outside the cell than inside. As a result of these concentration differences, which are maintained by a complex pumping machinery, there is a potential, V_m, across the membrane. The electrical activity of a neuron depends upon the movement of ions—primarily cations—across the cell membrane. This movement is made possible by the existence of *ion channels*, specialized proteins embedded in the membrane that allow the transport of ions down the concentration gradient across the otherwise impermeable cell membrane.

There are many kinds of channels, each permeable to a specific ion, for example, potassium. The channel exists in two basic conformations: open and closed. When the channel is closed, no ions can pass through, but when it is open, ions can flow across the membrane in or out of the cell. The likelihood of a channel being open or closed depends on a number of factors, the most important of which are the membrane potential, the concentrations of specific ions (particularly calcium) in the cell, and the presence of other neuromodulatory substances, such as acetylcholine. Different kinds of channels[1] display different dependences on membrane potential and ion concentrations, and also have different dynamics for their opening and closing. The regulation of channel opening by the potential provides a mechanism for feedback, either positive or negative. If, for example, the probability of a sodium channel being open increases with the membrane potential, and a flow into the cell of Na^+ raises V_m, then we have a positive feedback.

As a result of the different internal and external concentrations of ions, when it is not transmitting signals a typical neuron has a *resting potential* of -40 to -90 mV. By inserting a microelectrode into the cell through the membrane, it is possible to inject an external current, either positive or negative, into the cell.

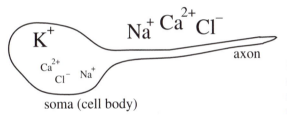

Figure 13.4 Crude representation of a neuron. Larger letters represent higher concentrations of ions.

[1]There may, for example, be three or four different kinds of potassium channels in a single cell.

Since, at rest, the cell membrane is polarized and has a negative potential, a decrease in V_m is referred to as a *hyperpolarization*, while an increase in membrane potential is a *depolarization*. Injected negative current causes the cell to hyperpolarize for the duration of the current injection by a voltage proportional to the magnitude of the current; when the current is turned off, the cell returns to its resting potential. Small injections of positive current produce a similar transitory *passive* depolarization. As a result of the positive feedback provided by the voltage-dependence of certain channels, the cell membrane is an excitable system, like the ones discussed in Chapter 6. If a large enough positive current is injected, the membrane undergoes a further large rapid depolarization; this *active* response, illustrated in Figure 13.5, is known as an *action potential*. It is by means of action potentials that neurons transmit information.

13.1.1 Modeling an Action Potential

The behavior of the membrane potential V can be understood, at least at a gross level, by applying some basic ideas of electrostatics to the current I_m that flows across the membrane. We assume, though this is a significant approximation, that the surface of the cell is homogeneous, namely, that the membrane is something like a well-stirred reactor, so that no spatial dependence need be considered. Since charge can build up on the two sides of the membrane, there is a capacitance C_m associated with the membrane and a corresponding current $C_m \, dV/dt$ that results from any potential change. There is a resistance to the flow of each type of ion. We shall employ the *conductance*, which is the reciprocal of the resistance, so that each ionic current is given by Ohm's law:

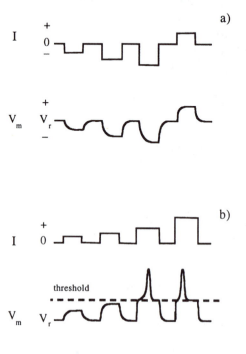

Figure 13.5 (a) Passive and (b) active (action potential) responses of a neuron to injected current I. (Adapted from Levitan and Kaczmarek, 1997.)

$$I_j = (V - V_j)/R_j = g_j(V - V_j) \qquad (13.1)$$

where I_j, R_j, and g_j are the current, resistance, and conductance, respectively, of the jth type of channel, and V_j is the rest potential for the ion that flows through that channel. The conductance, in general, depends upon time, the membrane potential, and possibly other quantities (such as $[Ca^{2+}]$), which we collect in a vector $\mathbf{x}$. The total membrane current is the sum of the capacitative current, the individual ionic currents, and any applied external current:

$$I_m(t) = C_m \, dV/dt + g_j(V, t, \mathbf{x})(V - V_j) + I_{ext}(t) \qquad (13.2)$$

By a variety of experimental techniques that make it possible to selectively block channels and record the time evolution of the current when a cell is brought to a chosen potential (voltage clamp, see section 13.1.3), it is possible to gain information about the conductances g_j in eq. (13.2). One of the most thoroughly studied systems is the squid giant axon, for which Hodgkin and Huxley (1952) developed a model that has served as the prototype for much of the neural modeling that has followed. The ionic currents are time- and voltage-dependent sodium and potassium currents and a nonspecific (primarily chloride) "leakage current," which has no time or voltage dependence. The model gives the equation for the evolution of the membrane potential [cf. eq. (13.2)] as

$$C_m \, dV/dt = -g_{Na}^0 m^3 h(V - V_{Na}) - g_K^0 n^4(V - V_K) - g_L(V - V_L) + I_{ext}(t) \quad (13.3)$$

where the quantities g_j^0 are the maximal conductances for the ion-specific channels. The variables m, h, and n, which take values between 0 and 1, specify the time and voltage dependences of the ion-specific conductances; m and n are *activation* variables that describe how the respective currents turn on; h is an *inactivation* variable that specifies how g_{Na} turns off. They are determined by the following auxiliary equations, in which the quantity $y(V)$ ($y = m$, h, or n) represents the steady-state voltage dependence and τ_y the corresponding time constant.

$$dm/dt = [m(V) - m]/\tau_m(V) \qquad (13.4a)$$
$$dh/dt = [h(V) - h]/\tau_h(V) \qquad (13.4b)$$
$$dn/dt = [n(V) - n]/\tau_n(V) \qquad (13.4c)$$

The steady-state activations, which are obtained from fits of experimental data, are s-shaped functions of V, something like $\{1 + \tanh[(V - E_y)/s_y]\}/2$, where E_y is the potential at which the variable y is half-activated and s_y specifies the width of the activation range below which y is essentially 0 (the channel is closed) and above which it is essentially 1 (the channel is open). The steady-state inactivation h is a mirror image of the steady-state activations; mathematically, we may think of s_h as being negative. Typical activation and inactivation curves are plotted in Figure 13.6.

The set of equations (13.3) and (13.4) leads to excitable dynamics. There is a stable steady state at the rest potential, which is near $V_K < 0$ V. At rest, the potassium conductance is significantly greater than the sodium conductance (the leak current is relatively small). If a large enough pulse of positive current is applied (i.e., if V is raised to a high enough level), m, which has a faster time scale than the other variables, increases rapidly, causing the sodium current to

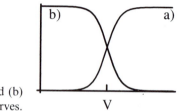

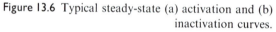

Figure 13.6 Typical steady-state (a) activation and (b) inactivation curves.

dominate and bringing the potential quickly up toward V_{Na}, which is about $+100$ mV. This fast spike in V is the action potential. When V rises, n increases and h decreases, but more slowly than the increase in m. The potential levels off and begins to decrease because the sodium channels shut down as a result of the inactivation h and the potassium channels open as n increases, bringing the system back toward V_K. The system eventually returns to rest, ready for another action potential.

A two-variable abstract model (Fitzhugh, 1961; Nagumo et al., 1962) can be derived from the Hodgkin–Huxley model by making the approximations that m changes much more rapidly than n and h, so that we can set $m = m(V)$, and that h is constant. The resulting Fitzhugh–Nagumo model contains a "voltage" variable v and a "recovery" variable w, which combines the functions of the activation and inactivation variables in the full model. The differential equations are given by

$$dv/dt = -f(v) - w + i_{ext} \tag{13.5a}$$
$$dw/dt = (v - cw)/\tau \tag{13.5b}$$

with

$$f(v) = v(a - v)(1 - v) \tag{13.6}$$

and $c, \tau > 0$, $0 < a < 1$.

Note the similarities between eqs. (13.5) and (13.6) and the Boissonade–De Kepper model, eqs. (4.2), which played such a key role in the design of chemical oscillators. Rinzel (1981) and Murray (1993) provide detailed phase-plane analyses of eqs. (13.5) and (13.6), showing how, depending on the parameters, they give rise not only to excitability (action potentials) but also to oscillation (beating or repetitive firing), which is also seen in actual neurons.

13.1.2 Bursting

The pattern of activity in many neurons consists not of single action potentials or of simple periodic firing, but of the more complex temporal mode known as bursting. We came across bursting, which consists of relatively quiescent, hyperpolarized periods alternating with periods in which a series or "burst" of action potentials occurs, in the previous chapter on coupled oscillators. Bursting oscillations have the potential to carry significantly more information than simple periodic (e.g., sinusoidal) oscillations, because, in addition to the frequency and amplitude, there is information content in the number of action potentials per cycle, the time between action potentials, and the "duty cycle," that is, the fraction of the period during which action potentials are being fired. Figure 12.18 shows an example of bursting behavior in the crustacean stomatogastric ganglion, a system that we shall have more to say about shortly.

Rinzel (1981) has studied a number of models for bursting behavior and has developed both an intuitive understanding of the origin of this phenomenon and a classification scheme that describes the different ways in which bursting can arise. The essence of his analysis, which is similar in concept to the development of Barkley's model for mixed-mode oscillations that we discussed in section 8.1, rests on identification of a set of slow and a set of fast processes. In neurons, the fast processes are associated with the generation of the action potentials, while the slow processes typically determine the period of the overall oscillation. The membrane potential provides the key link between the two sets of processes.

Imagine, to start with, that there is a single fast variable x and a single slow variable y. We think of y as a parameter on which the behavior of x depends, as pictured in Figure 13.7a. The x subsystem has three steady states, the upper and lower of which are stable, and on the upper branch dy/dt is positive, while y is decreasing when x is on the lower branch. This description should give the reader a sense of déjà vu, since it is closely related to both the Boissonade–De Kepper model of Chapter 4 and the Barkley model of Chapter 8. The system we have just described gives rise to the simple oscillation shown in Figure 13.7.

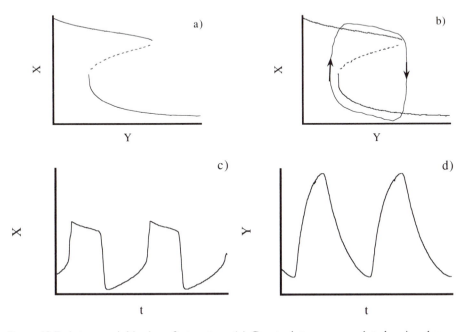

Figure 13.7 A two-variable slow–fast system. (a) Constraint–response plot showing dependence of steady states of x on y, treated as a parameter; (b) trajectory of oscillation in x–y plane; (c) x oscillations; (d) y oscillations. (Adapted from Rinzel, 1981.)

In order to turn the above system into a burster, we need to change the character of one of the x steady states shown in Figure 13.7a so that it becomes unstable and is surrounded by a stable limit cycle. This can only happen in a system where there are at least two fast variables (recall that it takes at least two variables to make an oscillator), so let us assume that this is the case. In Figure 13.8a, we show the new situation. The upper branch now has a Hopf bifurcation, marked HB, at which oscillation in x and the other fast variable begins. As y increases, we reach the point, labeled HC for homoclinic, where the limit cycle collides with the middle unstable state and ceases to exist, causing the system to jump down to the lower, stable branch of steady states. The resulting trajectory is shown in Figure 13.8b. It gives rise to the oscillatory behavior seen in Figures 13.8c and 13.8d. Note that only the fast variable x exhibits the rapid spiking characteristic of bursting. Comparison of Figures 13.7 and 13.8 shows that the spikes in x occur around the average value of the upper branch of the "slow wave" in Figure 13.7, and that the amplitude of oscillation of y is lower in the bursting mode because the system jumps to the lower x-branch on encountering the homoclinic point before reaching the end of the upper x-branch.

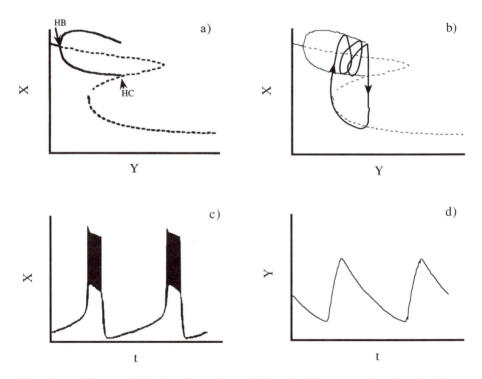

Figure 13.8 A three-variable slow–fast system. (a) Constraint–response plot showing dependence of steady states of x on y, treated as a parameter (curved lines surrounding upper dashed branch indicate limits of oscillation); (b) trajectory of bursting oscillation in x–y plane; (c) x oscillations; (d) y oscillations. (Adapted from Rinzel and Lee, 1987.)

A concrete example of a model for neuronal bursting (Plant, 1981) has been analyzed in considerable detail by Rinzel and Lee (1987). The model utilizes the Hodgkin–Huxley model,[2] eqs. (13.3) and (13.4), as the mechanism for action-potential generation and introduces two additional conductances, a calcium channel and a calcium-activated potassium channel, to produce the bursting behavior. The membrane potential is given by

$$C_m \, dV/dt = I_{HH} - g_{Ca}^0 x(V - V_{Ca}) - g_{K-Ca}^0 [Ca/(0.5 + Ca)](V - V_K) \qquad (13.7)$$

where I_{HH} represents the terms on the right-hand side of eq. (13.3) with n and h governed by eqs. (13.4), x is the activation variable for the calcium conductance, and Ca is a variable that is proportional to the calcium concentration in the cell. The behaviors of the slow variables x and Ca are determined by

$$dx/dt = [x(V) - x]/\tau_x(V) \qquad (13.8)$$
$$dCa/dt = \rho[K_c x(V_{Ca} - V) - Ca] \qquad (13.9)$$

Equation (13.7) constitutes the fast subsystem, with x and Ca treated as parameters and V, n, and h as variables; the slow subsystem consists of eqs. (13.8) and (13.9) with V treated not as a variable but as an instantaneous function of x, Ca, and t taken from the solution of the fast subsystem. Figure 13.9 shows a bursting solution with the spikes in the fast variable V and the slow wave solution for Ca.

13.1.3 An Experimentally Based Model

In most chemical oscillators, it is difficult or impossible to follow the concentrations of more than a small fraction of the reactants and intermediates. Experimental determination of all the rate constants is likewise a forbidding prospect. These difficulties in obtaining a complete set of measurements are even greater in biological systems, where the variety and complexity of relevant

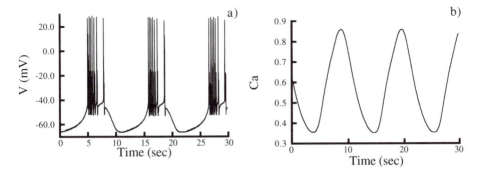

Figure 13.9 Bursting behavior in Plant's model for a neuron, eqs. (13.7)–(13.9). (Adapted from Rinzel and Lee, 1987.)

[2]With the minor modification that the sodium activation variable m is assumed to relax instantaneously to m (V), so that m is a voltage-dependent parameter rather than a variable.

molecules and reactions tends to be significantly greater than in inorganic chemical systems.

Perhaps the most detailed effort to construct a model for a neural oscillator based on experimental data has been carried out by Buchholtz et al. (1992). These investigators studied the lateral pyloric neuron in the rock crab *Cancer borealis*. By utilizing a variety of pharmacological techniques, they were able to block selectively the currents in this cell, thereby isolating a single type of channel for study (Golowasch and Marder, 1992). They then applied the *voltage clamp* technique, in which a cell at rest is brought to a desired membrane potential by rapid injection of current and then held, or "clamped," at that potential by injecting smaller amounts of current to compensate for time-dependent activation and/or inactivation of channels at the new potential. By recording the magnitude of the current necessary f· the clamping at a series of voltages, one gains information about the time and voltage dependences of the channels being studied. The data can then be fitted to a conductance model of the Hodgkin–Huxley type, extracting parameters that describe the maximal conductance and the activation and inactivation of each channel.

The model contains six voltage-dependent conductances: three for potassium, one each for sodium and calcium, and a mixed conductance, as well as a voltage-independent leak conductance. In addition, the model includes a calcium buffering system; one of the potassium conductances is calcium-activated. Obtaining the experimental data needed to describe seven different conductances is a painstaking task, which required nearly four years (Golowasch, 1990). The nonlinear fitting by which the parameters were extracted was also time-consuming, but the quality of the fit is impressive. Figure 13.10 shows examples of experimental voltage clamp data and the corresponding simulations.

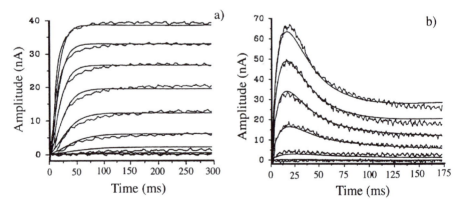

Figure 13.10 Voltage clamp data (current vs. time) for (a) Hodgkin–Huxley ("delayed rectifier")-type potassium current i_d, and (b) calcium-activated potassium current i_o. Noisy traces are experimental data, smooth curves are model simulations. Note that i_d has no inactivation, so current remains at higher level, while i_o contains inactivation, so that current reaches a peak and declines. (Adapted from Buchholtz et al., 1992.)

13.2 Microtubule Polymerization

Microtubules are polymeric fibers that play a key role in cell division and in cell shape and motility. They are formed from subunits of the protein tubulin (molecular weight about 100 kd), which assemble into protofilaments. Each microtubule is composed of thirteen linked protofilaments. Intermediate in size are oligomers, which are short stretches of protofilaments (Correia and Williams, 1983). The assembly of microtubules is driven by the hydrolysis of the nucleotide guanosine triphosphate (GTP), which binds to the tubulin subunits and hydrolyzes to the diphosphate (GDP) on polymerization. The GDP can also bind to tubulin, but it inhibits microtubule assembly. The nucleotides are rapidly exchangeable when bound to tubulin, but become nonexchangeable when attached either to oligomers or to microtubules.

It was originally thought that microtubule assembly was a simple polymerization process in which a slow nucleation step is followed by endwise addition of tubulin subunits to the growing polymer (Johnson and Borisy, 1977). Such a mechanism would be governed by pseudo-first-order kinetics and would result in a steady state. However, experiments involving light (Carlier et al., 1987) and x-ray (Mandelkow et al., 1988) scattering reveal that the process can occur in an oscillatory fashion. Figure 13.11 shows x-ray scattering data that indicate the existence of large-amplitude oscillations both in the extent of polymerization and the concentration of microtubules.

The microtubule oscillations gradually damp out to a low-concentration steady state, while the extent of polymerization, that is, the oligomer concentration, slowly grows. Higher levels of initial GTP extend the period of oscillation. Parallel experiments to determine the amounts of GTP and GDP bound to tubulin show that the amount of protein-bound GDP reaches a maximum at or just after the polymerization peaks, while bound GTP peaks at about the time of the troughs in polymerization, that is, when the free tubulin concentration is highest. Consumption of GTP via binding to tubulin and subsequent hydrolysis appears to occur in bursts coupled to assembly cycles (Mandelkow et al., 1988). The damping of oscillations is likely to result from depletion of the available GTP.

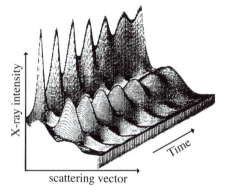

Figure 13.11 Time-resolved x-ray scattering data from solutions of oscillating microtubules. Large peak at left shows total extent of polymerization (oligomers and microtubules); smaller, central peak arises from microtubules. (Reprinted with permission from Mandelkow, E.-M.; Lange, G.; Jagla, A.; Spann, U.; Mandelkow, E. 1988. "Dynamics of the Microtubule Oscillator: Role of Nucleotides and Tubulin–MAP Interactions," *EMBO J. 7*, 357–365. © 1988 Oxford University Press.)

Mandelkow et al. (1988) suggest a model for the oscillatory assembly of microtubules, which is shown in Figure 13.12. In step 1, microtubules (MtGTP) are formed from active subunits (TuGTP), and the GTP is hydrolyzed (step 2) to give MtGDP. The GDP-bearing microtubules can disassemble in step 3 to produce oligomers (OlGDP), which can further dissociate in step 4 to regenerate tubulin, now bound to GDP. The tubulin subunits can exchange GDP for GTP in step 5 to become active again. Autocatalysis is provided by the enhancement of the rate of MtGTP production by higher concentrations of microtubules. The oscillations will naturally damp out as the GTP concentration decreases owing to hydrolysis to GDP. Side reactions consist of the reversible formation of rings from oligomers and the nucleation that occurs in the first round of microtubule assembly.

In addition to temporal oscillations, solutions of tubulin and GTP can generate spatial patterns, including traveling waves of microtubule assembly and polygonal networks (Mandelkow et al., 1989). This system may provide a useful experimental model for understanding pattern formation in cells.

13.3 Dynamical diseases

In any healthy human being, a multitude of rhythmic processes take place over a wide range of time scales. Some of these are listed in Table 13.1. Occasionally, one or more of these processes changes its character: the period may lengthen or decrease, a periodic oscillation may become chaotic (or vice versa), an oscillatory process may cease to oscillate, or a previously constant quantity may become oscillatory. Such changes often have deleterious effects on the organism. Mackey and Glass (1977) call pathological conditions characterized by abnormal temporal organization *dynamical diseases*. Cardiac arrhythmias are an obvious example, but there are many more, some of them quite exotic.

Glass and Mackey (1988) note that the clinical literature contains relatively little data of use in understanding dynamical diseases and that the medical profession has been reluctant to appreciate the importance of dynamical considerations in pathology. They point out that an understanding of this aspect of disease

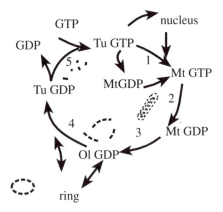

Figure 13.12 A model for oscillations in microtubule assembly. (Adapted from Mandelkow et al., 1988.)

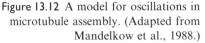

would open the way to the development of therapies that exploit our knowledge of dynamical systems. A relatively trivial example is the attempt to analyze how circadian clocks reset themselves in relation to the phenomenon of jet lag (Winfree, 1986).

A number of biological systems lend themselves to modeling by differential delay equations of the sort described in Chapter 10, where we looked at the respiratory disorder Cheyne–Stokes syndrome. Introduction of delays is often more natural in biological than in chemical systems. Another example occurs in the production of white blood cells (Mackey and Glass, 1977; Glass and Mackey, 1979). The cells of interest, called neutrophils, are produced from stem cells in the bone marrow at a rate that depends on the population of circulating neutrophils in the blood. When this population goes down, the hormone granulopoietin is released and this acts on the stem cells to speed up the rate at which new neutrophil precursors are released. Although the production of new cells increases with decreasing cell concentration over most of the range, when there are very few cells circulating, the rate of production begins to decrease, so the production rate has a maximum as a function of the circulating concentration and decreases toward zero at both very high and very low concentrations. Precursor cells take a considerable amount of time to mature, about 6 days. Neutrophils also have a finite lifetime, approximately 7 h. These features can be summarized in a mathematical model as a single differential delay equation:

$$dx(t)/dt = -\gamma x(t) + \beta \theta^n x(t - \tau)^n / [\theta^n + x(t - \tau)^n] \qquad (13.10)$$

where x is the population of circulating neutrophils; γ is the rate at which neutrophils are destroyed; τ is the time between release of a neutrophil precursor and its maturation to a circulating cell; and β, n, and θ are parameters that determine the shape of the production vs. concentration curve for neutrophils.

With parameters corresponding to normal humans, linear stability analysis reveals two steady states for eq. (13.10), both unstable, and numerical integration yields a mild oscillation in x with a period of about 20 days, consistent with observations that white counts in healthy people go up and down with periods in the range of 12–24 days. It has been suggested that in chronic myologenous leukemia, there is an increase in the time of maturation of neutrophils. Figure 13.13b illustrates the results of increasing τ in eq. (13.10) from 6 to 20 days (Mackey and Glass, 1977). The large-amplitude chaotic oscillations that result bear a striking resemblance to those shown in Figure 13.13a, which are white blood cell counts from a twelve-year-old girl with chronic granulocytic leukemia (Gatti et al., 1973).

13.4 Conclusions

The number and variety of biological oscillators is overwhelming. Nearly every process in every living system has an "exotic" dynamical component associated with it. Why this should be so is not well understood at this time, though Richter and Ross (1980) have suggested that certain oscillatory processes may be more efficient than the corresponding steady-state processes and may thus confer an

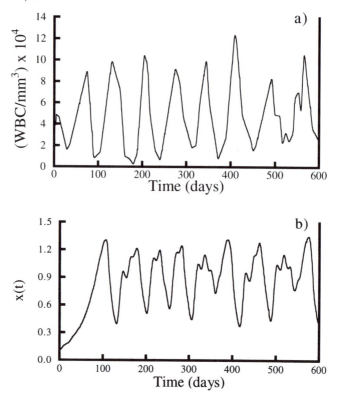

Figure 13.13 (a) White blood cell counts from a leukemia patient. (b) Results of numerically integrating eq. (13.10) with the abnormally long maturation time $\tau = 20$ days. (Adapted from Glass and Mackey, 1988.)

evolutionary advantage, while Boiteux et al. (1980) propose a role for oscillations in generating and maintaining the temporal and spatial organization of metabolic systems. As we warned you at the outset, we have only been able to scratch the surface in this chapter. We have looked at a few examples and perhaps developed an appreciation for how much is to be gained by improving our current state of understanding of these important systems. Further study promises great rewards!

14

Turing Patterns

In the first chapter of this book, we noted the "dark age" of nearly forty years separating the work of Bray and Lotka in the early 1920s and the discovery of the BZ reaction in the late 1950s. Remarkably, the history of nonlinear chemical dynamics contains another gap of almost the same length. In 1952, the British mathematician Alan Turing wrote a paper in which he suggested that chemical reactions with appropriate nonlinear kinetics coupled to diffusion could lead to the formation of stationary patterns of the type encountered in living organisms. It took until 1990 for the first conclusive experimental evidence of Turing patterns to appear (Castets et al., 1990).

Turing was a formidable figure (Hodges, 1983). He was responsible for much of the fundamental work that underlies the formal theory of computation, and the notion of a "Turing machine" is essential for anyone who wishes to understand computing and computers. During World War II, Turing was a key figure in the successful effort to break the Axis "Enigma" code, an accomplishment that almost certainly saved many lives and shortened the war in Europe. His 1952 paper, entitled "The Chemical Basis of Morphogenesis," was his only published venture into chemistry, but its impact has been enormous. Recently, this classic paper has been reprinted along with some of Turing's unpublished notes on the origins of phyllotaxis, the arrangement of leaves on the stems of plants (Saunders, 1992).

In this chapter, we shall describe the nature of Turing patterns and some of the systems in which they may play a role, explore why they have been so elusive, examine the experimental systems in which they have been demonstrated, and consider other systems and other methods for generating them. Much of our discussion will focus on the chlorite–iodide–malonic acid (CIMA) reaction in which the patterns were first seen. In the study of Turing patterns, the CIMA system and its relatives play much the same role today that the BZ reaction played during the 1960s and 1970s in the study of chemical oscillation.

14.1 What Are Turing Patterns?

When we think of diffusion acting on a system in which there are concentration inhomogeneities, our intuition suggests that diffusion should act to lessen, and eventually eliminate, the inhomogeneities, leaving a stable pattern with concentrations that are equal everywhere in space. As in the case of temporal oscillation, for a closed system the laws of thermodynamics require that this intuition be valid and that the eventual concentration distribution of the system be constant, both in time and in space. In an open system, however, just as appropriate nonlinear rate laws can lead to temporal structure, like oscillations and chaos, the interaction of nonlinear chemical kinetics and diffusion can produce nonuniform spatial structure, as suggested schematically in Figure 14.1.

In Chapter 6, we examined one sort of spatial pattern formation in reaction–diffusion systems, but the trigger waves, target patterns, and spirals considered there arise when the temporally uniform steady state becomes unstable. Here, we wish to examine, as Turing did, whether diffusion can lead to spatial pattern formation *even when the temporally uniform steady state is stable in the absence of diffusion.* Thus, the first criterion for a pattern to qualify as a Turing pattern is that the system have a stable spatially homogeneous steady state. Second, we shall require that only reaction and diffusion occur in the system. Mass transport, such as convection, is not permitted. This specification rules out certain patterns in chemical systems (Möckel, 1977) that were initially identified as Turing patterns, but were later shown to be generated by convection. Next, in order for the structures to arise, there must be at least one spatially inhomogeneous infinitesimal perturbation to the uniform steady state that will grow rather than decay. That is, the steady state is stable to all *spatially homogeneous* infinitesimal perturbations, but it may be unstable to inhomogeneous perturbations of a particular symmetry. Turing pattern formation is thus a *symmetry-breaking* phenomenon. Finally, as the perturbation grows, it eventually evolves into a temporally constant, spatially nonuniform pattern—for example, stripes or spots. This last condition is not necessary; Turing did, in fact, consider nonconstant patterns as well,

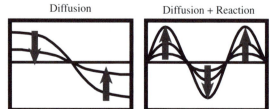

Diffusion   Diffusion + Reaction

Figure 14.1 Schematic comparison of effects on spatial structure of pure diffusion and diffusion coupled to nonlinear chemical reaction kinetics.

and there has been recent work in this area, as we shall mention briefly toward the end of this chapter. Our focus here, however, like the focus of nearly all the theoretical and experimental studies of the past four decades, is on temporally stationary Turing patterns. One additional feature of Turing patterns deserves mention. The spatial structure possesses an intrinsic wavelength that is determined by the rate and diffusion constants and the concentrations. This wavelength is independent of the size of the system, so long as the system dimensions exceed the wavelength. If the system length is smaller than the wavelength in all directions, then no Turing patterns can arise, and only the uniform steady state will be stable.

To make these conditions more precise, and to derive some important insights into what sorts of systems can exhibit Turing patterns, we consider the simplest possible class of systems capable of supporting this type of behavior, systems in one spatial dimension with two concentration variables. Let $x(t, t)$ and $y(r, t)$ be the two concentrations whose time evolution is governed by

$$\partial x/\partial t = f(x, y) + D_x \, \partial^2 x/\partial r^2 \tag{14.1}$$

$$\partial y/\partial t = g(x, y) + d_y \, \partial^2 y/\partial r^2 \tag{14.2}$$

We assume that the system has a spatially uniform stable steady state (x_s, y_s) such that

$$f(x_s, y_s) = g(x_s, y_s) = 0 \tag{14.3}$$

We define the elements a_{ij} of the Jacobian matrix $\mathbf{J}$, where all the partial derivatives are evaluated at (x_s, y_s):

$$a_{11} = \partial f/\partial x, \qquad a_{12} = \partial f/\partial y, \qquad a_{21} = \partial g/\partial x, \qquad a_{22} = \partial g/\partial y \tag{14.4}$$

The linear stability analysis developed in Chapter 2 implies that the steady state will be stable to spatially uniform perturbations if, and only if,

$$\text{tr}\,(\mathbf{J}) = a_{11} + a_{22} < 0 \tag{14.5}$$

and

$$\det\,(\mathbf{J}) = a_{11}a_{22} - a_{12}a_{21} > 0 \tag{14.6}$$

Equations (14.5) and (14.6) are derived by requiring the decay of all infinitesimal perturbations (α, β) such that

$$x(r, t) = x_s + \alpha \, \exp\,(\lambda t) \tag{14.7}$$

$$y(r, t) = y_s + \beta \, \exp\,(\lambda t) \tag{14.8}$$

Let us now consider what happens if the perturbation is allowed to be spatially, as well as temporally, dependent. If we write, instead of eqs. (14.7) and (14.8),

$$x(r, t) = x_s + \alpha \, \exp\,(\lambda t)\,\cos\,(qr) \tag{14.9}$$

$$y(r, t) = y_s + \beta \, \exp\,(\lambda t)\,\cos\,(qr) \tag{14.10}$$

and carry out the linear stability analysis as before, we obtain an extra term in each of the diagonal elements of the Jacobian matrix as a result of the spatial derivatives in eqs. (14.1) and (14.2). The terms a_{11} and a_{22} in eq. (14.4) must be replaced by

$$a'_{11} = a_{11} - q^2 D_x \tag{14.11}$$

$$a'_{22} = a_{22} - q^2 D_y \tag{14.12}$$

When these matrix elements are inserted into the stability conditions (14.5) and (14.6), we obtain the following inequalities that must be satisfied for an inhomogeneous perturbation of the form described by eqs. (14.10) and (14.11) to be stable:

$$\text{tr } (\mathbf{J}) = a_{11} + a_{22} - q^2 (D_x + D_y) < 0 \tag{14.13}$$

and

$$\det (\mathbf{J}) = (a_{11} - q^2 D_x)(a_{22} - q^2 D_y) - a_{12}a_{21} > 0 \tag{14.14}$$

If the homogeneous steady state is stable, inequality (14.5) must hold, and since the diffusion constants are positive, condition (14.13) must hold as well. Thus, the only possibility for the steady state to be unstable to an inhomogeneous perturbation is for condition (14.14) to be violated. We multiply out the terms in eq. (14.14) and rewrite the resulting condition for instability as an inequality on a polynomial in q^2:

$$H(q^2) = D_x D_y q^4 - (a_{11}D_y + a_{22}D_x)q^2 + a_{11}a_{22} - a_{12}a_{21} < 0 \tag{14.15}$$

Note that, as a function of q^2, H is a quadratic that opens upward, that is, it has a minimum. Since eq. (14.6) tells us that $a_{11}a_{22} - a_{12}a_{21} > 0$, we know that H cannot be less than or equal to zero for any positive value of q^2 unless

$$(a_{11}D_y + a_{22}D_x) > 0 \tag{14.16}$$

because if eqs. (14.16) did not hold, all the terms in eq. (14.15) would be positive for any $q^2 > 0$. Thus, eq. (14.16) is a necessary condition for a Turing-type instability.

Equation (14.16) provides an important physical insight into the requirements for Turing pattern formation. We know from eq. (14.5) that the sum of the diagonal Jacobian elements, $a_{11} + a_{22}$, is negative. If both elements are negative, however, eq. (14.16) cannot be true. Therefore, for both eqs. (14.5) and (14.16) to hold, we must have one negative and one positive diagonal element. That is, one species enhances the rate of its own production, while the other decreases its rate of production as its concentration grows. For specificity, let us suppose that

$$a_{11} > 0 \tag{14.17}$$

We shall then call species x the *activator*, since making more of it activates the production of still more. Species y is referred to as the *inhibitor*. A model of the form of eq. (14.1), with one positive and one negative Jacobian element, is called an *activator–inhibitor model*. Models of this type have played a key role in the investigation of Turing patterns.

Note that, in order for eq. (14.5) to hold, we must have

$$|a_{11}| < |a_{22}| \tag{14.18}$$

If we rearrange eq. (14.16), making use of eqs. (14.17) and (14.18), we obtain

$$D_x/D_y < a_{11}/(-a_{22}) < 1 \qquad (14.19)$$

Equation (14.19) provides an important restriction on the diffusion constants of the activator and inhibitor in any system that permits Turing structure formation. *The inhibitor must diffuse more rapidly than the activator.* This condition is sometimes referred to as "local activation and lateral inhibition," since the activation, which often takes the form of autocatalysis, tends to be confined to the neighborhood of its initiation site, in contrast to the inhibition, which can spread more rapidly throughout the medium. Murray (1993) gives the intuitively attractive analogy of a dry forest in which there are firefighters with helicopters equipped with a fire-retardant spray dispersed throughout the forest. The fire is the activator and the firefighters are the inhibitor. If the helicopters travel faster than the fire can spread, then when fire begins to break out at random points in the forest, the firefighters will be able to outrun the fire fronts, spraying enough trees to stop the line of advance, and eventually a stable pattern of burnt black and unburnt green trees will be established. If the fire moves more quickly than the firefighters, then the outcome will be a homogeneous state of black.

Although eq. (14.19) is important, and we shall return to it below, it is not a sufficient condition for instability, that is, for eq. (14.15) to hold. If we think of eq. (14.15) as a quadratic equation in q^2 of the form $a(q^2)^2 + bq^2 + c$, then in order for there to be real roots, the discriminant, $b^2 - 4ac$, must be nonnegative. Otherwise, the roots are imaginary. Applying this condition to eq. (14.15), we obtain

$$(a_{11}D_y + a_{22}D_x)^2 - 4D_xD_y(a_{11}a_{22} - a_{12}a_{21}) > 0 \qquad (14.20)$$

or, using eqs. (14.6) and (14.16) to keep track of the signs,

$$a_{11}D_y + a_{22}D_x > 2[D_xD_y(a_{11}a_{22} - a_{12}a_{21})]^{1/2} > 0 \qquad (14.21)$$

If a system of the form of eq. (14.1) satisfies conditions (14.5), (14.6), (14.17), (14.19), and (14.21), then it can give rise to Turing pattern formation when the homogeneously stable steady state (x_s, y_s) is subject to inhomogeneous perturbations whose spatial scale, q^{-1}, is such that q satisfies eq. (14.15). In such a system, the initial, infinitesimal perturbation will grow, and the system will ultimately evolve to a stable, spatially inhomogeneous structure, the Turing pattern.

14.2 The Search for Turing Patterns

Turing's 1952 paper had a profound effect on the way scientists thought about pattern formation. A search of the literature reveals over 2000 papers that cite this seminal work. Interestingly, while most of the studies that refer to Turing pattern formation are in the fields of chemistry and biology, several investigations range as far afield as economics (Maruyama, 1963), semiconductor physics (Röhrlich et al., 1986), and star formation (Nozakura and Ikeuchi, 1984). More conventional applications of Turing's ideas have focused on the formation of hair fibers (Nagorcka and Mooney, 1982), spiral patterns in sunflowers (Berding et al.,

1983), tissue regeneration in hydra (Bode and Bode, 1987), and pattern formation in the BZ reaction (Rovinsky, 1987). What is striking is that until 1990, essentially every paper dealing with Turing patterns refers to them in theoretical or mechanistic terms. There is no experimental evidence for the existence of Turing patterns in any real system!

Here we have the gap of nearly forty years referred to at the beginning of this chapter. Why, if Turing patterns provide such an attractive mechanism of pattern formation in such a variety of environments, should it have been so hard to find one experimentally? Part of the answer in biological (or economic or astrophysical) systems lies in the extreme difficulty in identifying the activator and inhibitor species associated with the pattern formation and in demonstrating that the homogeneous steady state is stable. Chemical systems, as is often the case, offer far better prospects for the study of pattern formation. To appreciate why it took nearly four decades for chemists to find real Turing patterns, we must go back to the criteria for pattern formation that we developed in the previous section. Two problems, in particular, stand out.

First, true Turing patterns can only be found in an open system. Turing's models maintained an effective flow by holding the concentrations of certain key reactants constant. Turing also assumed perfect mixing, so that these reactant concentrations were constant not only in time but also in space. In real chemical systems, it is not possible to achieve these idealized conditions. We have seen earlier how the introduction of stirred flow reactors made it possible to generate chemical oscillation in essentially homogeneous open systems. We have also looked at how transient nonstationary patterns can arise in closed systems like the BZ reaction in a Petri dish. By the late 1980s, chemists were looking for ways to study target patterns and similar phenomena in open unstirred systems so that they could observe true far-from-equilibrium structures instead of transients. A number of ideas were developed (Tam et al., 1988a), one of the most fruitful of which proved to be the use of a gel as the reaction medium. In this scheme, illustrated in Figure 14.2, the two broad faces of a slab of gel, composed, for example, of polyacrylamide or agar, are in contact with two solutions of different compositions. The solutions, which are circulated through CSTRs, contain the reactants for the reaction that is to give rise to the pattern formation. These reactants diffuse into the gel, encountering each other at significant concentrations in a region near the middle of the gel, where the pattern formation can occur. The CSTRs provide a continuous flow of fresh reactants, which diffuse into the gel, thereby maintaining the system as open.

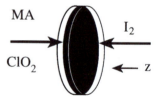

Figure 14.2 Schematic diagram of reactor for studying Turing pattern formation.

The second problem that needed to be solved in order to produce Turing structures in a real chemical system is summarized in eq. (14.19): the diffusion constant of the inhibitor must exceed that of the activator. In fact, for the concentrations and rate constants encountered in typical chemical systems, the ratio of D_y to D_x must be significantly greater than 1, perhaps as large as 8–10. As students of physical chemistry will recall, diffusion constants of molecular and ionic species in aqueous solution are depressingly predictable. Essentially all species have diffusion constants within a factor of two of 2×10^{-5} cm^2 s^{-1}. How can we "tune" a diffusion constant enough to produce the disparity between activator and inhibitor diffusion rates required for Turing pattern formation? To answer this question, we turn to the chemical system, the CIMA reaction, in which Turing patterns were first produced experimentally.

14.3 The CIMA Reaction

As we have seen, the success of a systematic approach to designing new chemical oscillators led to the discovery of many new oscillating reactions in the 1980s. Since the method required that the experiments be conducted in a CSTR, the reactions obtained, unlike the earlier accidentally discovered oscillators, tended to oscillate only under flow conditions and not in a closed, batch system. By far the richest vein of new oscillators was found in the reactions of the chlorite ion, particularly with iodine-containing species like iodide, iodate, and iodine. Sensing the symbolic importance of finding some new batch oscillators to complement their many new flow systems, the Brandeis group, which pioneered the development of the cross-shaped diagram approach to oscillator design in a CSTR, turned its attention to adapting one or more flow oscillators to function under batch conditions. To accomplish this goal, they started with two chlorite oscillators that showed a wide range of interesting dynamical behavior in a CSTR: chlorite–thiosulfate and chlorite–iodide (De Kepper et al., 1982). They obtained two new batch oscillators, one of which, the chlorite–iodide–malonic acid (CIMA) reaction, gave rise not only to temporal oscillation but also to target and spiral patterns. Soluble starch is used as an indicator in the CIMA system because it forms a deep blue complex with the triiodide ion, and this color, which is found when the system is in its reduced (high-iodide) state, provides a sharp contrast with the pale yellow or colorless oxidized state.

The CIMA reaction proved to be useful in a number of studies of pattern formation in an imaginative reactor discussed in Chapter 3 (Figure 3.10) (Ouyang et al., 1989), in which the reactants are fed from two CSTRs into the region between a pair of concentric counter-rotating cylinders. The relative rotation rate of the two cylinders serves as a bifurcation parameter, analogous to an effective one-dimensional (axial) diffusion constant (Tam and Swinney, 1987). As the rotation rate is increased, the system goes through a series of steady and oscillatory spatiotemporal states; on decreasing the rotation rate, one observes hysteresis between some of these states.

The chemistry of the CIMA reaction is complex. The inorganic part, the chlorite–iodide reaction, has been thoroughly studied and is relatively well understood, as we have discussed in earlier chapters. As in the case of the BZ reaction, adding the malonic acid makes it exceedingly difficult to formulate an elementary step mechanism for the CIMA reaction, because the number of possible organic intermediates is enormous. Far less, in fact, is known about the organic chemistry of the CIMA reaction than that of the BZ reaction. Simply extending the most widely accepted elementary step mechanism for the chlorite–iodide reaction by adding the reaction of malonic acid with iodine to regenerate iodide fails to account for the batch oscillation in the CIMA system. A key feature that makes the CIMA reaction more difficult to characterize than the chlorite–iodide system is that in the CIMA reaction the radical species ClO_2, which can be neglected in describing the oscillatory behavior of the chlorite–iodide reaction, plays a major role.

An alternative approach to seeking an elementary step description of the CIMA reaction is to characterize the system in terms of a few overall stoichiometric processes and their empirical rate laws (Rábai et al., 1979). Such an approach works well if there is no significant interaction among the intermediates of the component net processes, that is, no "cross talk," and if none of the intermediates builds up to significant levels. The proof of the pudding lies, of course, in comparing the predictions of this empirical rate law approach with the experimental results for the system of interest.

Spectrophotometric analysis of the CIMA reaction showed conclusively that ClO_2 is a key species in the batch oscillator. After an initial rapid consumption of most of the chlorite and iodide to generate ClO_2 and I_2, oscillation results from the reactions of these latter species with the reactant malonic acid. To demonstrate this hypothesis, Lengyel et al. (1990a, 1990b) showed that if we start from ClO_2, I_2, and MA, we observe oscillations with the same waveform and frequency as in the original CIMA system, but without any induction period. The oscillatory behavior of the CIMA reaction and of the derived chlorine dioxide–iodine–malonic acid (CDIMA) reaction is shown in Figure 14.3.

The simplest empirical rate law description of the CIMA and CDIMA systems consists of three component processes: the reaction between MA and I_2 to produce I^-, the reaction between ClO_2 and I^-, and the reaction between ClO_2^- and I^-:

$$MA + I_2 \rightarrow IMA + I^- + H^+$$
$$r_1 = k_{1a}[MA][I_2]/(k_{1b} + I_2) \tag{14.22}$$

$$ClO_2 + I^- \rightarrow ClO_2^- + \tfrac{1}{2}I_2$$
$$r_2 = k_2[ClO_2][I^-] \tag{14.23}$$

$$ClO_2^- + 4I^- + 4H^+ \rightarrow Cl^- + 2I_2 + 2H_2O$$
$$r_3 = k_{3a}[ClO_2^-][I^-][H^+] + k_{3b}[ClO_2^-][I^-][I_2]/(\alpha + [I^-]^2) \tag{14.24}$$

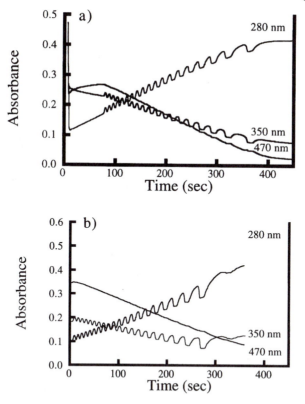

Figure 14.3 Oscillations in the (a) CIMA and (b) CDIMA reactions. Initial concentrations: (a) $[ClO_2^-]$ = 5.71×10^{-3} M, $[I^-]$ = 4.29×10^{-3} M, $[MA]$ = 1.2×10^{-3} M, $[H_2SO_4]$ = 5×10^{-3} M; (b) $[ClO_2]$ = 1.98×10^{-4} M, $[I_2]$ = 5.14×10^{-4} M, $[MA]$ = 1.0×10^{-3} M, $[H_2SO_4]$ = 5×10^{-3} M. Absorbance at 280 nm due to iodomalonic acid and triiodide, at 350 nm due to iodine and triiodide, and at 470 nm due to total iodine concentration. (Adapted from Lengyel and Epstein, 1995.)

Here, IMA is iodomalonic acid, $CHI(COOH)_2$, and α is an empirical parameter that determines the iodide concentration above which the last term in the rate law of reaction (14.24) becomes self-inhibitory in the reactant I^-. The values of the kinetic parameters are given in Table 14.1.

Reaction (14.24) is the crucial feedback process; it is autocatalytic in iodine and inhibited by the reactant iodide. In the CDIMA reaction, (I_2) is high and almost constant. Thus, the autocatalytic feature plays no significant role in the dynamics. However, (I^-) can change by several orders of magnitude, and its inhibitory role is crucial, providing the destabilization of the steady state necessary for oscillation or other exotic modes of dynamical behavior. The direct reaction of chlorite with iodine is not included in the model, but this omission, fortunately, does not introduce significant error at the concentrations present during the oscillations.

Table 14.1 Parameters of the Model for the CIMA and CDIMA Reactions

Rate or Diffusion Constant[a]	Value at 25 °C	EA (kJ mol^{-1})	Value at 4 °C	Reference
k_{1a} (s^{-1})	7.5×10^{-3}	8.15	6.2×10^{-4}	Lengyel and Epstein, 1991
k_{1b} (M)	5×10^{-5}		5×10^{-5}	Lengyel et al., 1990a
k_2 (M^{-1} s^{-1})	6×10^{3}	62.0	9.0×10^{2}	Lengyel and Epstein, 1991
k_{3a} (M^{-2} s^{-1})	460	51.4	100	Kern and Kim, 1965
k_{3b} (s^{-1})	2.65×10^{-3}	110.0	9.2×10^{-5}	Kern and Kim, 1965
α (M^2)	1×10^{-14}		1×10^{-14}	Lengyel et al., 1990a
$10^5 \times D_{ClO_2}$[b]	1.5		0.75	Estimated from mobility data
$10^5 \times D_{I^-}$	1.4		0.7	Ruff et al., 1972
$10^5 \times D_{ClO_2}$	1.5		0.75	As ClO_2^-
$10^5 \times D_{I_2}$	1.2		0.6	Estimated values for I$^-$ and I$_3^-$ (Ruff et al., 1972)
$10^5 \times D_{MA}$	0.8		0.4	Estimated from mobility data
D_{starch}	$0\,(< 10^{-7})$		$0\,(< 10^{-7})$	Lengyel et al., 1990a

[a]These rate constants were set to values in the calculations with the six-variable model that ensure they are not rate limiting steps. These values are smaller than anticipated from T-jump relaxation data. Choosing higher values will not modify any calculations.
[b]All diffusion coefficients are in cm^2 s^{-1}.

The chlorite–iodine reaction in a flow reactor at relatively high (10^{-3} M) input concentrations of ClO_2^- and I$^-$ is, of course, a chemical oscillator. However, these oscillations do not occur at the lower concentrations used in the CIMA and CDIMA systems because the key intermediates cannot accumulate to sufficiently high levels.

The differential equations derived from reactions (14.22)–(14.24) contain six variables: the concentrations of H$^+$, MA, I$_2$, ClO$_2$, I$^-$ and ClO_2^-. If we integrate these equations numerically, we obtain oscillations that are remarkably similar to the experimental results, particularly in view of the simplicity of the model (Lengyel et al., 1990b). The numerical integration also reveals that during each cycle of oscillation, four of the concentrations, [H$^+$], [MA], [I$_2$], and [ClO$_2$], change very little, while the other two, [I$^-$] and [ClO_2^-], vary by several orders of magnitude. We shall make use of this observation a bit later on to simplify the model still further.

The model suggests that the malonic acid serves only to generate iodide via reaction (14.22). It should therefore be possible to replace MA by other species that produce iodide from iodine at a similar rate to malonic acid. In fact, oscillatory behavior persists if ethyl acetoacetate is substituted for malonic acid (Lengyel et al., 1990a, 1990b). An alternative test of the model is to delete the malonic acid and supply iodide ion from an external input flow rather than by a chemical reaction. Chlorine dioxide and iodide ion do react with oscillatory kinetics in a flow reactor over a broad range of input concentrations and residence time, even in the absence of malonic acid (Lengyel et al., 1992b). In Figure 14.4, we compare an experimental two-parameter bifurcation diagram in the [I$^-$]$_0$ vs. flow rate parameter plane with the corresponding plot calculated from reactions (14.23) and (14.24). The agreement is excellent.

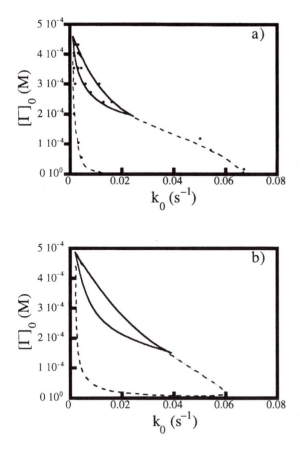

Figure 14.4 (a) Experimental and (b) calculated two-parameter bifurcation diagrams in the $[I^-]_0$ $- k_0$ phase plane for the $ClO_2^- - I^-$ CSTR oscillator. Dashed line encloses a region of oscillation, solid line bounds a region of bistability. $[ClO_2^-]_0 = 1.0 \times 10^{-4}$, $[H_2SO_4] = 5.0 \times 10^{-3}$ M. (Adapted from Lengyel and Epstein, 1995.)

14.4 The Role of Starch

Although the above discussion suggests why the CIMA reaction is an activator–inhibitor system that might be capable of producing some interesting spatiotemporal behavior, it does not seem to say much about the key criterion that must be satisfied for Turing patterns to arise: a large difference in the diffusion constants of the activator and the inhibitor. We recall that in the chlorite–iodide and CIMA reactions, starch is often used as a color indicator to enhance the visibility of the oscillations and spatial structures. In the first open spatial gel reactors developed in Bordeaux and Texas, like that shown in Figure 14.2, polyacrylamide (PAA) gel was used as the reaction medium to prevent convective motion. The starch is introduced into the acrylamide monomer solution before polymerization to the gel because the huge starch molecules cannot diffuse into the gel from the reservoirs. The diffusivity of starch in the gel is effectively zero, and any small molecules, for example, triiodide ion, bound to the starch diffuse at a rate negligible in comparison with their free diffusion. The complex formation, shown schematically in Figure 14.5, may be written as

$$S + I_2 + I^- \leftrightarrow SI_3^- \qquad K = [SI_3^-]/[S][I_2][I^-] \qquad K' = K[S][I_2] \qquad (14.25)$$

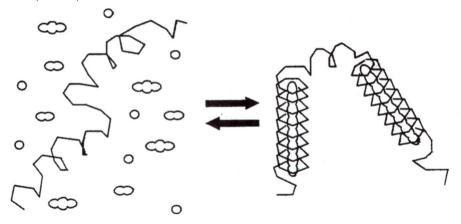

Figure 14.5 Schematic depiction of the formation of the starch–triiodide complex.

where S represents starch, and K' is roughly constant throughout the reaction because, under the conditions of interest,

$$[S] \gg [I_2] \gg [I^-] \tag{14.26}$$

Representing the complex formation by eq. (14.25) is a slight oversimplification of the actual processes involved. The starch–triiodide complex does not have a well-defined stoichiometry. The actual stoichiometry depends on the starch polymer chain length, the iodide and salt concentrations, and the physical state of the complex. Cesaro et al. (1986) proposed a theory of the amylose–iodine–triiodide complex, based on a generalization of a matrix model for the statistical mechanics of the helix–coil phase transition of polypeptides (Zimm and Bragg, 1959). The model describes well many properties of the complex and their dependence on the chain length of the amylose molecules, including the fact that the binding is cooperative, so that additional triiodide ions bind to a complex more rapidly than the first one. The starch–triiodide complex has been known for centuries. The presence of iodide, iodine, and a sufficient amount of water (Murdoch, 1992) is all that is required for the formation of the deep blue complex. The structure has been shown by x-ray diffraction to be a sixfold symmetrical helix (Rundle, 1943). Starch forms helical complexes not only with triiodide but also with many organics, such as butanol or fatty acids, and this property can be used to separate amylose, which forms the helical complex, from other polycarbohydrates (amylopectins), which do not. Despite these complexities, eq. (14.25) is an excellent description of the complex formation when conditions (14.26) hold, so that, in effect, the complex concentration is determined by the level of iodide.

The answer is now at hand. The starch, by binding iodine and iodide molecules, slows their effective diffusivity relative to that of chlorite and chlorine dioxide. It is as if the chlorine-containing inhibitor molecules undergo a random walk through an open field, but the iodine-containing activator molecules must make their way through a field filled with traps (the starch molecules) that capture them for a while before releasing them to continue on their way. The activator thus diffuses more slowly than the inhibitor, and if the starch concentration is high enough, the necessary condition for Turing pattern formation is fulfilled (Lengyel and Epstein, 1991). Although they did not realize it at the time, it was the fortuitous choice of starch as the indicator that led to the success of Castets et al. (1990) in producing the first experimental Turing structures. These workers were not merely lucky, however. While they did not add the starch deliberately to slow the diffusion of the activator, they instantly recognized that they had generated a Turing pattern. As ever, chance favors the prepared mind!

Turing pattern experiments with the CIMA reaction in a capillary reactor in the absence of gel (Agladze et al., 1992) support the idea that complexation by the indicator plays an essential role. Turing patterns occur at high starch concentrations, while at lower starch concentrations only waves are generated. Not only starch but also other organic polymers, such as polyvinylalcohol (PVA) and even polyacrylamide, are capable of binding iodine and iodide. Both PVA (Noszticzius et al., 1992) and polyacrylamide gel of high concentration (Lee et al., 1992) have sufficient capacity to bind iodine and iodide to produce Turing patterns, even in the absence of starch. Because starch is the most widely used color indicator of iodine and triiodide and the structures of other iodine(triiodide)–polymer complexes are similar to that of the starch complex, we have focused our description on starch. The notion of generalizing starch's role as a complexing agent to generate a difference in diffusion rates is an attractive one, which we shall exploit a bit further on.

14.5 A Simpler Model

Historically, it was the CIMA reaction in which Turing patterns were first found. Under the conditions of these experiments, however, our analysis suggests that, after a relatively brief initial period, it is really the CDIMA reaction that governs the formation of the patterns. Even when the input feeds consist of chlorite and iodide, chlorine dioxide and iodine soon build up within the gel and play the role of reactants whose concentrations vary relatively slowly compared with those of ClO_2^- and I^-. We have therefore found it more practical to work with the CDIMA system, using chlorine dioxide and iodine along with malonic acid as the input species, since in this way the relevant parameters can more easily be measured and controlled. Working with the CDIMA system also leads us naturally toward a simpler version of the model described by eqs. (14.22)–(14.24).

Our goal is a model with only two or three variables that will provide insight into the CIMA and CDIMA reactions in much the same way that the Oregonator has illuminated the dynamics of the BZ reaction. We recall that numerical calculations show that during oscillation in the CIMA and CDIMA systems, only two concentrations, $[I^-]$ and $[ClO_2^-]$, vary significantly. Letting $X = I^-$ and $Y = ClO_2^-$, and treating all other concentrations in eqs. (14.22)–(14.24) as constants, we can simplify those equations to reflect the behavior only of the two variable species X and Y:

$$\rightarrow X \qquad r_1 = k_1' \qquad\qquad\qquad k_1' = k_{1a}[MA] \qquad (14.27)$$

$$X \rightarrow Y \qquad r_2 = k_2'[X] \qquad\qquad\qquad k_2' = k_2[ClO_2] \qquad (14.28)$$

$$4X + Y \rightarrow \qquad r_3 = k_3'[X][Y]/(\alpha + [X]^2) \qquad k_3' = k_{3b}[I_2] \qquad (14.29)$$

where we have neglected the first term in the rate law of eq. (14.24), since, under the conditions of interest, it is much smaller than the second term.

The resulting differential equations for $[X]$ and $[Y]$ are

$$d[X]/dt = k_1' - k_2'[X] - 4k_3'[X][Y]/(\alpha + [X]^2) \qquad (14.30)$$

$$d[Y]/dt = k_2'[X] - k_3'[X][Y]/(\alpha + [X]^2) \qquad (14.31)$$

To simplify things still further, we define dimensionless variables $u = [X]/\alpha^{1/2}$, $v = k_3'[Y]/k_2'\alpha$, $\tau = k_2't$, and parameters $a = k_1'/k_3'\alpha^{1/2}$, $b = k_2'/k_3'\alpha^{1/2}$. Equations (14.30) and (14.31) can then be written as

$$du/d\tau = a - u - 4uv/(1 + u^2) \qquad (14.32)$$

$$dv/d\tau = b[u - uv/(1 + u^2)] \qquad (14.33)$$

It is not difficult to show that the unique steady state of this system is given by $u_{ss} = a/5$, $v_{ss} = 1 + a^2/25$. Straightforward linear stability analysis reveals that this state is unstable if

$$b < 3a/5 - 25/a \qquad (14.34)$$

By the Poincaré–Bendixson theorem, the system will have a periodic limit cycle solution when eq. (14.34) holds. This inequality defines a surface in the $[ClO_2]$–$[I_2]$–$[MA]$ plane that separates the regions of stable oscillatory and steady states. The experimental range of oscillation is well described by this equation, except when the initial concentration of ClO_2 is so low that it is nearly consumed in one oscillatory period, violating the assumption that (ClO_2) is a constant.

Our simple model thus describes well the temporal behavior of the CDIMA system. What about the Turing patterns? Obviously, to obtain Turing patterns, we need to add diffusion. If this is all we do, with roughly equal diffusion coefficients for X and Y, no Turing patterns result. What is missing? The difference in diffusion constants between activator and inhibitor is characterized mathematically by eq. (14.19). If we cheat and simply make D_X a factor of 10 or so less than D_Y, we can, in fact, find conditions where the steady state is stable to homogeneous perturbations [eq. (14.34) is violated] and where certain inhomogeneous perturbations cause the system to evolve away from the homogeneous state to a stationary patterned state.

An approach that more accurately reflects the experiments is to take into account the effects of starch by adding eq. (14.25) to our description of the system. We see how this can be done in a more general context in the following section.

14.6 A Design Scheme

We wish to show how, by adding diffusion terms and the complexation of the activator species (iodide in this case) by an immobile agent, such as starch, Turing patterns can be generated from an activator–inhibitor model like eqs. (14.32) and (14.33). We shall analyze a more general version of the problem, because its solution will lead us to a systematic approach to designing new systems that exhibit Turing pattern formation.

We start from a general reaction–diffusion system with activator–inhibitor kinetics in one spatial dimension with coordinate r:

$$\partial x/\partial t = f(x, y) + D_x\, \partial^2 x/\partial r^2 \tag{14.35}$$

$$\partial y/\partial t = g(x, y) + d_y\, \partial^2 y/\partial r^2 \tag{14.36}$$

where x is the activator and y is the inhibitor and the functions f and g specify the kinetics. Let us also suppose that x forms an unreactive and immobile complex, sx, with an immobile complexing agent s, and that this complexation process and the corresponding dissociation occur rapidly compared with the processes that contribute to the rate functions f and g. Let us also assume that s is present in large excess over x, so that the concentration of s is effectively constant.

$$s + x \underset{k_b}{\overset{k_f}{\longleftrightarrow}} sx \qquad K = k_f/k_b \tag{14.37}$$

We must modify eq. (14.35) to take account of the complexation, and we must add a new rate equation for sx. We define $k = k_f s$ and $K' = k/k_b = Ks$. Then,

$$\partial x/\partial t = f(x, y) + D_x\, \partial^2 x/\partial r^2 + k_b(sx) - kx \tag{14.38}$$

$$\partial(sx)/\partial t = -k_b(sx) + kx \tag{14.39}$$

If, as we have assumed, the formation and dissociation of the complex are rapid, we can replace x by $K'(sx)$. This substitution requires a bit more analysis to be mathematically rigorous, but it can be justified using singular perturbation theory (Lengyel and Epstein, 1992). If we now add eqs. (14.38) and (14.39) and eliminate the concentration of the complex, we obtain

$$\partial(x + sx)/\partial t = (1 + K')\, \partial x/\partial t = f(x, y) + D_x\, \partial^2 x/\partial r^2 \tag{14.40}$$

To obtain the final form of our reaction–diffusion equations, we define $\sigma = 1 + K'$, $c = D_y/D_x$, $t' = \sigma t$, and $z = D_x^{-1/2} r$. Equations (14.40) and (14.36) become

$$\partial x/\partial t' = f(x, y) + \partial^2 x/\partial z^2 \tag{14.41}$$

$$\partial y/\partial t' = \sigma[g(x, y) + c\, \partial^2 y/\partial z^2] \tag{14.42}$$

In effect, the complexation separates the time scales for the evolution of the activator and inhibitor by a factor σ, which increases with the stability of the complex and with the concentration of the complexing agent. The equations for the steady states of the system remain unchanged, so the steady states occur at the same composition as in the absence of s. However, *the stability* of the steady state(s) may change as a result of complexation. Typically, if there is a single steady state that loses stability via Hopf bifurcation and becomes oscillatory as some parameter is varied, adding the complexing agent (i.e., increasing the value of σ from 1) broadens the range of stability of the steady state. Although the result may seem counterintuitive, it is *by stabilizing the homogeneous steady state* that the complexation reaction facilitates the formation of Turing patterns. How can this be?

Figure 14.6 shows a bifurcation diagram for the model of the CDIMA reaction given by eqs. (14.32) and (14.33), with c set at a physically reasonable value of 1.5. The Turing bifurcation curve is the dashed line, given by

$$(3ca^2 - 5ab - 125c)^2 = 100abc(25 + a^2) \qquad (14.43)$$

This curve is independent of the value of σ, and below it the homogeneous steady state is unstable to inhomogeneous perturbation. The Hopf curve, below which the stable homogeneous steady state gives way to homogeneous oscillation, is given by

$$\sigma b = 3a/5 - 25/a \qquad (14.44)$$

If we increase σ, the Hopf curve shifts downward proportionally; for example, if we double σ, the Hopf bifurcation occurs at a b value lower than the previous one by a factor of 2. Note that for $\sigma = 1$ (i.e., no complexation), the Hopf curve lies well above the Turing curve. Therefore, if we start from the stable steady state at relatively high b value (large $[ClO_2]/[I_2]$ ratio) and then decrease b, before the Turing bifurcation can occur, the system begins to oscillate; no Turing pattern can be observed because the system undergoes bulk oscillation. As we increase s by adding complexing agent, the Hopf curve moves down. At about $\sigma = 8$, it falls below the Turing curve. We see in Figure 14.6 for $\sigma = 10$ that the Turing bifurcation can occur; Turing patterns arise in the region below the dashed Hopf curve and above the solid $\sigma = 10$ Turing curve.

We are now ready to generalize our analysis to a simple design or search algorithm for new chemical systems that show Turing patterns. We start with a system that oscillates in a CSTR, preferably at relatively low flow rates, since high flow rates are not practical in a gel reactor. All model and experimental systems

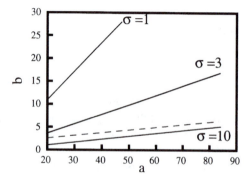

Figure 14.6 Bifurcation curves in a model of the CDIMA reaction in the presence of an immobile complexing agent with $c = 1.5$. Solid line: Hopf bifurcation curve for several values of σ; steady state is stable above the curve, oscillatory state is stable below the curve. Dashed line: Turing bifurcation curve; homogeneous steady state is stable to inhomogeneous perturbations above, unstable below the curve. (Adapted from Lengyel and Epstein, 1992.)

that have been found to exhibit Turing patterns also show oscillatory behavior for some other set of parameters. We choose our oscillatory system so that the activator species is known and is capable of forming unreactive complexes with large, immobile molecules. Ideally, for our analysis to be accurate, the complexation and dissociation reactions should be very rapid compared with the kinetics of the unperturbed chemical oscillator. We run the oscillatory system in a CSTR and find a range of conditions that lead to oscillation. Now we begin to add complexing agent to the input flow. We increase the amount of complexing agent until the oscillations disappear, that is, until the homogeneous steady state is stabilized. We now switch to a gel reactor, incorporating the same concentration of complexing agent into the gel that led to stabilization of the steady state in the CSTR. Finally, we flow into the gel reactor the concentrations of reactants that gave oscillations in the CSTR without complexing agent and a stable steady state in the presence of complexation. If we are fortunate, Turing patterns will appear. If not, we must vary the concentrations of reactants and complexing agent, or perhaps the flow rate, and try again.

Although one might argue that the discovery of Turing patterns was based, at least in part, on the above algorithm, no new Turing systems have been found by this route starting from scratch. To date, the obstacles have been the difficulty of building a suitable gel reactor (at the time of this writing, only three groups in the world possess working gel reactors!) and the challenge of finding a reaction in which an appropriate complexing agent exists for the activator species. We expect these challenges to be met in the very near future.

14.7 Why Are the Patterns Two-Dimensional?

One puzzling but fortuitous aspect of the experiments that resulted in the discovery of Turing patterns in the CIMA reaction, and in the experiments that followed, is the fact that the patterns observed appear to occur in a single plane parallel to the faces of the gel at which the reactor is fed. The patterns are essentially two-dimensional. This observation is puzzling because the gel is typically about 2 mm thick, while the wavelength of the patterns is generally of the order of 0.2 mm. Thus, there would be room for up to ten layers of patterns if the whole gel were susceptible to pattern formation. That only a single layer is found is fortuitous because, in most experiments, one observes the pattern by shining light through the gel along the direction of the concentration gradient (z in Figure 14.2). If there were multiple layers of stripes or spots, their absorption of light would be additive, and, unless each layer was perfectly aligned with all the others, they would add up to a pattern that was indistinguishable from a homogeneous concentration distribution. In fact, model calculations (Borckmans et al., 1992) suggest that if three-dimensional Turing patterns did occur, the spots or stripes would arrange themselves in a lattice, much like the atoms in a crystal, in which the spots or stripes in one layer would be found in the interstices of the next layer. Is it possible to understand why Turing patterns tend to be two- rather than three-dimensional and under what conditions might one obtain more than a single layer of patterns?

Intuitively, we can see the answer immediately. Formation of Turing patterns requires that concentrations of all reactants lie within ranges that allow the system to satisfy a set of conditions: in the case of a two-variable activator–inhibitor system, eqs. (14.5), (14.6), (14.17), (14.19), and (14.21). Because of the way the experiment is done (recall Figure 14.2), each reactant concentration is position-dependent, ranging from its input feed value at the end of the gel where it enters to essentially zero at the other end. Clearly, the conditions for Turing pattern formation can be satisfied only in a portion of the gel, if at all.

Some algebraic manipulation (Lengyel et al., 1992a), reduces the set of conditions for our prototype activator–inhibitor system in the presence of a complexing agent to the pair of inequalities

$$K' > H_1 > H_2 \tag{14.45}$$

where K' is the effective equilibrium constant for complex formation, Ks, defined in the previous section, and in terms of the Jacobian matrix elements defined in eqs. (14.4),

$$H_1 = -a_{11}/a_{22} - 1 \tag{14.46}$$

$$H_2 = -a_{11}/\{2[c(a_{11}a_{22} - a_{12}a_{21})]^{1/2} - ca_{11}\} - 1 \tag{14.47}$$

Because of the geometry of the reaction, H_1 and H_2, which contain the reactant concentrations, will be position-dependent. In the simple case of binary complex formation, $s + x \leftrightarrow sx$, K' is position-independent, since s is assumed to be uniformly distributed throughout the gel. If the complexation reaction involves more species, as in the case of the CIMA and CDIMA reactions, where I_2 is also required for complex formation, K' can be position-dependent as well. In the CIMA and CDIMA systems, K' is proportional to the concentration of iodine, which is one of the feed species, and thus varies with z.

The simplest assumption we can make, and one that turns out to be in good agreement with the results of more elaborate calculations, is that the concentration of each feed reactant varies linearly from its input value at the end where it enters the gel to zero at the opposite end. The assumed concentration-dependence in the CDIMA reaction is shown in Figure 14.7.

Using our model of eqs. (14.27)–(14.29), supplemented with the complexation reaction (14.25), we can derive the position-dependence of the quantities K', H_1, and H_2. These are plotted in Figure 14.8.

The two inequalities (14.45) are satisfied only within the dashed lines of Figure 14.8. Thus, Turing pattern formation can occur in at most about 20% of the thickness of the gel. If the gel were much thicker, perhaps 20–50 times the wavelength of the structures, we would almost surely find multiple layers of patterns. Mechanical problems with thick slabs of gel put a practical upper limit of a few millimeters on the usable thickness of cylindrical slabs of polyacrylamide or agar gel.

Lengyel et al. (1992a) undertook a set of experiments to test predictions of the sort shown in Figure 14.8. By focusing their optical system at different depths in the gel, they were able to estimate, to within about ±0.02 mm, the position of the

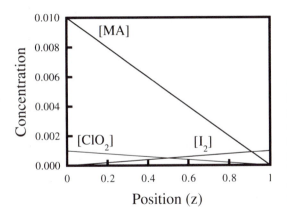
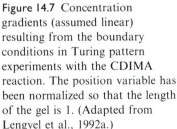

Figure 14.7 Concentration gradients (assumed linear) resulting from the boundary conditions in Turing pattern experiments with the CDIMA reaction. The position variable has been normalized so that the length of the gel is 1. (Adapted from Lengyel et al., 1992a.)

pattern in the gel. They did two series of experiments, one in which they varied the input concentration of malonic acid, the other in which they varied the concentration of starch embedded in the gel. Both of these parameters affect the location of the patterns and the width of the region in which patterns can arise. A comparison of the calculated and observed positions of the structures is shown in Figure 14.9. Figure 14.10 demonstrates the excellent agreement between the patterns calculated from our simple mechanism for the CDIMA–starch system and the patterns actually observed.

Numerical simulations (Borckmans et al., 1992) do predict Turing patterns with interesting three-dimensional structure in sufficiently thick media. These calculations were carried out assuming a uniform distribution of parameters, that is, no gradients resulting from flow into the gel, a condition that can be realized in an open system only in theory. In this case, the entire medium can support Turing pattern formation. However, Figure 14.8 suggests that even a 2-mm gel is close to being able to contain two layers of 0.2-mm wavelength patterns. It is worth considering how one might be able to detect and analyze three-dimensional Turing patterns. Confocal microscopy, a tool rapidly being perfected for biological studies, offers one possibility. A second, used very effectively in a recent study of traveling waves in the BZ reaction (Tzalmona et al., 1990) is nuclear

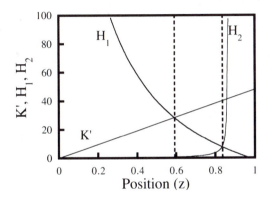

Figure 14.8 Position-dependent functions in Turing pattern experiment on the CDIMA system derived from the concentration profiles of Figure 14.7. $D_{ClO_2} = D_{I_2} = 7 \times 10^{-6}$ cm^2 s^{-1}. $K[S] = [SI_3^-]/[I^-][I_2] = 6 \times 10^4$ M^{-1} at 4 °C. (Adapted from Lengyel et al., 1992a.)

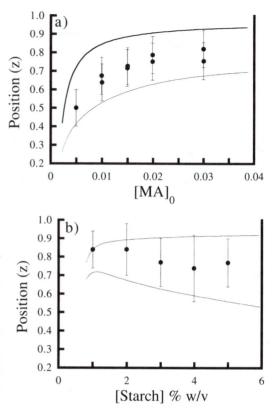

Figure 14.9 Dependence of the position of Turing structures in the CDIMA reaction on (a) input concentration of malonic acid, and (b) starch concentration. $[ClO_2]_0 = 1 \times 10^{-3}$ M, $[I_2]_0 = 8 \times 10^{-3}$ M. $[S] = 3\%$ w/v in (a) and $[MA]_0 = 2 \times 10^{-2}$ M in (b). Lines are calculated limits, solid circles are experimental points. (Adapted from Lengyel, et al., 1992a.)

Figure 14.10 (a) Experimental and (b) calculated Turing patterns in the CDIMA reaction. Bright pixels represent low $[I^-]$. (a) $[ClO_2]_0 = 1 \times 10^{-3}$ M, $[I_2]_0 = 8 \times 10^{-4}$ M, $[MA]_0 = 1 \times 10^{-2}$ M. $[S] = 3\%$ w/v; (b) $[ClO_2]_0 = 1.5 \times 10^{-4}$ M, $[I_2]_0 = 7 \times 10^{-4}$ M, $[MA]_0 = 2 \times 10^{-3}$ M, $K[S] = 1.83 \times 10^{-4}$ M^{-1}. (Reprinted with permission from Lengyel, I.; Kádár, S.; Epstein, I. R. 1992. "Quasi-Two-Dimensional Turing Patterns in an Imposed Gradient," *Phys. Rev. Lett.* **69**, 2729–2732. © 1991 American Institute of Physics.)

magnetic resonance (NMR) imaging. The trick here is to find a nucleus that gives sufficient contrast to make NMR imaging feasible. This has not yet proven possible in the CDIMA reaction, but there is no doubt that in the not too distant future three-dimensional Turing patterns will be found experimentally.

14.8 Turing Patterns in a Dish

If one were to ask what single factor contributed the most to the emerging interest of chemists in nonlinear dynamics during the 1970s and 1980s, a convincing answer would be that it was not any theoretical breakthrough or brilliantly conceived experiment, but rather the visually attractive and thought-provoking lecture demonstrations of temporal oscillation and spatial pattern formation in the BZ reaction. The availability of a system that produces striking color changes with a period of a minute or so and can be set up at room temperature in a beaker or a Petri dish with relatively inexpensive and harmless chemicals was a major factor in creating an awareness of chemical oscillators and reaction–diffusion systems among audiences of all levels of scientific sophistication. For this reason, it seems worth exploring whether it might be possible to design a reaction that would produce Turing patterns under easily accessible conditions without the requirement of complex gel reactors.

Like the temporal and spatial demonstrations involving the BZ reaction, a Turing pattern demonstration would be most convenient if it ran in a closed system, for example, a Petri dish. Using a closed system, of course, means that we are willing to forego the possibility of true, sustained patterns. We are willing to settle for a transient pattern. This limitation to transient behavior proved to be no obstacle to the acceptance of oscillations and pattern formation in the BZ system. So long as the behavior persists for many (about ten or more) cycles of oscillation, or long enough for the patterns to develop and be followed, perhaps a quarter of an hour or so, our goal will have been fulfilled. For Turing patterns, we would like to find a system that will generate patterns that will form and remain stationary for at least 15 min after an induction time not much longer than the lifetime of the patterns. The transient patterns should resemble the true Turing patterns seen in open gel reactors in terms of their shape and their wavelength.

The studies described in the previous section provide a route toward accomplishing this goal, and, with the minor (?) flaw that the demonstration works only at temperatures in the neighborhood of 5 °C, the problem of generating Turing structures in a dish has been solved. In attempting to predict where in a gel reactor Turing patterns could arise, Lengyel et al. (1992a) used the model of eqs. (14.41) and (14.42) for the CDIMA reaction to calculate not only the position-dependent functions H_1, H_2, and K' that determine whether a Turing instability will occur, but also the concentrations of the reactants and intermediates in the region where the patterns can arise. If one wishes to generate transient Turing patterns in a closed system, it would seem logical to start that system with reactant concentrations corresponding to those that prevail in the region of the open system where the patterns emerge. This is exactly the approach that was taken by Lengyel et al.

(1993), who mixed, in a Petri dish at 4 °C (the temperature at which the gel reactor experiments were done), solutions of chlorine dioxide, iodine, malonic acid, and starch at concentrations predicted by the model equations to lead to Turing pattern formation. No gel was used. We note that although this experiment violates the spirit of Turing's calculations by employing a closed rather than an open system, it is more faithful to Turing's approach in another way, since it starts with a spatially uniform distribution of reactant concentrations—that is, unlike the gel reactor experiments, this system has no parameter gradients.

The experiments were remarkably successful. For several sets of concentrations, patterns developed after an induction period of about 25 min and then remained stationary for 10–30 min, eventually giving way to bulk oscillation or a homogeneous steady state, depending on the initial concentrations. During the induction period, spots or stripes form spontaneously and move around until they reach the pattern that remains stationary. The patterns observed consist of mixed spots and stripes or networklike structures, as illustrated in Figure 14.11. They have wavelengths of a few tenths of a millimeter, like the patterns found in gel reactors. The relatively brief induction period, which is 4–10 times shorter than the time required for stationary structures to form in gel reactors, arises from the fact that, in the closed system configuration, it is not necessary to wait for a stationary concentration gradient to become established via the flow into the reactor and diffusion.

Computer simulations (Kádár et al., 1995) of the evolution of transient Turing patterns have been carried out using the six-variable model of the CDIMA system given by eqs. (14.22)–(14.25). Figure 14.12 shows a one-dimensional simulation that illustrates how the initial concentration of malonic acid influences the beha-

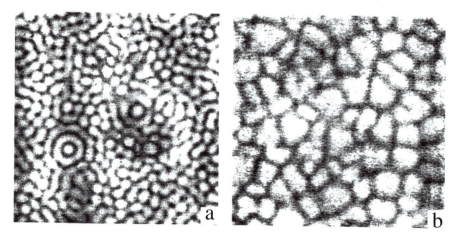

Figure 14.11 Transient Turing patterns in the CDIMA reaction in a Petri dish. (a) Hexagonal pattern of (mostly) spots and (a few) stripes, (b) networklike structure. $[ClO_2]_0 = 5 \times 10^{-4}$ M, $[I_2]_0 = 8 \times 10^{-4}$ M, $[MA]_0 = 1 \times 10^{-3}$ M in (a) and 3×10^{-3} M in (b), [starch] = 1 g per 100 mL, [acetic acid] = 10%, $T = 4$ °C. (Reprinted with permission from Lengyel, I.; Kádár, S.; Epstein, I. R. 1993. "Transient Turing Structures in a Gradient-Free Closed System," *Science 259*, 493–495. © 1993 American Association for the Advancement of Science.)

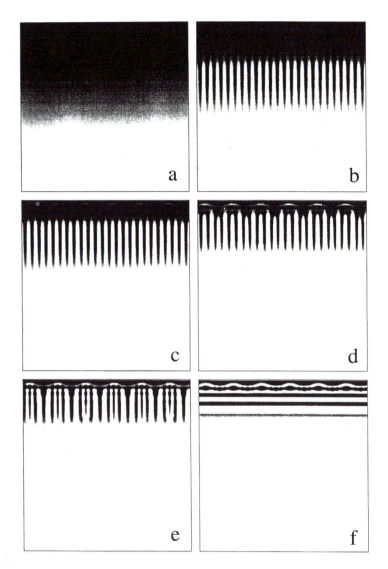

Figure 14.12 Simulated time development of closed CDIMA reaction in one spatial dimension as a function of initial malonic acid concentration. $[ClO_2]_0 = 6 \times 10^{-3}$ M, $[I_2]_0 = 8.2 \times 10^{-3}$ M, $[H^+] = 10^{-3}$ M. $[MA]_0 = 2 \times 10^{-3}$ M (a), 2.5×10^{-3} M (b), 3×10^{-3} M (c), 4×10^{-3} M (d), 5×10^{-3} M (e), 7×10^{-3} M (f). Horizontal coordinate is space; each frame is 5 mm. Vertical coordinate is time; each frame is 2000 s. (Reprinted with permission from Kádár, S.; Lengyel, I.; Epstein, I. R. 1995. "Modeling of Transient Turing-Type Patterns in the Closed Chlorine–Dioxide–Iodine–Malonic Acid–Starch Reaction System," *J. Phys. Chem. 99*, 4054–4058. © 1995 American Chemical Society.)

vior of the system. As we increase $(MA)_0$, we move from (a) no pattern formation to (b–e) pattern formation that gives way to a homogeneous steady state to (f) pattern formation that persists only very briefly before evolving to homogeneous bulk oscillation. Note the development of more complex spatial structure in (d) and (e) as the system approaches the condition where the Hopf and Turing bifurcations collide.

Two-dimensional simulations, shown in Figure 14.13, give a clearer picture of how the structures evolve and emphasize the resemblance between the computed and experimental behaviors.

The success of these efforts bodes well for the goal of developing a simple, easily shown demonstration of Turing patterns. For the moment, because of the temperature-dependence of the CDIMA system, one can only view the demonstration in a cold room or outdoors during the depths of winter. However, once other reactions that show Turing patterns are developed, the approach described here should make it a straightforward task to convert at least some of them into more convenient vehicles for displaying the realization of Alan Turing's vision.

14.9 Future Directions

Especially because of the wide range of applicability of ideas behind Turing pattern formation, we may expect rapid developments in this area during the coming years. Here, we suggest a few directions that seem particularly promising.

Our discussion of experimental results has basically been limited to one system, the CIMA reaction in a gel reactor, and its variants. We expect a considerable expansion in the repertoire of systems that exhibit Turing patterns. Turing structures have already been found experimentally (Falta et al., 1990) and modeled (Imbihl et al., 1991) in the catalytic oxidation of carbon monoxide on single crystals of platinum. It seems likely that other heterogeneous systems will exhibit Turing patterns as well, though the elaborate high-vacuum technology required for these experiments may delay the discovery of new systems. Less expensive alternatives for chemical media that may support Turing pattern formation include membranes (Winston et al., 1991) and sol–gel glasses (Epstein et al., 1992). An even larger potential source of Turing patterns lies in the biological world. Here, the issue is one of identifying and monitoring the activator and inhibitor. One very promising development has been the rapid advance in techniques for imaging calcium concentrations with dyes that are sensitive to (Ca^{2+}). Recent experiments, for example, have revealed spiral waves of calcium concentration moving across the surface of a developing frog oocyte (Lechleiter et al., 1991). As methods evolve for obtaining spatially resolved images of other metabolically important species, we can expect experiments to reveal the existence of Turing-like patterns.

We have already alluded to three other expected developments: the design, by appropriate complexation of the activator species, of new reactions that give rise to Turing patterns; the development of systems that show transient patterns under conveniently obtainable conditions; and the generation and analysis of three-

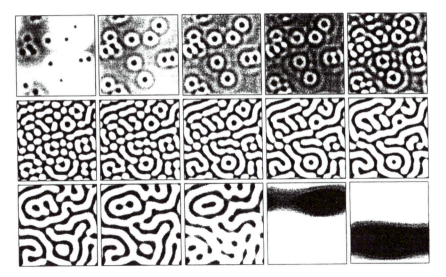

Figure 14.13 Simulated time development of closed CDIMA reaction in two spatial dimensions. $[ClO_2]_0 = 6 \times 10^{-3}$ M, $[I_2]_0 = 8.2 \times 10^{-3}$ M, $[H^+] = 10^{-3}$ M. $[MA]_0 = 3.5 \times 10^{-3}$ M. Horizontal coordinate is space; each frame is 5 mm.

dimensional Turing patterns by microscopy or magnetic resonance imaging techniques.

Finally, we expect a considerable expansion in knowledge about more "exotic" forms of behavior in reaction–diffusion systems. Turing patterns are, in a sense, only the tip of the iceberg. When a Turing bifurcation occurs in the neighborhood of another bifurcation, such as a Hopf bifurcation, new phenomena can arise. Figure 14.14 shows a "chemical flip-flop," (Perraud et al., 1992) in which the interaction between a Turing and a Hopf bifurcation in the CIMA reaction leads to formation of a stationary droplet that periodically sends out traveling waves, first to the left and then to the right.

Recent model calculations (Pearson, 1993), supported by experiments (Lee et al., 1993) on the EOE reaction in an open unstirred reactor, show that even relatively simple systems are capable of exhibiting an enormous variety of steady and time-dependent, periodic and aperiodic, spatiotemporal structures. To date, it is not clear whether all of the possibilities are known. The task of cataloging them, and of characterizing the bifurcations that may occur as one varies the constraints on a system, will provide exciting work for chemists and mathematicians for many years.

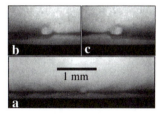

Figure 14.14 A "chemical flip-flop." (a) Snapshot of wave source of the surrounding wave trains. (b) and (c) Snapshots of the wave source taken at 16-s intervals. Notice that they are mirror images. Waves are produced antisymmetrically. (Courtesy of P. De Kepper.)

15

Stirring and Mixing Effects

In almost everything that we have discussed so far, we have assumed, explicitly or implicitly, either that the systems we are looking at are perfectly mixed or that they are not mixed at all. In the former case, concentrations are the same everywhere in the system, so that ordinary differential equations for the evolution of the concentrations in time provide an appropriate description for the system. There are no spatial variables; in terms of geometry, the system is effectively zero-dimensional. At the other extreme, we have unstirred systems. Here, concentrations can vary throughout the system, position is a key independent variable, and diffusion plays an essential role, leading to the development of waves and patterns. Geometrically, the system is three-dimensional, though for mathematical convenience, or because one length is very different from the other two, we may be able to approximate it as one- or two-dimensional.

In reality, we hardly ever find either extreme—that of perfect mixing or that of pure, unmixed diffusion. In the laboratory, where experiments in beakers or CSTRs are typically stirred at hundreds of revolutions per minute, we shall see that there is overwhelming evidence that, even if efforts are made to improve the mixing efficiency, significant concentration gradients arise and persist. Increasing the stirring rate helps somewhat, but beyond about 2000 rpm, cavitation (the formation of stirring-induced bubbles in the solution) begins to set in. Even close to this limit, mixing is not perfect. In unstirred aqueous systems, as we have seen in Chapter 9, it is difficult to avoid convective mixing. Preventing small amounts of mechanically induced mixing requires considerable effort in isolating the system from external vibrations, even those caused by the occasional truck making a delivery to the laboratory stockroom. It is possible to suppress the effects of convection and mechanical motion in highly viscous media, such as the gels used in the experiments on Turing patterns as discussed in the previous chapter. There, we can finally study a pure reaction–diffusion system.

Systems in nature—the oceans, the atmosphere, a living cell—are important examples in which chemical reactions with nonlinear kinetics occur under conditions of imperfect mixing. Industrial reactors, in which enormous quantities of chemicals worth billions of dollars—fibers, fertilizers, pharmaceuticals—are produced every year, provide additional instances in which we need to comprehend how partial mixing influences the outcome of a chemical reaction. Our present understanding of such systems is limited, but much progress has been made in the past decade and a half. In this chapter, we will present some of the key ideas and most significant experimental results on the effects of stirring and mixing on nonlinear chemical dynamics.

How important stirring and mixing effects are in any given system depends on the nonlinearity of that system. The more nonlinear the kinetics, the greater the consequences of changes in the quality of mixing. In Table 15.1, we consider some simple examples of reactions governed by different rate laws and with different spatial distributions of the reactant x. We fix the spatially averaged concentration of x at 1 and consider three cases: "x constant" (i.e., $x = 1$ throughout the system); "smooth gradient" (i.e., x increases linearly from 0 to 2 across the system); and "step function" (i.e., $x = 0$ in the left half and $x = 2$ in the right half). We then calculate the average reaction rate for each of the five rate laws and three spatial distributions. All the rate laws are normalized so that the average rate is unity in the "constant x" case.

For linear kinetics, $dx/dt = x$, the average rate is independent of the distribution of the reactant; the average rate is affected only by the average concentration. For all of the other rate laws, and for *any* nonlinear rate law, the average rate does change with the spatial distribution of the reactants. If we move reactant from one part of the vessel to another, the increased rate in one region is not exactly balanced (as it is in the case of linear kinetics) by the decreased rate in the other region. For example, in the step function case with $dx/dt = x^2$, the rate increases to 4 in half the vessel and drops to zero in the other half. The average, however, doubles, going from 1 to 2, because with the nonlinear rate law the decrease in the depleted region is more than made up for by the increase in the enriched region.

Table 15.1 Effect of Kinetic Nonlinearity and Spatial Distribution of Reactant on Average Rate of a Chemical Reaction

Rate Law, dx/dt	x Constant	Smooth Gradient	Step Function
x	1	1	1
x^2	1	1.33	2
x^3	1	2	4
$e(x - 1)$	1	1.18	1.54
$e(1 - 1/x)$	1	0.89	0.821

As the examples of the power law rates $(dx/dt = x, x^2, x^3)$ illustrate, the more nonlinear the rate law, the greater the difference between the actual average rate and that calculated from the spatially averaged concentration (i.e., by assuming perfect mixing). Nonlinear kinetics will amplify departures from spatial homogeneity. Note that the inverse exponential rate law $dx/dt = \exp(1 - 1/x)$ is the only case in which the spatially homogeneous system has a *higher* average reaction rate than a system with spatial inhomogeneities. This rate law, with appropriate rescaling, is equivalent to the Arrhenius temperature dependence, $A \cdot \exp(-E_a/RT)$, of chemical rate constants. The role of the "concentration" x is now played by the temperature T.

What Table 15.1 demonstrates is that, unless the kinetics of a reaction are extremely simple, the reaction rates calculated by assuming spatial homogeneity (i.e., by using average concentrations) will be in error if the concentrations in an experiment are not uniform. This observation casts suspicion on many of the results of kinetics experiments now in the literature, or at least on the accuracy of their error estimates. Of course, the spatial distributions used in Table 15.1 are extreme cases, chosen to exaggerate the effects of imperfect mixing. Nevertheless, real systems show measurable and occasionally remarkable effects as a result of imperfect mixing, even when great pains are taken to ensure that mixing is complete.

15.1 Mixing

A complete, or even just a convincing, description of the mixing process is well beyond the scope of this book. To understand mixing in detail has long been a goal of hydrodynamicists and chemical engineers, among others. They have learned a great deal, some of which we attempt to summarize here, but there is still a long way to go. A real understanding of the mixing process would be a valuable thing to have, since so much important chemistry occurs in imperfectly mixed systems. Rather than pursue the quest for the elusive (perhaps unattainable) grail of perfect mixing, it may be more useful to recognize the extent of imperfection that exists in systems of interest and to assess the consequences that such imperfection brings. In the prototype BZ reaction, for example, more rapid stirring will tend to enhance the rate at which bromine gas generated by the reaction leaves the solution, but it will increase the rate at which oxygen, which can react witl some of the organic intermediates, is taken up by the reaction mixture. Since both Br_2 and O_2 have significant effects on the chemistry of the BZ reaction, one may expect that changes in the mixing efficiency will affect the oscillatory behavior. More generally, we may anticipate large mixing effects in any reaction that involves more than one phase. If a solid surface is involved, the mixing efficiency will determine the rate at which fresh material from the solution phase reaches the surface. However, even if there is only a single phase, mixing can play a major role in determining the behavior of the system.

Chemical engineers have devoted much attention to the issue of mixing in flow reactors because of the industrial importance of reactions that occur in CSTRs.

They have learned a great deal, and readers seeking a more thorough treatment than we are about to provide should seek it in the chemical engineering literature (Curl, 1963; Villermaux, 1983; Zwietering, 1984; Ulbrecht and Pearson, 1985). Here, we give a brief summary of some of the most important features of the mixing process in a CSTR.

To assess the effects of mixing (Villermaux, 1991), one must consider a number of simultaneously occurring processes and the coupling between them. These processes include the reactants coming into contact with each other, the hydro-dynamics of mixing of the various fluid streams, the mass and heat transfer within the reactor, and the kinetics of the chemical reactions. It is useful to think in terms of two sorts of mixing, *macromixing* and *micromixing*, which occur on two different time scales. In the faster process, macromixing, lumps of fluid having different compositions and/or ages (time spent in the reactor) come together and become mixed in such a way that the system becomes homogeneous on a macroscopic scale. After macromixing, a sample of macroscopic dimensions, having a volume of a significant fraction of a cubic centimeter or more, will have the same average composition as another sample taken from elsewhere in the reactor. The characteristic time for macromixing can be estimated by experiments with tracers or with differently colored unreactive feedstreams. Micromixing, which generally requires significantly more time than macromixing, is the process whereby the fluid becomes mixed on a *molecular* scale, so that in a solution consisting of *A* and *B*, for example, any *A* molecule has the same probability of being next to a *B* molecule as does any other *A* molecule. Note that this situation does not necessarily hold for a fluid that is homogeneous on the macroscopic scale.

The rate of macromixing is largely controlled by the average velocity of fluid in the reactor, which can be enhanced not only by more rapid stirring, but also by clever reactor design, for example, by introducing baffles and by using an appropriately shaped stirrer. If turbulent flow can be achieved, energy from the stirrer is dissipated relatively efficiently to large and then to small eddies, aiding the mixing process. Micromixing occurs through molecular diffusion. It is enhanced by turbulent flow and by increased power dissipation within the fluid. It is not well understood, but its effectiveness plays a key role in determining the efficiency and product distribution of many industrial processes. Model calculations (Puhl and Nicolis, 1987) suggest that changing the characteristic time for micromixing can cause a system to undergo a transition from one steady state to another. These calculations also offer an explanation for the observation that in a number of systems, very different results are obtained if reactants are *premixed* in a small mixing chamber before entering the CSTR.

One of the clearest demonstrations of how difficult it is to obtain perfect mixing in a CSTR is found in the studies of Menzinger and Dutt (1990) on the chlorite–iodide reaction. These authors placed a platinum microelectrode at different distances from the central stirring rod in their cylindrical flow reactor. In one experiment, at a stirring rate of 880 rpm, consistent with many "well-stirred" CSTR experiments, they observed a difference in electrode potential of about 50 mV when the electrode was moved radially from a position at the axis to a position 12 mm away in the radial direction. This difference corresponds to a factor of roughly 8 in the iodide concentration! Even at a stirring rate of 2820 rpm, far above the rates normally employed in CSTRs, and very near the threshold for cavitation, $[I^-]$ differed by 17% between the axis of the reactor and the tip of the stirrer. As we see in Figure 15.1, increasing the stirring rate does make the iodide concentration more homogeneous and does lower the amplitude of fluctuations in the electrode potential, but despite the experimenters' best efforts, a significant inhomogeneity remains.

A more striking demonstration of the effects of stirring in this system can be seen in Figure 15.2, which shows absorption spectra measured at several positions in a CSTR containing a chlorite–iodide mixture that has reached a steady state. We see that at 340 rpm, the spectra differ significantly at the two ends of the reactor ($x = \pm 10$ mm) particularly at 288 and 352 nm, where the triiodide ion has its peak absorption. At a stirring rate of 1100 rpm, the spectra become nearly, though not quite, independent of position.

15.2 Examples

The effects of imperfect mixing on chemical systems make themselves felt in a variety of ways, depending on the kinetics of the reactions involved and on the physical configuration of the system. We consider now a few examples of these effects, ranging from experiments in which tuning the mixing by varying the stirring rate can be used to sweep through the dynamical states of a system to studies in which imperfectly mixed systems generate chaotic or even stochastic behavior.

15.2.1 Mixing Quality as a Bifurcation Parameter

The studies of Menzinger and coworkers described above were inspired by a pioneering set of experiments by Roux, De Kepper, and Boissonade (Roux et al., 1983a) on the chlorite–iodide reaction in a flow reactor, in which they measured, at several different stirring rates, the hysteresis curve as a function of flow

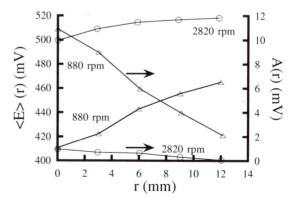

Figure 15.1 Dependence of the mean electrode potential $\langle E(r) \rangle$ and of the average fluctuation amplitude $A(r)$ on r, the distance from the stirrer axis in the chlorite–iodide reaction without premixing. $[I^-]_0 = 2.96 \times 10^{-4}$ M, $[H_2SO_4]_0 = 1.36 \times 10^{-3}$ M, $[Na_2SO_4]_0 = 6.8 \times 10^{-3}$ M, $[ClO_2^-]_0 = 1.15 \times 10^{-4}$ M, $k_0 = 1.57 \times 10^{-2}$ s^{-1}, T = 20 ± 0.5 °C. (Adapted from Menzinger and Dutt, 1990.)

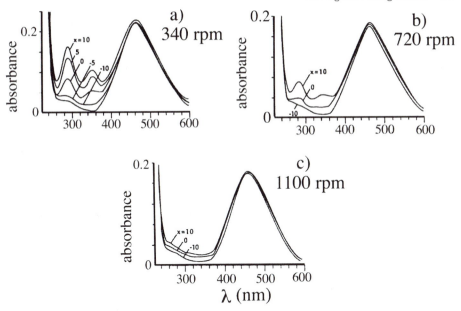

Figure 15.2 Spatial dependence of absorption spectrum at different positions in a CSTR. x indicates distance in mm between the light beam (diameter 1.5 mm) and the stirrer axis. pH $= 1.75$, $[I^-]_0 = 8.9 \times 10^{-4}$ M, $[ClO_2^-]_0 = 3.0 \times 10^{-4}$ M, $k_0 = 5.0 \times 10^{-2}$ s^{-1} S = stirring rate in rpm. (Adapted from Ochiai and Menzinger, 1990.)

rate in the bistable region. Typical results are shown in Figure 15.3. Note that changing the stirring rate has almost no effect on the transition from the flow branch to the thermodynamic branch, but increasing the rate of stirring shifts the flow rate for transition from the thermodynamic to the flow branch to higher k_0. Thus, better stirring stabilizes the thermodynamic branch. Although the bistability and hysteresis in this system had previously been characterized, both experimentally and theoretically, it came as a shock to most investigators that the critical flow rate for transition from the upper to the lower branch of steady states was strongly dependent on the stirring rate, even at relatively rapid rates. The fact

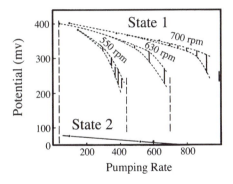

Figure 15.3 Effects of stirring rate on bistability in the chlorite–iodide reaction. Dashed line indicates steady-state potential in state I, solid line shows steady-state potential in state II. Note how the bistable region shifts and broadens as the stirring rate is increased from 550 to 700 rpm. (Adapted from Roux et al., 1983a.)

that only one of the two transitions was sensitive to stirring also cried out for explanation.

Luo and Epstein (1986) later studied the same reaction over a wider range of input concentrations, and demonstrated conclusively that the stirring rate can be treated as a bifurcation parameter, like the flow rate or the input concentration of a reactant, which shifts the dynamical behavior of the system among its steady and oscillatory states. They carried out a series of experiments at several different flow rates in the steady state, bistable, and oscillatory regions of the concentration–flow rate parameter space. They also investigated the effects of premixing the input streams in a small chamber, about one-twentieth the volume of the main reactor, before the reagents entered the CSTR. The phase portraits derived from their experiments are shown schematically in Figure 15.4. Premixing or high stirring rates tend to stabilize the steady state(s) at the expense of the oscillatory state.

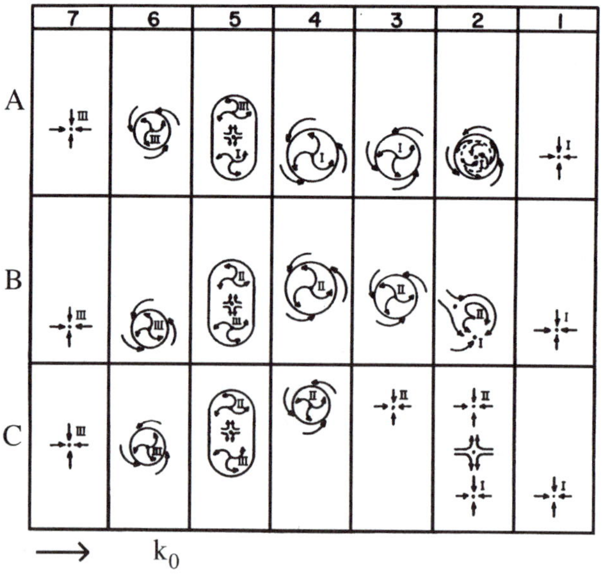

Figure 15.4 Phase portraits of the chlorite–iodide system with pH = 1.56, $[I^-]_0 = 1.8 \times 10^{-3}$ M, $[ClO_2^-]_0 = 5.0 \times 10^{-4}$ M. Points with Roman numerals are steady states, closed curves are limit cycles. Arrows indicate how the system evolves in concentration space. Row A shows evolution of the system without premixing at low stirring rate (< 550 rpm) as flow rate increases; Row B shows intermediate stirring rate without premixing; Row C show high stirring rate or premixing. (Reprinted with permission from Luo, Y.; Epstein, I. R. "Stirring and Premixing Effects in the Oscillatory Chlorite–Iodide Reaction," *J. Chem. Phys. 85*, 5733–5740. © 1991 American Institute of Physics.)

15.2.2 Modeling Mixing Effects

As we have stated, mixing is an extremely complicated phenomenon, but so is the BZ reaction, and a relatively simple model, the Oregonator, has provided a great deal of insight into that system. Is it possible to construct a simple model that will tell us something about the nature of mixing effects in chemical systems with interesting nonlinear kinetics? The model shown in Figure 15.5 was proposed by Kumpinsky and Epstein (1985) to provide a qualitative understanding of some of the stirring effects seen in a CSTR.

Inevitably, some regions of any CSTR—for example, those closest to the stirrer—are better mixed than others. As the stirring rate increases, the relative size of the poorly mixed volume shrinks, though it never reaches zero. As a first approximation, we represent the actual reactor by two coupled reactors: one, the well-mixed region with volume V_a, in contact with the input and output flows; the other, the poorly mixed region with volume V_d, linked only to the well-mixed region but not to the external flows. The parameters Q_x and V_a/V_d, the cross flow and volume ratio, respectively, between the two regions, are used to characterize the quality of the mixing, which improves as the stirring rate increases. The system is then described by the equations

$$dC_{i,a}/dt = k_a(C_{i,0} - C_{i,a}) + [x/(1-x)]k_d(C_{i,d} - C_{i,a})$$
$$+ \sum_{j=1}^{m} \alpha_{ij}R_{j,a} \qquad i = 1, 2, \ldots, n \qquad (15.1)$$

in the active zone and

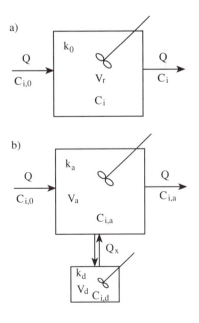

a)

b)

Figure 15.5 Two-reactor model of imperfect mixing in a CSTR. (a) Idealized CSTR in which mixing is perfect and concentrations are homogeneous throughout the reactor. (b) Stirred tank with cross flow between active (well-mixed) and dead (poorly mixed) zones. Q represents volume flow, C is concentration, k is flow rate (reciprocal residence time), and V is volume. Subscripts: i specifies chemical species, 0 signifies input from reservoirs, r is homogeneous reactor, and a and d are active and dead zones, respectively. (Adapted from Kumpinsky and Epstein, 1985.)

$$dC_{i,d}/dt = k_d(C_{i,a} - C_{i,d}) + \sum_{j=1}^{m} \alpha_{ij} R_{j,d} \qquad i = 1, 2, \ldots, n \qquad (15.2)$$

in the dead zone. Here, the n chemical species are denoted by the subscript i and the m reactions are denoted by the subscript j. Subscripts a, d, and 0 refer to the active and dead zones and the input flow, respectively. Species concentrations are represented by C and rates of mass transfer by k. The volume ratio x is given by $V_d/(V_d + V_a)$, and $k_a = k_0/(1 - x)$, where $k_0 = Q/(V_d + V_a)$ is the total flow rate through the reactor. The rates of the chemical reactions in the two zones are represented by R, and the reaction stoichiometries are specified by the coefficients α_{ij}. We are thus able to describe approximately the effects of mixing on the imperfectly stirred system with $2n$ ordinary differential equations with the introduction of two new free parameters, which may be taken to be the volume ratio x and the ratio of residence times in the active and dead zones, $z = k_d/k_a$.

This crude model, which may be thought of as the first in a hierarchy of models that characterize the mixing process in increasing detail, gives surprisingly good agreement with the experimental results, not only in the chlorite–iodide reaction, but also in systems, such as the minimal bromate oscillator, whose chemistry causes the mixing quality to affect primarily the high flow rate rather than the low flow rate branch. Analysis suggests that the ratio of the transition time between states to the residence time in the reactor is the key factor that determines how changing the degree of mixing will influence the stability of the two bistable branches of steady states. If the transition time (induction period) is relatively short, then poorly mixed regions will be found on the thermodynamic (low-flow-rate) branch, that is, with concentrations similar to those of the completed reaction. Decreasing the stirring rate will tend to stabilize this state at the expense of the high-flow-rate branch. If, in contrast, the induction period is relatively long, then isolated regions will tend to be found in flow branch states, whose concentrations resemble those of the unreacted input flow. Material will be "washed out" of the reactor before it has a chance to react to any significant degree. Decreasing the stirring rate under these conditions will tend to decrease the range of stability of the low-flow-rate branch.

15.2.3 Chaos and Mixing

The two-reactor model shown in Figure 15.5 played a major role in a recent controversy about the origins of chemical chaos in the Belousov–Zhabotinsky (BZ) reaction. In Chapter 8, we discussed the fact that chaotic behavior was found experimentally, under a variety of conditions, by several groups in the 1970s and early 1980s. A decade later, however, no chemically realistic model of the BZ reaction had been shown to generate chaos under conditions similar to those of the experimental chaotic regime, and some investigators were questioning whether the chaotic behavior observed in the BZ and other chemical systems was truly a consequence of the homogeneous chemistry. Faint echoes of earlier skepticism about the legitimacy of periodic chemical oscillation could be detected. In 1989, Györgyi and Field demonstrated that if a four-variable partially reversible Oregonator-type model for the BZ chemistry was incorporated into a two-reactor model for the mixing, chaos was easily obtained. They then argued that the experimentally observed chaos might arise, not from the inherent (homogeneous) chemical kinetics of the BZ system, but from the coupling of those nonlinear rate laws with the imperfect mixing in the experiments.

The Györgyi–Field proposal called into question the source of chaotic behavior in the best characterized chaotic chemical system, the BZ reaction. After several months of occasionally heated discussion with experimentalists, Györgyi and Field (1992) developed several chemically plausible variants of the Oregonator model that produce chaotic behavior with perfect mixing under conditions resembling those in the experiments. Their models are derived from an initial eleven-variable Mass Action model shown in Table 15.2. The four-variable versions contain the three variables—Br⁻, BrO₂, and Ce(IV)—of the standard Oregonator, supplemented with the bromomalonic acid concentration as a fourth variable. Because of the steady-state approximations used for such species as cerous ion, malonyl radical, and bromine dioxide, these models contain some nonpolynomial (non-Mass Action) terms, but they are nonetheless chemically based. It is even possible to simplify further to obtain a still chaotic, but somewhat less faithful, three-variable model.

While it remains extremely difficult to pinpoint the origin of chaos in any particular experiment, it now appears that chemical chaos in the BZ reaction can result *either* from the homogeneous chemistry *or* from perturbation of that chemistry under periodic oscillating conditions by the effects of imperfect mixing.

Table 15.2 The Eleven-Variable Györgyi–Field Model for the BZ Reaction[a]

Number	Reaction	Rate Constant	
A1	$HOBr + Br^- + \{H^+\} \to Br_2 + \{H_2O\}$	6.0×10^8	$M^{-1} s^{-1}$
A2	$Br_2 + \{H_2O\} \to HOBr + Br^- + \{H^+\}$	2.0	s^{-1}
A3	$Br^- + HBrO_2 + \{H^+\} \to 2HOBr$	5.2×10^5	$M^{-1} s^{-1}$
A4	$Br^- + \{BrO_3^-\} + \{2H^+\} \to HOBr + HBrO_2$	0.01352	s^{-1}
A5	$HOBr + HBrO_2 \to Br^- + \{BrO_3^-\} + \{2H^+\}$	3.2	$M^{-1} s^{-1}$
A6	$2HBrO_2 \to \{BrO_3^-\} + HOBr + \{H^+\}$	3.0×10^3	$M^{-1} s^{-1}$
A7	$\{BrO_3^-\} + HBrO_2 + \{H^+\} \to 2BrO_2 + \{H_2O\}$	0.858	s^{-1}
A8	$2BrO_2 + \{H_2O\} \to \{BrO_3^-\} + HBrO_2 + \{H^+\}$	4.2×10^7	$M^{-1} s^{-1}$
A9	$Ce(III) + BrO_2 + \{H^+\} \to HBrO_2 + Ce(IV)$	1.612×10^4	$M^{-1} s^{-1}$
A10	$HBrO_2 + Ce(IV) \to Ce(III) + BrO_2 + \{H^+\}$	7.0×10^3	$M^{-1} s^{-1}$
A11	$MA + Br_2 \to BrMA + Br^- + \{H^+\}$	40.0	$M^{-1} s^{-1}$
A12	$MA + HOBr \to BrMA + \{H_2O\}$	8.2	$M^{-1} s^{-1}$
A13	$MA \cdot + Ce(IV) \to MA \cdot + Ce(III) + \{H^+\}$	0.3	$M^{-1} s^{-1}$
A14	$BrMA + Ce(IV) \to Ce(III) + Br^- + \{products\}$	30.0	$M^{-1} s^{-1}$
A15	$MA \cdot + BrMA \to MA + Br^- + \{products\}$	2.4×10^4	$M^{-1} s^{-1}$
A16	$MA \cdot + Br_2 \to BrMA + Br \cdot$	1.5×10^8	$M^{-1} s^{-1}$
A17	$MA \cdot + HOBr \to Br \cdot + \{products\}$	1.0×10^7	$M^{-1} s^{-1}$
A18	$2MA \cdot \to MA + \{products\}$	3.0×10^9	$M^{-1} s^{-1}$
A19	$Br \cdot + MA \cdot \to Br^- + MA \cdot + \{products\}$	1.0×10^5	$M^{-1} s^{-1}$

[a] Species in {} are assumed to be at fixed concentrations, which are incorporated into the rate constants. $[H_2O] = 55$ M, $(BrO_3^-) = 0.1$ M, $(H^+) = 0.26$ M. MA $\equiv CH_2(COOH)_2$, MA $\cdot \equiv \cdot CH(COOH)_2$, BrMA $\cdot \equiv BrCH(COOH)_2$.

15.2.4 Chiral Symmetry Breaking

Questions about origins, whether of the universe, of the Earth, or of life, have always held great fascination. One intriguing problem of this type is to explain chiral symmetry-breaking in nature—how, for example, it came to be that the optically active amino acids in living systems are nearly all levo- rather than dextrorotatory. In an experiment that rivals Pasteur's work on optical activity for its simplicity and elegance, Kondepudi et al. (1990) obtained some remarkable data on the distribution of optical activity among crystals obtained from a super-saturated solution of sodium chlorate under different mixing conditions. The handedness of the crystals was established visually, using crossed polarizers.

The results are summarized in Figure 15.6. When the solution was unstirred, of 1000 crystals collected from 17 different crystallizations, 525 were of levo and 475 of dextro chirality; that is, the total numbers were statistically equal, as were the numbers in each individual experiment, as seen in Figures 15.6a and 15.6c. When the solution was stirred at 100 rpm during the crystallization process, the total numbers for 11,829 crystals were consistent with a 50/50 distribution of L- and D-crystals. However, with only one exception, each of the 32 individual crystalliza-

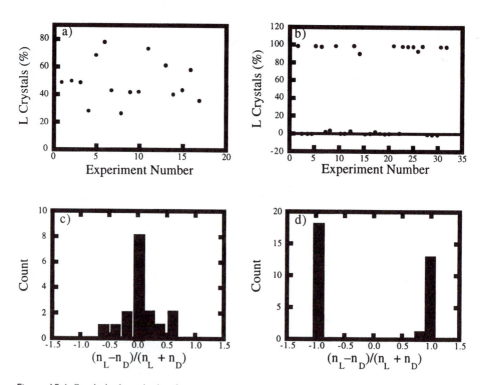

Figure 15.6 Statistical analysis of crystallization from stirred and unstirred sodium chlorate solutions. Upper: Scatter plots of the percentage of L-crystals for (a) unstirred and (b) stirred experiments. Each point represents a single experiment. Lower: Enantiomeric excess of L- over D-crystals in (c) unstirred and (d) stirred experiments. (Adapted from Kondepudi et al., 1990.)

tions (Figures 15.6b and 15.6d) gave rise to a set of crystals that was more than 99% levo- or 99% dextrorotatory! Even relatively weak stirring somehow tips the balance in any given beaker all the way to one handedness or the other.

The authors attribute the observed selectivity to an autocatalytic effect arising from the rapid production of secondary nuclei from a single primary nucleus in the stirred system. In essence, they propose that the first crystal to form will be either an L-crystal or D-crystal, each with 50% probability. When the solution is stirred, this ur-crystal (i.e., original crystal) serves as the template or the source for the formation of other crystals from solute brought to it by the stirring-induced transport in the supersaturated solution. In the absence of stirring, crystals form independently—there is no communication between crystals already formed and the rest of the solution, except for that provided by the relatively slow process of diffusion.

In a recent experiment, Kondepudi and Sabanayagam (1994) followed the crystallization process by scanning electron microscopy. When the degree of supersaturation is high, they observe the formation of needlelike structures, as shown in Figure 15.7 on the surface of a crystal. These structures do not appear at low levels of supersaturation. They postulate that the breakage of needles as the result of contact with the stirrer or by liquid shear provides a vehicle for formation of secondary nuclei of the same chirality as the initial sodium chlorate crystal. This mechanism requires that the supersaturation, the size of the primary nucleus, and the stirring rate all exceed their threshold values, leading to an expression for the rate of secondary nucleation:

$$R_s = (C - C_s)sA\theta(C - C_{min})\theta(\rho - \rho_{min})\theta(s - s_{min}) \tag{15.3}$$

where C is the concentration of solute, C_s is the saturation concentration, C_{min} is the threshold concentration for the formation of needles, s is the stirring rate, ρ is the size of the crystal, A is the surface area of the crystal (which is proportional to ρ^2), and $\theta(x)$ is 0 if $x \leq 0$ and is 1 if $x > 0$. This expression differs significantly from the traditional rate expression for secondary nucleation:

$$R_s = K(C - C_s)^\alpha sA \tag{15.4}$$

where K and α are empirical parameters.

The chiral crystallization experiments we have just described would seem to defy simple modeling, but Metcalfe and Ottino (1994) used a simple autocatalytic reaction scheme combined with a model of chaotic mixing to simulate many of the features of Kondepudi's experiments. The "chemistry" is mimicked by a pair of autocatalytic reactions involving white (W), red (R) and green (G) particles:

$$W + G \rightarrow 2G \tag{15.5}$$
$$W + R \rightarrow 2R \tag{15.6}$$

The mixing is approximated by a bounded, two-dimensional, well-characterized Stokes flow typical of a fluid residing between two parallel but nonconcentric cylinders (Swanson and Ottino, 1990). The simulation is started with a large number of white particles and a single autocatalytic seed particle of each color. Although the simulations do not lead to "ultra-high-purity product" of the sort found in the experiments on sodium chlorate crystallization, many initial condi-

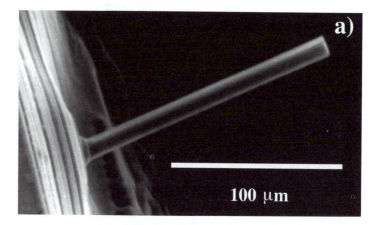

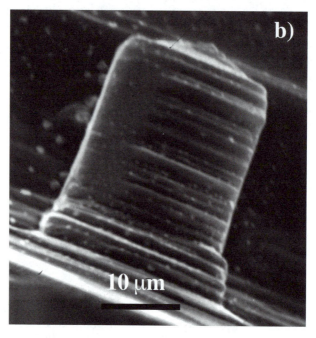

Figure 15.7 (a) Electron micrograph of a typical needle found on crystals grown under high supersaturation. (b) Micrograph of an isolated needle under higher magnification showing banded growth structure. (Reprinted from Kondepudi, D. K.; Sabanayagam, C. 1994. "Secondary Nucleation that Leads to Chiral Symmetry Breaking in Stirred Crystallization," *Chem. Phys. Lett. 217*, 364–368, with kind permission from Elsevier Science, The Netherlands.)

tions quickly yield states in which over 90% of the colored particles are of the same color. Perhaps more suggestive is the fact that very small changes in the initial location of one of the colored seeds can turn a red-dominated state into a green-dominated one. The authors point to the detailed topology of the mixing, the "highly interwoven nature" of the manifold along which the dominant color propagates as the source of this surprising behavior. Just as in the case of temporal chaos, tiny variations in the initial conditions result in major changes in the outcome.

15.2.5 "Crazy Clocks"

We noted earlier that one fruitful source of autocatalytic reactions is the section in the library containing books on lecture demonstrations. Autocatalytic reactions that occur in aqueous solution and that give rise to a color change serve as the basis for the lecture demonstrations known as "clock reactions." In a clock reaction, reactant solutions of known concentration are mixed, and there is apparently no reaction until, suddenly, perhaps accompanied by a wave of the hand or magic words from the demonstrator, the solution changes, for example, from red to blue. The time at which the color change takes place is predictable to very good accuracy (the "clock" aspect) from a knowledge of the reactant concentrations and the temperature. Typically, the reaction is left unstirred or, perhaps, swirled by the demonstrator or a volunteer from the audience. Occasionally, one might employ a magnetic stirrer for mixing.

In some clock reactions, however, there is a narrow range of concentrations in which the quality of the mixing becomes critically important. Under these conditions, the time of the sharp transition from initial to final state becomes essentially unpredictable. The prototype system of this type is the chlorite–thiosulfate reaction (Nagypál and Epstein, 1986). Measurements of pH vs. time for five replicate experiments starting from the same initial concentrations are shown in Figure 15.8. For the first several minutes, all the curves are identical. The pH increases smoothly. In three of the curves, we observe a sharp drop in pH at approximately 3, 5, and 9 min; in the other two, this decrease occurs at times greater than 20 min. When an acid–base indicator like phenolphthalein is added to the solution, the pH

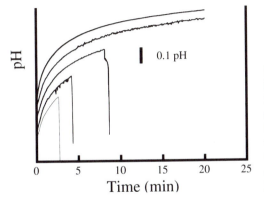

Figure 15.8 Representative pH traces in the chlorite–thiosulfate reaction at 20.0 °C measured under identical conditions. Successive curves have been shifted by 0.07 pH units for easier viewing, since in the absence of a shift, the initial portions of the curves coincide. (Adapted from Nagypál and Epstein, 19861.)

change is accompanied by a striking color change. In an "ordinary" clock reaction or in the chlorite–thiosulfate reaction in a different range of reactant concentrations, this sharp pH drop would occur at the same time, to within a few seconds, in all of the curves. What, other than experimental error, might lead to such irreproducibility, sometimes as much as several orders of magnitude, in the reaction times?

The investigators who first observed this behavior (Nagypál and Epstein, 1986) started with the sensible assumption that there was some hidden source of error in their experimental design. However, their painstaking efforts to remove all possible causes of variability among experiments met with total failure. Despite elaborate schemes to ensure that all experiments were the same with regard to temperature, initial concentrations, exposure to light, vessel surface, age of solutions, and mixing procedure, the reaction times still varied over a wide range. They did notice, however, that the *distribution* of reaction times for, say, a set of 100 replicate experiments under the same conditions, was statistically reproducible. If the conditions, e.g., reactant concentrations, temperature, stirring rate or, most remarkably, the volume of the vessel, were varied, the distribution of reaction times, as characterized, for example, by its mean and variance, changed significantly. Figure 15.9 shows a cumulative probability distribution of reaction times for three sets of experiments carried out with three different stirring rates. In these experiments, both the mean and the standard deviation of the reaction time distribution increase as the stirring rate increases or as the reaction volume decreases.

Mechanistic analysis of the chlorite–thiosulfate reaction, as well as consideration of the reaction time distributions, leads to the following qualitative explanation for the observed behavior. The pH rise and subsequent drop result from competition between two reaction pathways starting from chlorite and thiosulfate:

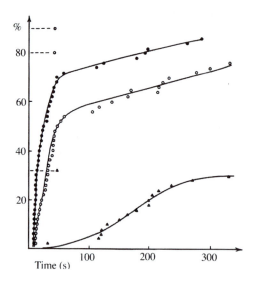

Figure 15.9 Percentage of samples reacted as a function of time in the chlorite–thiosulfate reaction for three different stirring rates. Volume = 4.00 cm^3, temperature = 30.0 °C, $[Na_2S_2O_3]_0$ = 0.100 M, $[NaClO_2]_0$ = 0.300 M, $[NaOH]_0$ = 0.100 M. Stirring rate: filled circles, 500 rpm; open circles, 620 rpm; triangles, 700 rpm. Note that the points are equidistant vertically, being separated by 2%, since each distribution is based on 50 measurements. (Adapted from Nagypál and Epstein, 1986.)

$$4S_2O_3^{2-} + ClO_2^- + 4H^+ \rightarrow 2S_4O_6^{2-} + 2H_2O + Cl^- \qquad (15.7)$$

$$S_2O_3^{2-} + 2ClO_2^- + H_2O \rightarrow 2SO_4^{2-} + 2H^+ + 2Cl^- \qquad (15.8)$$

Reaction (15.7) is responsible for the initial pH rise; its rate is proportional to $[H^+]$. Reaction (15.8) accounts for the pH drop. It is *supercatalytic*, that is, autocatalytic with an order greater than 1 in the autocatalyst; its rate is proportional to $[H^+]^2[Cl^-]$. In the bulk of the solution, reaction (15.7) is the dominant pathway. However, if in some small region a fluctuation leads to a sufficiently high concentration of hydrogen ions, the autocatalysis in reaction (15.8) can produce a rapid local buildup of $[H^+]$, which can spread through the solution, causing the observed rapid pH drop. The key assumption is that there is a threshold for the size of a drop that will grow rather than be dissipated by diffusion or other sources of mixing.

The stochastic nature of the reaction time distribution results from the fact that these supercritical concentration fluctuations are rare, random events. The observed stirring and volume effects on the distribution support this interpretation. By increasing the effectiveness of the mixing, one decreases the likelihood that any given fluctuation will grow to critical size before it is destroyed by mixing with the high pH bulk. Note that if the stirring is stopped at any time after the first few seconds, all samples react completely within 3 s. The apparently more puzzling volume effect can be understood simply in terms of the number of possible microvolumes within the sample in which a supercritical fluctuation can occur. The larger the volume of the system, the more likely such a fluctuation and the shorter the mean reaction time.

Although the behavior described above is rare, it is by no means unique to the chlorite–thiosulfate reaction. Similar distributions of reaction times are found in the chlorite–iodide reaction (Nagypál and Epstein, 1988) in a concentration range where it, too, shows high-order autocatalysis. More interesting is the fact that strikingly similar behavior occurs in two very different systems: the polymerization of the mutant hemoglobin S molecules (Hb S) that cause sickle cell anemia in their unfortunate hosts, and the combustion of simple hydrocarbons in a cool flame.

Ferrone et al. (1985a, b) found a stochastic variation in the distribution of times at which polymerizing sickle hemoglobin molecules undergo a sharp transition from the monomeric to the polymeric state. The experiment involves laser photolysis of a thin layer of a concentrated solution of hemoglobin molecules to which carbon monoxide is bound. The conditions employed are in the physiological range of hemoglobin concentration, pH, and temperature. The strong complexation of hemoglobin by CO (which accounts for the lethal effects of carbon monoxide) prevents the polymerization of Hb S, the reaction that leads to the symptoms associated with sickle cell disease. The laser flash causes the dissociation of the CO molecules from Hb S, which then begins to polymerize. The polymerization is monitored by light scattering, since the polymer scatters light much more effectively than the monomers. No polymerization is detected if the experiment is carried out with normal hemoglobin, Hb A. The rate of the reaction increases sharply with increasing Hb S concentration in the solution. When (Hb S) is between 0.2 and 0.3 g cm^{-3}, the rate varies as the *thirty-sixth* power of the concentration, an extraordinary degree of supercatalysis!

When the Hb S concentration is high, the polymerization proceeds rapidly, and the results obtained are quite reproducible. However, when the Hb S concentration is low enough that the reaction time is of the order of half a minute or more, stochastic behavior sets in, and it shows many of the features found in the inorganic clock reactions discussed above. In Figure 15.10, we see that the relative deviation in the characteristic time of reaction (specified here as the tenth time, i.e., the time to reach 1/10 of the maximum light-scattering intensity) increases sharply with the mean time, suggesting that when the reaction proceeds rapidly enough, nucleation events occur independently throughout the solution, but when reaction is slow, a single primary nucleus governs the fate of the entire sample. Also, as we saw in Figure 15.8, the shapes of the progress curves for different experiments are identical. They are simply shifted along the time axis because the primary nucleation event occurs at a different, stochastically varying time in each sample.

The high-order catalysis results (Hofrichter, 1986) from the fact that the growing Hb S polymer is thermodynamically unstable to dissociation until it reaches a

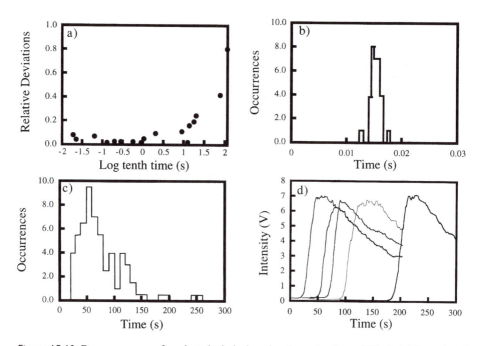

Figure 15.10 Progress curves for photolysis-induced polymerization of Hb S. (a) Fractional deviation of reaction time as a function of mean reaction time for sets of replicate experiments. Mean times were varied by changing sample concentration and/or temperature. (b) and (c) Histograms showing distribution of reaction times. (b) [Hb S] = 5.82 mM; 40 experiments are recorded and binned in 1-ms intervals. (c) [Hb S] = 4.17 mM; 109 experiments are recorded and binned in 10-s intervals. (d) Sample progress curves from experiments shown in part c. (Adapted from Ferrone et al., 1985a.)

threshold length. The rate-determining step is thus the accumulation of the critical number of monomers; this number is the order of the autocatalysis. The analogy to the need for a critical nucleus of reacted material in the chlorite–thiosulfate system is an attractive one. This first supercritical polymer then serves as a template for the formation of secondary strands, a feature bearing some similarity to suggested routes for secondary nucleation in chiral crystallization.

In a rather different set of experiments, Lemarchand et al. (1989) studied the thermal ignition of cool flames in mixtures of n-pentane and air. This phenomenon is characterized by an ignition time t_i that depends on the oven temperature and the initial reactant composition and pressure. In the experiments of interest, t_i decreases as the three-halves power of the pressure, suggesting supercatalysis. Again, there is a parameter range in which the ignition times vary stochastically, and, again, the reaction progress curves for a given set of replicate experiments are identical if one shifts them along the time axis.

Several rather elegant and sophisticated theoretical techniques have been used to analyze these stochastic reaction time distributions (Nicolis and Baras, 1987; Sagués et al., 1990). In keeping with our inclination toward simple, intuitive models, we describe here an approach developed by Szabo (1988) to calculate the distribution of reaction times in the experiments of Ferrone et al. using the theory of first passage times (Weiss, 1967). If one assumes, in agreement with the Hb S polymerization experiments, that monomer is incorporated into polymer at a constant rate B and that nucleation occurs at a constant and smaller rate c, then the time rate of change of the average normalized concentration of polymerized monomers $\langle n(t)\rangle$ is given by

$$d\langle n(t)\rangle/dt = B\langle n(t)\rangle + c \tag{15.9}$$

This equation actually describes the system only during the first, autocatalytic phase of the reaction, but that, fortunately, is the period we are interested in.

If we interpret $\langle n(t)\rangle$ as the average number of bacteria in some volume, then eq. (15.9) describes a process in which bacteria divide at a rate B and are added to that volume at a rate c. In the stochastic interpretation, we view the actual number of bacteria as a random variable, so that the probability of finding n bacteria at time t is given by the probability distribution $P_n(t)$. We then have

$$\langle n(t)\rangle = \sum_{n=0}^{\infty} nP_n(t) \tag{15.10}$$

By considering how the probability distribution evolves between times t and $t + dt$, where dt is sufficiently small that, at most, one bacterium will be born or will die during the interval, one can derive the following equation:

$$P_n(t + dt) = P_{n-1}(t)[B(n - 1) + c]dt + P_n(t)[1 - (Bn + c)]dt \tag{15.11}$$

If we let dt go to 0, we obtain

$$dP_n(t)/dt = [B(n - 1) + c]P_{n-1}(t) + [Bn + c]P_n(t) \tag{15.12}$$

Multiplying eq. (15.12) by n and summing over n confirms that the average number of bacteria $\langle n(t)\rangle$ satisfies eq. (15.9).

The theory of first passage times enables one to take a master equation like eq. (15.12) that describes the evolution of a probability distribution $P_n(t)$ for a population, and derive from it an equation for the probability distribution of times $T_n^m(t)$ for reaching a population m at time t when we start out with n at time zero. For the Szabo model of eq. (15.12), we find

$$dT_n^m(t)/dt = (Bn + c)[T_{n+1}^m(t) - T_n^m(t)] \tag{15.13}$$

with $T_n^m(0) = 0$ for $m \neq n$ and $T_n^n(t) = \delta(t)$. Equation (15.13) can be solved exactly by Laplace transform techniques. The final result of interest, the probability distribution for reaching m bacteria (or polymerized monomers in the case of Hb S polymerization), is given by

$$T_0^m(t) = B\Gamma(m + c/B)e^{-ct}(1 - e^{-Bt})^{m-1}[\Gamma(m)\Gamma(c/B)] \tag{15.14}$$

From this distribution, one can calculate, for example, the mean tenth time $t_{1/10}$, in terms of $m_{1/10}$, the number of polymerized molecules when the material is one tenth polymerized:

$$\langle t_{1/10} \rangle = \sum_0^{m_{1/10}} tT_0^m(t)\,dt = 1/c + 1/(B + c) + 1/(2B + c)$$
$$+ \ldots + 1/[(m_{1/10} - 1)B + c] \tag{15.15}$$

Note that in this oversimplified model, all the details of the mixing are lumped into the nucleation parameter c, while the chemistry is summed up by the polymerization addition rate B.

15.3 Implications for Population Dynamics

As our bacterial growth analogy in the Szabo model for stochastic polymerization dynamics suggests, there are many similarities between the laws that govern the growth of biological populations and those that determine the time evolution of chemical reactions. In this last section, we consider some of the consequences of imperfect mixing in living systems. Population growth is an inherently autocatalytic process, as Reverend Thomas Malthus recognized two centuries ago when he pointed out that the population tends to increase geometrically, while the food supply grows only arithmetically (Malthus, 1798).

We shall examine some simple models of population growth to see what consequences imperfect mixing might have. Consider first a set of species with populations $p_1, p_2, \ldots$ competing under perfectly mixed conditions for an unlimited food supply. Assume that each species grows at a rate $g_i(p_i)$, which depends upon its population. It has a death rate $d_i p_i$ proportional to its population. The growth rate g_i might be also be proportional to p_i, so that

$$dp_i/dt = (b_i - d_i)p_i = r_i p_i \tag{15.16}$$

where b_i is the per capita birth rate and r_i is the net per capita growth rate of the ith species. If there are no interactions between species (e.g., predator–prey), then these equations provide a reasonable description of the system. They can be solved exactly to yield

$$p_i(t) = p_{i0}(t) \exp(r_i t) \tag{15.17}$$

The species with the largest growth rate r_i will eventually come to dominate the population, regardless of the initial populations p_{i0}. A similar conclusion holds for any other growth law if the growth is unconstrained. Thus, with no limits on the supply of food, all species will survive, and the fittest (fastest growing one) will survive best.

Now let us consider the more realistic case that the food supply is fixed. One way to take into account what happens when one limits the food supply is to modify the growth equation (15.16) by introducing a finite *carrying capacity* K:

$$dp_i/dt = r_i(1 - p_i/K)p_i \tag{15.18}$$

The resulting equation (15.18) is known as the *logistic equation* and has played a major role in efforts to understand population dynamics. It is analyzed in detail in many books on mathematical biology (e.g., Edelstein-Keshet, 1988). The key result is that for any initial population, the system approaches the single stable steady-state solution $p_i = K$.

An alternative, perhaps more realistic approach to representing the effects of finite resources when there is direct competition among species for the food supply is to include these effects by constraining the total population. Eigen and Schuster (1979) have taken such an approach in their models of molecular evolution. We can maintain the total population, $P = \Sigma p_i$, constant by adding a flux term to our population growth equation. The particular form of the constraint used in eq. (15.19) is referred to by Eigen and Schuster as the *constraint of constant organization*:

$$dp_i/dt = r_i p_i - (p_i/P)\Sigma r_j p_j \tag{15.19}$$

If we sum up the growth equations (15.19), we see that the dp_i/dt terms add up to zero; the total population P is a constant.

Now consider such a system with just two species whose initial populations are p_{10} and $p_{20} = P - p_{10}$, where P is the fixed total population. The growth equations are

$$dp_1/dt = r_1 p_1 - (p_1/P)(r_1 p_1 + r_2 p_2) \tag{15.20}$$
$$dp_2/dt = r_2 p_2 - (p_2/P)(r_2 p_2 + r_1 p_1) \tag{15.21}$$

Since $p_2 = P - p_1$, we can eliminate p_2 from eq. (15.20) to obtain a single growth equation for p_1:

$$dp_1/dt = r_1 p_1 - (p_1/P)[r_1 p_1 + r_2(P - p_1)] = [(r_2 - r_1)/P][p_1(p_1 - P)] \tag{15.22}$$

Equation (15.22) can be solved exactly for $p_1(t)$ to yield

$$p_1(t) = P/\{1 - (1 - P/p_{10}) \exp[(r_2 - r_1)t]\} \tag{15.23}$$

As $t \to \infty$, our expression for $p_1(t)$ approaches zero if $r_2 > r_1$; it approaches P if $r_2 < r_1$. Solving for $p_2(t)$ shows that this population approaches zero if r_1 is the larger rate and P if r_2 is larger. We thus have *all-or-none* selection of the faster growing species, independent of the initial populations so long as neither starts at zero.

Now consider a pair of species evolving with a nonlinear growth rate, say, quadratic, being again under the constraint of constant organization. The evolution equations for this system take the form

$$dp_1/dt = r_1 p_1^2 - (p_1/P)(r_1 p_1^2 + r_2 p_2^2) \tag{15.24}$$

$$dp_2/dt = r_2 p_2^2 - (p_2/P)(r_1 p_1^2 + r_2 p_2^2) \tag{15.25}$$

Substitution of $P - p_1$ for p_2 in eq. (15.24) yields, after some algebraic manipulation,

$$dp_1/dt = -1/[P(r_1 + r_2)]p_1[p_1 - r_2 P/(r_1 + r_2)](p_1 - P) \tag{15.26}$$

Equation (15.26) can be solved exactly, but a graphical solution is more revealing. In Figure 15.11, we plot dp_1/dt vs. p_1. From the sign of the time derivative, we see that if p_1 is initially less than $p' = r_2 P/(r_1 + r_2)$ it will decrease until it reaches a final value of zero, that is, species 1 will die out. If $p_1 > p'$, then it will grow until it reaches P. Again, we have all-or-none selection, but now the winner of the battle depends, not simply on the per capita growth rate, but on whether or not p_1 initially exceeds p'. Note that

$$p' = r_2 P/(r_1 + r_2) = r_2(p_1 + p_2)/(r_1 + r_2) \tag{15.27}$$

so that the condition $p_1 > p'$ is equivalent to

$$p_1(r_1 + r_2) > r_2(p_1 + p_2) \tag{15.28}$$

or, equivalently,

$$r_1 p_1 > r_2 p_2 \tag{15.29}$$

Although our analysis has been carried out for one simple growth law and a model containing only two competing species, it can be shown that, under the constraint of constant organization or other similar constraints, selection that is dependent on the initial conditions occurs for any growth law that exceeds linearity in its dependence on the populations. Similar conclusions can be reached with a bit more effort for an arbitrary number of species (Eigen and Schuster, 1979). All-or-none selection is a general feature of a perfectly mixed system of species competing for a common, fixed food supply. The situation resembles somewhat the L- and D-crystals competing for the dissolved sodium chlorate in the experiments of Kondepudi et al.

The "fittest" species ultimately wins out. As we have seen, the definition of fitness varies somewhat with the rate law for reproduction and the environment in which the competition takes place. If the food supply is unlimited or if the growth

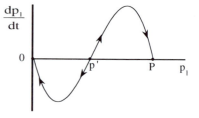

Figure 15.11 Plot of dp_1/dt vs. p_1 in eq. (15.26). Arrows show direction in which the system evolves.

rate is linear, even a single reproductively capable individual will give rise to descendants who will dominate an established population of lower fecundity (or survivability). If the food supply is limited and the growth rate is nonlinear, then history plays a role. In order to win the competition, a new species must first reach a certain population threshold at which its absolute rate, not its relative rate, of reproduction is the largest in the system. In effect, a favorable mutation cannot survive unless there is a way for its population to reach a critical level before it is swamped by competition from its less able but more populous neighbors.

All-or-none selection is not a particularly desirable outcome (unless yours is the species to be selected). It throws away genetic diversity that may prove valuable at some future time when the environment has changed. One way in which real entities avoid the wastefulness of all-or-none selection is through less-than-perfect mixing. Mobile organisms can isolate themselves from direct competition with their better adapted fellows by spatially segregating themselves—by choosing a niche. Darwin recognized that more exotic, though perhaps "less fit," species were likely to be found in isolated habitats such as islands. To the extent that different molecules can find different puddles of primeval broth in which to grow, or that compartmentalization can arise from the development of primitive cell membranes, the mixture will be able to support a richer variety of complex species.

15.4 Conclusions

The past decade has witnessed a growing understanding of the rich variety of behaviors of which nonlinear chemical reactions are capable. The experiments and models presented in this chapter suggest that mixing effectiveness, difficult though it may be to quantify, can serve as an additional, more subtle control parameter. In addition, the interaction of mixing with nonlinearity, especially autocatalysis, can give rise to new phenomena, such as chiral symmetry-breaking and stochasticity in both open and closed systems. The utility of even very simple models in providing us with intuitions about these complex effects is encouraging. From a theoretical point of view, the extreme cases of perfectly mixed, homogeneous systems and of unmixed, diffusive systems are far better understood than the intermediate case of partially mixed systems, though the latter are far more common in nature.

Further study of the behavior that can arise in imperfectly mixed autocatalytic systems should prove rewarding. There may well be new phenomena to be discovered. The notion of tuning the mixing efficiency to control the composition of the output from an industrial reactor is an appealing potential extension of the experiments that produce chirally pure sodium chlorate crystals. Whether the sorts of experiments and models described here will give us further insight into controlling the outputs of vats of industrially important chemicals or will allow us to understand better the mysteries of molecular evolution remains to be seen. What is already clear is that the interaction of nonlinear chemical dynamics with imperfect mixing leads to a rich variety of phenomena and that there is still much to be learned about the behavior of such systems.

APPENDIX I
Demonstrations

One of the great appeals of nonlinear chemical dynamics lies in the striking visual demonstrations that can be created. It is a rare person who is not impressed the first time he or she sees a clear solution repeatedly turn brown, then blue, then back to clear! In this appendix, we explain how to perform demonstrations of oscillating reactions, fronts, and waves. In the next section, we provide information on how some of these systems can be systematically studied in the upper-level undergraduate laboratory.

In addition to the demonstrations and experiments we present, there have been many pedagogical articles in the *Journal of Chemical Education* (Winfree, 1984; Epstein, 1989; Field, 1989; Field and Schneider, 1989; Fink et al., 1989; Glasser, 1989; Noyes, 1989; Sharbaugh and Sharbaugh, 1989; Soltzberg, 1989; Pojman, 1990; Vemaulpalli, 1990; Hollinger and Zenzen, 1991; Jahnke and Winfree, 1991; Rosenthal, 1991; Melka et al., 1992; Talanquer, 1994; Weimer and Smith, 1994; Benini et al., 1996; Pojman and West, 1996; Strizhak and Menzinger, 1996). An excellent collection of demonstrations of oscillating reactions and waves appears in *Chemical Demonstrations: A Handbook for Teachers* (Shakhashiri, 1985).

AI.I The Briggs–Rauscher Reaction

Although the BZ reaction is the most famous oscillating reaction, the best demonstration is the Briggs–Rauscher reaction (Briggs and Rauscher, 1973). It oscillates dependably, colorfully, and rapidly, making it an ideal classroom demonstration.

Two recipes are often used, one with 30% hydrogen peroxide and one with 3%. We will only present the latter, because the concentrated hydrogen peroxide is not readily available; both recipes work equally well.

We will describe how to prepare a total of 3 L of solution.

- Solution A is just drugstore (3%) hydrogen peroxide.
- Solution B is prepared by placing 29 g of potassium iodate (KIO_3) (26 g of $NaIO_3$ can be substituted) and about 400 mL of distilled water in a 1-L beaker. Add 8.6 mL of 6.0 M H_2SO_4. (To prepare a stock solution of 6.0 M sulfuric acid, carefully pour 330 mL of concentrated [18 M] sulfuric acid into 500 mL of distilled water and dilute to 1.0 L.) Stir the solution until the iodate dissolves (this may require some heating—sodium iodate will dissolve more readily).
- Solution C is prepared by dissolving 10.4 g of malonic acid and 2.2 g of manganese(II) sulfate monohydrate ($MnSO_4 \cdot H_2O$) in about 400 mL of distilled water in another 1-L beaker. In a 100-mL beaker, heat 50-mL of distilled water to boiling. In a 50-mL beaker mix, 0.2 g of soluble starch with about 5 mL of distilled water and stir the mixture to form a slurry. Pour the slurry into the boiling water and continue heating and stirring the mixture until the starch is dissolved. (NB: This step is crucial, because if the starch is not dissolved, the oscillations will not be visible.) Pour the starch solution into the malonic acid/manganese sulfate solution. Dilute it to 500 mL.

Pour equal volumes of solutions B and C into a beaker equipped with a magnetic stir bar. (The demonstration can be performed without mixing, but the transitions will not be as sharp as with mixing.) Add a volume of solution A equal to the sum of the volumes of solutions B and C. An amber color will appear quickly, then turn to deep blue. The blue will disappear, leaving a clear solution. Copious bubble production will accompany these changes. The sequence will repeat; the period of oscillation is initially about 15 s, but it will increase until, after several cycles of oscillation, the reaction remains deep blue.

Safety Information

A large amount of I_2 will be produced. The beaker should be kept covered, for example, with parafilm, to keep the vapors from escaping. The solid iodine that remains at the end of the reaction should be reduced to iodide by the addition of 5 g L^{-1} of sodium thiosulfate. This reduction can be quite exothermic! The cold solution should be flushed down the drain with large amounts of water.

AI.2 The Belousov–Zhabotinsky Reaction

Prepare the following solutions in deionized water.

- Solution A: 19 g $KBrO_3$ (or 17 g $NaBrO_3$) dissolved in enough water to make 500 mL of solution. $[BrO_3^-] = 0.23$ M.
- Solution B: 16 g of malonic acid and 3.5 g of KBr (3.0 g $NaBr$) dissolved in enough water to make 500 mL of solution. $[MA] = 0.31$ M, $[Br^-] = 0.059$ M.

- Solution C: 5.3 g cerium(IV) ammonium nitrate dissolved in enough 2.7 M sulfuric acid to prepare 500 mL. $[Ce(NH_4)_2(NO_3)_6] = 0.019$ M · (To prepare a stock solution of 2.7 M sulfuric acid, carefully pour 149 mL of concentrated (18 M) sulfuric acid into 500 mL of distilled water and dilute to 1.0 L.)
- Solution D: 3.6 g of methylmalonic acid (Aldrich), instead of malonic acid in solution B and 0.70 g KBr in 100 mL water.

Pour solutions A and B into a 2-L flask equipped with a magnetic stir bar. The solution will become brown from bromine formation. After the solution clears, add solution C and 30 mL of 25 mM ferroin (Fisher). The solution will change from green to blue to violet and then to red over a period of about 1 min. These oscillations will persist for about 20 min.

A smaller version can be prepared using appropriately scaled volumes.

Substituting solution D for C will produce an oscillator with an initial period of 3–5 min that will continue oscillating for 8 h! The period will gradually lengthen to 15 min.

Safety Information

Bromates are strong oxidizing agents and should be treated with caution. Spills of sulfuric acid should be neutralized with bicarbonate.

The completed reaction can be neutralized with bicarbonate and flushed down the drain with copious amounts of water.

AI.3 BZ Waves

Prepare the following solutions.

- Solution A: Dissolve 84 g of $KBrO_3$ (or 69 g of $NaBrO_3$) in 750 mL of 0.6 M sulfuric acid. (To prepare the acid solution, carefully pour 33 mL of concentrated (18 M) sulfuric acid into 500 mL of distilled water and dilute to 1.0 L.) Dilute to 1.0 L with more acid. $[BrO_3^-] = 0.50$ M.
- Solution B: Dissolve 52 g of malonic acid in 750 mL of distilled water and dilute to 1.0 L. [Malonic acid] = 0.50 M.
- Solution C: Dissolve 1g of NaBr in 10 mL of distilled water. [NaBr] = 0.97 M.
- Solution D: 25 mM ferroin (Fisher). Make sure the ferroin is the sulfate version and not the chloride salt.

Into a small Erlenmeyer flask, introduce 7 mL of solution A, 3.5 mL of solution B, and 1 mL of solution C. Stopper the flask and allow to stir on a magnetic stirrer. The brown color is bromine, which forms from the oxidation of bromide by bromate. The bromine slowly disappears as it reacts with the malonic acid to form bromomalonic acid. When the solution has cleared, add 1.0 mL of ferroin (solution D) and stir.

The reaction may oscillate between red and blue, but ignore this. Use a pipette to transfer sufficient solution to form a thin (1–2-mm) layer in a clean Petri dish. Cover the dish and wait. To demonstrate the waves to the class using an overhead projector, it is essential to have a thin layer. Wetting the surface of a Petri dish can be achieved by adding a surfactant (Field and Winfree, 1979). Sodium dodecyl sulfate works well. Adding a small amount of ultrafine silica gel (CAB-O-SIL, Cabot Corp.) eliminates convection, but the solution remains sufficiently translucent to allow projection.

Rings will appear spontaneously after a few minutes (see Figure 6.6). To create spirals, gently break a wave using a pipette. Two counter-rotating spirals will slowly form (see Figure 6.7).

Safety Information

Safety and disposal information are the same as for the BZ oscillating reaction.

AI.4 A Propagating pH Front

This is a very simple reaction to perform in order to show how an autocatalytic front propagates and how convection affects such a front. The reaction is also described in Summerlin et al. (1988) and was studied by Nagy and Pojman (1993).

A solution initially contains sulfite–bisulfite and chlorate with the pH indicator bromophenol blue. No reaction occurs until some 3 M sulfuric acid is added. The solution rapidly turns yellow as the pH drops.

$$ClO_3^- + 3HSO_3^- \rightarrow Cl^- + 3SO_4^{2-} + 3H^+$$

If the solution initially contains 0.88 M SO_3^{-2} and 0.12 M HSO_3^-, the reaction rate is effectively zero at 298 K. If no sulfite is present, then the bisulfite can disproportionate:

$$2HSO_3^- \Leftrightarrow H_2SO_3 + SO_3^{2-} \qquad K_2/K_1 = K = 5.56 \times 10^{-6}$$

The presence of the sulfite suppresses the reaction, and the pH remains above 7.

If a strong acid is added, then H^+ reacts with sulfite:

$$SO_3^- + H^+ \Leftrightarrow HSO_3^- \qquad 1/K_2 = 1.0 \times 10^7$$

Given the value of the equilibrium constant, and the excess of sulfite, all the H^+ is converted to bisulfite. The increased bisulfite and decreased sulfite shifts the bisulfite–H_2SO_3 equilibrium to form H_2SO_3. The H_2SO_3 can react with the chlorate to produce more H^+. This H^+ will protonate sulfite, replacing the bisulfite and, in turn, the H_2SO_3 consumed. The key is that as the concentration of sulfite decreases during the reaction, the bisulfite disproportionation equilibrium will shift farther to the right, increasing the concentration of H_2SO_3. Thus, the pH remains nearly constant until enough sulfite is reacted to leave the system unbuffered. Then, all the H^+ produced will protonate the bisulfite, and the reaction can "explode."

The recipe is simple: 4 g potassium chlorate, $KClO_3$, (or 3.3 g sodium chlorate) and 12.5 g sodium sulfite, are dissolved in about 75 mL of distilled or deionized water. Add a few milligrams of bromophenol blue. (The solution should be a dark blue.) Slowly add 4 mL of 3 M sulfuric acid while the solution is stirred, and dilute the solution to a final volume of 100 mL with water. The solution should be dark blue (basic). Pour the solution into a graduated cylinder or test tube and add a few drops of 3 M sulfuric acid to the top, without stirring. A yellow solution will appear at the top, and the reaction will propagate down the tube. Convective fingering can be observed (see Figure 9.9).

Safety Information

The reaction is extremely exothermic, and the temperature difference between reacted and unreacted regions is about 50 °C. After the reaction cools down, it can be neutralized with bicarbonate and poured down the drain.

APPENDIX 2
Experiments for the Undergraduate Lab

A2.I Frontal Polymerization

We will describe a simple and inexpensive experiment with propagating fronts of addition polymerization. The method can be used to determine the front velocity dependence on the initiator concentration and to determine the effects of heat loss and gravity on a chemical reaction. This experiment is also described in Pojman, et al. (1997b).

Propagating fronts can be created with many monomers, including acrylamide, *n*-butyl acrylate, methacrylic acid, methyl methacrylate, and triethyleneglycol dimethacrylate (TGDMA) (Pojman, 1991; Pojman et al., 1992a, 1993, 1995a, b). We chose TGDMA because of its high boiling point, which reduces the risk of explosion, and the fact that it forms a crosslinked product that will not undergo fingering (see Figure 11.9).

A2.1.1 Experimental

Six reactions are employed in this experiment. Four are used to determine the front velocity dependence on the initiator concentration. The other two demonstrate the effects of convection and heat loss on the reaction.

Safety Information

- Benzoyl peroxide (BPO) should not be heated.
- Triethylene glycol dimethacrylate (TGDMA) should be kept cool.

- Dimethyl aniline (DMA) should be kept in the hood. Do not get this compound on your skin.
- Do not heat solutions of monomer and initiator as they will polymerize very rapidly.
- Do not expose skin to either BPO or TGDMA. Gloves should be worn at all times. Safety glasses must be worn during this experiment. All reactions should be performed in a working fume hood behind shatterproof glass.

The test tubes containing the traveling front will be very hot—above 200 °C, so they should be allowed to cool for at least an hour before handling.

Any unused monomer–initiator solution can be stabilized by adding a few grams of hydroquinone and disposed of as any other organic waste.

Propagating Front Velocity

Three 100 mL solutions of BPO in TGDMA are prepared with concentrations between 1 g per 100 mL and 4 g per 100 mL. The solutions are stable at room temperature for a few hours and should be prepared and used in the same lab period. (Initiator–monomer solutions may be refrigerated for a day or two if necessary.) Benzoyl peroxide does not dissolve quickly in TGDMA above 4 g per100 mL. (NB: The solution should **not** be heated to speed dissolution, because polymerization will occur.)

Students should record the position of the front as a function of time. This can be done by either making centimeter markings on the test tube or placing a ruler beside the test tube. We suggest using five 16- × 150-mm and one 22- × 150-mm test tubes. The size of the tube makes a difference, as it changes the amount of heat lost to the surroundings. If the tube is too small, then the heat losses will quench the front.

A 16- × 150-mm test tube is arranged in a hood so that the markings (or ruler) can be easily seen. Then, one of the solutions is poured into the test tube. The tube is clamped to a ring stand and a small aliquot (∼ 1–2 mL) of DMA is added to the top of the tube. (DMA reacts with BPO via a redox process and produces free radicals that initiate polymerization, releasing heat.) The DMA is mixed with the first centimeter of solution, and a reaction should be observed almost immediately. A soldering iron can be substituted for DMA. The hot soldering iron is inserted into the first centimeter of solution to initiate the reaction. Those using a soldering iron should take care that the iron is not too hot, as the solution can splatter out of the test tube.

Once the reaction has begun, it is allowed to proceed 2–3 cm down the tube. This is done to allow enough time for the front to stabilize. Then, the time at which the front reaches each centimeter mark is recorded, and a plot of position vs. time is made for each of the three solutions. (A video camera can be used to obtain more precise data.)

Heat Loss

The effect of heat transfer on the front is investigated in two ways. First, the tube size is changed from the 16- × 150-mm tube to the larger 22- × 150-mm tube.

Second, the reaction is run with the smaller tube in a water bath. The first experiment changes the surface-to-volume ratio; the second changes the heat transfer coefficient between the tube and its surroundings.

Because the polymer formed is crosslinked, the effect of initiator concentration on the molecular weight distribution cannot be studied. Other monomers (methacrylic acid, butyl acrylate) can be used for such a study, but they require additional techniques and equipment. See references (Pojman et al., 1995b, 1996d) for more information.

A2.1.2 Typical Results

The front velocity is the slope of a plot of front position vs. time. This plot is prepared for each concentration, and the data are compared with the predictions of the steady-state model. Students observe that as the initiator concentration increases, the velocity of the front increases according to a power function.

Heat losses in a propagating front have a significant effect on the front velocity. This is related to two factors: the size of the tube that affects the surface-to-volume ratio, and the medium outside the tube that affects the rate of heat conduction. If the tube diameter is decreased, the front travels slower. In tubes of the size we have used, the polymerization front stops when the vessel is placed in room-temperature water, because water conducts heat more effectively than air.

Changing the orientation of the test tube demonstrates the effects of gravity on the front. If a well-established front is inverted, the hot reaction zone rises quickly to the top of the mixture, and the front may be extinguished, depending on the tube diameter and viscosity. If the tube is tilted, the front reorients so that is perpendicular to the gravitational vector (see Figure 11.8). These effects occur because the hot, liquid reaction zone is trapped between cold, dense monomer below and solid polymer above. Buoyancy forces the reaction zone to remain above the monomer layer.

A2.2 Oscillations in the Homogeneous Belousov–Zhabotinsky Reaction

The goals of the experiment are:

1. To observe spontaneous temporal self-organization.
2. To observe and measure the induction period and determine how it is affected by the organic substrate used.
3. To understand the use of electrodes and the physical chemistry behind their operation.
4. To understand what a redox indicator is and to determine if ferroin acts purely as an indicator or if it affects the reaction dynamics.
5. To determine the effects of chloride ion and oxygen.

A2.2.1 Experimental

The electrical potential of the reaction mixture is measured with a platinum electrode and a Hg/Hg$_2$SO$_4$ reference (available from Rainin). The output can be recorded on a single-channel strip chart recorder, with the platinum electrode attached to the positive terminal and the reference to the negative, or the recording can be made via a high-impedance input to an A/D board (analog-to-digital converter) on a computer (see Chapter 3).

The recipe has been obtained from Shakhashiri (1985).

Prepare the following solutions in deionized water using volumetric glassware.

- Solution A: 3.4 g NaBrO$_3$ dissolved in enough water to make 100 mL of solution.
    — Solution A1: [NaBrO$_3$] = 0.23 M.
    — Solution A2: 50 mL of solution A1 diluted to 100 mL.
    — Solution A3: 50 mL of A2 diluted to 100 mL.
- Solution B: 3.2 g of malonic acid and 0.70 g of KBr dissolved in enough water to make 100 mL of solution. [MA] = 0.31 M, [KBr] = 0.059 M.
- Solution C: 0.14 g cerium(IV) ammonium nitrate dissolved in enough 2.7 M sulfuric acid to prepare 100 mL. [Ce(NH$_4$)$_2$(NO$_3$)$_6$] = 0.019 M.
- Solution D: 3.6 g of methylmalonic acid (Aldrich), instead of the malonic acid in solution B.

Now add 15 mL each of solutions A1 and B to a 100-mL beaker equipped with a stir bar. A brown color will appear because of the production of bromine, which disappears as it reacts with the malonic acid. When the solution clears, add 15 mL of solution C, and position the electrodes immediately so that the induction period may be accurately observed and measured. Stir at a low rate, or oscillations may be inhibited. The recorder may be set at any desired speed (typically about 1 cm min^{-1}). The voltage scale is usually set at either 250 or 500 mV.

Note the time that the oscillations commence, that is, the induction period. Also record the period of the oscillations and the amplitude (number of millivolts during one oscillation).

Students should be asked to consider these questions and perform the following procedures:

1. After several periods of oscillations, add 2 mL of a 25 mM ferroin solution (available from Fisher). Ferroin (tris(1,10-phenanthroline)iron(II) sulfate) is a redox indicator. As the (Ce(IV)) increases, it can oxidize the iron in ferroin from iron(II) to iron(III). The iron(II) complex is red and the iron(III) complex is blue; consequently, the color changes as the potential changes. What effect does the ferroin have on the period and amplitude of the oscillations? Is it really acting just as an indicator?
2. Test the effect of Cl$^-$ by adding a small grain of NaCl. What happens? How long does it take the system to recover (if at all)? Add a pinch of NaCl. What happens?
3. Test the effect of stirring on the system by turning the stirrer to a high speed. Turn it up to full speed. Are the oscillations affected?
4. Field, Körös, and Noyes (Field et al., 1972) proposed a mechanism in which radicals play important roles (BrO$_2$ ·, MA ·) Add a drop of acrylonitrile to an oscillating reaction and explain the relevance of the white precipitate (Váradi and Beck, 1973).

5. Observe the oscillations for half an hour. Note changes in the period as a function of time. What is the relevance of this change to the Second Law of Thermodynamics?
6. Prepare a fresh reaction mixture using solution A2. Does it oscillate? If it does, what is the period? Try with solution A3.
7. Finally, set up an experiment with methylmalonic acid (solution D) and solution A1 and leave it running for several hours.

A2.2.2 Typical Results

Figure A.1 shows digitized data for the platinum and bromide electrodes.

The amplitudes of solutions A1, A2, and A3 are approximately the same. The periods of oscillation become increasingly longer with decreased bromate ion concentration.

The addition of ferroin reduces the amplitude and the period. Because the Fe(II) (ferroin) is available to reduce the Ce(IV) back to Ce(III), the oscillations may occur faster. Ferroin itself can also catalyze the BZ reaction.

Adding a grain of salt slightly decreases the amplitude and period. Adding more salt can suppress oscillations. The added chloride ions are oxidized to chlorous acid ($HClO_2$), which is then able to reduce the Ce(IV) back to Ce(III), inhibiting oscillations. Oscillations return when the chlorous acid is completely oxidized to inert ClO_3^- (Jacobs and Epstein, 1976).

Increasing the stirring rate until vortexing occurs entrains oxygen into the system (Farage and Janjic, 1980, 1981). The oscillations cease because the oxygen molecules are able to react with malonyl radicals that are present in the reaction via an autocatalytic process that increases the radical concentration. The malonyl radicals react with bromomalonic acid, liberating Br. The net effect of oxygen is to increase the bromide concentration beyond the range allowable for oscillations (Field and Burger, 1985). When the stirring rate is reduced, the oscillations resume.

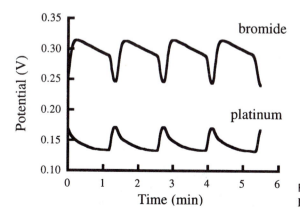

Figure A.1 Typical time series for BZ reaction.

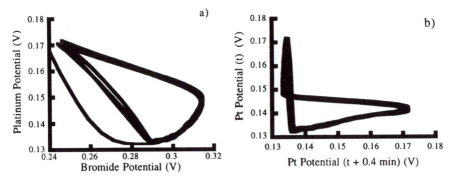

Figure A.2 (a) Phase plot prepared by plotting platinum vs. bromide electrode potentials. (b) Phase plot prepared by using time-delay technique with only platinum electrode.

Addition of large amounts of acrylonitrile stops oscillations for about 20 minutes and a white precipitate (insoluble polymer) forms. See Chapter 11 for a discussion.

With methylmalonic acid as the substrate, the period increases by a factor of about 9 and the induction period by a factor of about 13.

If data are digitized, then it is possible to create phase plots such as those shown in Figure 4.5. If a bromide electrode is unavailable, a pseudo phase plot can be constructed by using the time-delay method in which the platinum potential is plotted vs. the potential at a fixed later time interval (see section 8.2.2). Figure A.2 compares phase plots prepared by both methods.

A2.3 Unstirred BZ System: "Measuring Rate Constants with a Ruler"

In this experiment, we determine the rate constant of an important autocatalytic reaction in the BZ reaction, where Red is the reduced form and Ox is the oxidized form of the metal catalyst:

$$3H^+ + BrO_3^- + HBrO_2 + 2Red \rightarrow 2HBrO_2 + 2Ox + H_2O$$

$$\frac{d[HBrO_2]}{dt} = k[H^+][BrO_3^-][HBrO_2] \tag{A.1}$$

However, measuring k directly is difficult because we cannot run an experiment in a beaker in which we mix bromate and $HBrO_2$ in acid solution ($HBrO_2$ is not a stable species). In the BZ system, the formation of $HBrO_2$ is accompanied by the oxidation of the metal ion, which we observe as a propagating color change. Tyson (1985) proposed the following expression for the wave velocity in terms of the $HBrO_2$ diffusion coefficient and the rate constant of the autocatalytic reaction:

$$\text{Velocity} = 2(kD[H^+][BrO_3^-])^{1/2} \tag{A.2}$$

where D is the diffusion coefficient of $HBrO_2$ (2×10^{-5} cm^2 s^{-1}).

The goals of this experiment are:

1. To test this model by determining the sensitivity of the front velocity to the concentrations of ferroin, $[BrO_3^-]_0$, and $[H^+]_0$.
2. To use these data to calculate a value for k.
3. To observe the phenomenology of waves and spirals.

A2.3.1 Experimental

The following solutions are prepared in advance (see Appendix 1 for recipes).

- Solution A: 100 mL of 0.6 M $NaBrO_3$ / 0.6 M H_2SO_4.
- Solution B: 40 mL of 0.48 M malonic acid.
- Solution C: 25 mM ferroin (Fisher).
- Solution D: 0.97 M NaBr (1 g per 10 mL).

Students should prepare the following additional solutions.

- Solution A1: Take 40 mL of solution A and dilute to 50 mL. (0.48 M $NaBrO_3$ / 0.48 M H_2SO_4.)
- Solution A2: Take 20 mL of solution A1 and dilute to 25 mL. (0.38 M $NaBrO_3$ / 0.38 M H_2SO_4.)
- Solution A3: Take 10 mL of solution A and dilute to 20 mL. (0.3 M $NaBrO_3$ / 0.3 M H_2SO_4.)

The following procedure is from Epstein (1987).

Into a small Erlenmeyer flask, introduce 7 mL of solution A, 3.5 mL of solution B, and 1 mL of solution D. Stopper the flask and stir with a magnetic stirrer. The orange-brown color is bromine, which forms from the oxidation of bromide by bromate. The bromine slowly disappears as it reacts with the malonic acid to form bromomalonic acid. When the solution has cleared, add 0.5 mL of the 25 mM ferroin solution and stir.

The reaction may oscillate between red and blue, but ignore this. Use a pipette to transfer sufficient solution to a clean Petri dish. Cover the dish and wait.

Students should be asked to consider these questions and perform the following procedures:

1. You will notice small rings of blue forming in the red solution. Notice how rapidly the blue color spreads. Calculate how long a molecule would take to diffuse just 1 cm ($D \approx 10^{-5}$ cm^2 s^{-1}, distance $(Dt)^{1/2}$). Clearly, simple diffusion cannot account for the speed of color propagation.
2. You will see bubbles form. What are they? Recent work indicates that they are not CO_2, but CO (Ouyang et al., 1987). If too many bubbles form to be able to see the waves clearly, swirl the dish around. New waves will form.
3. What happens when two waves collide? How is this different from what happens with water waves?
4. What happens when a wave encounters a barrier? Does it reflect?
5. To make a spiral, slowly move a pipette tip through the center of a target pattern. In which direction does the spiral rotate?
6. Repeat the above experiment. Use graph paper to measure the change in the radial distance of the wave from the center of a target pattern as a function of

time. The slope of a line drawn through these data will provide the wave speed. Also measure the wavelength (the distance between fronts).

7. Does the velocity remain constant with time? If not, why?
8. Repeat these measurements for the following solutions:
   (a) Use 0.2 mL ferroin instead of 0.5 mL with solution A. Is there any effect?
   (b) Using 1.0 mL ferroin and the same volumes of other reagents, make velocity and wavelength measurements for solutions A1, A2, and A3.
9. Plot the velocity as a function of the square root of the product of $[H^+]$ and $[BrO_3^-]$. Fit a line. Using the slope and the diffusion coefficient of $HBrO_2$, calculate the rate constant of the autocatalytic reaction. Compare it to the reported value of $20 \text{ M}^{-2} \text{ s}^{-1}$.
10. Does the velocity depend on the ferroin concentration?

A2.3.2 Typical Results

As the concentrations of BrO_3^- and sulfuric acid are decreased, the wave velocity decreases. A plot of the square of the velocity vs. $[H^+][BrO_3^-]$ was also constructed to solve for k from eq. (A.2) because the slope $= 4kD$. Using nothing more than a ruler and a stopwatch, physical chemistry lab students at the University of Southern Mississippi prepared Figure 6.12, whose slope is $4.1 \text{ cm}^2 \text{ min}^{-2} \text{ M}^{-2}$. A rate constant of $14 \text{ M}^{-2} \text{ s}^{-1}$ was calculated using a diffusion coefficient for $HBrO_2$ of $2 \times 10^{-5} \text{ cm}^2 \text{ s}^{-1}$ (Tyson, 1985). The percentage deviation from the literature value of $20 \text{ M}^{-2} \text{ s}^{-1}$ is 30%, based on the experiments of Field and Noyes (1974a) and Showalter (1981).

For each concentration of bromate ion–acid, several velocity measurements were recorded. Data collected from the same samples do not indicate substantial change of the velocity with time. Keeping $[BrO_3][H_2SO_4]$ constant, the ferroin concentration was increased from 0.5 mL to 2 mL. The average velocity of the waves decreased from 0.65 cm min^{-1} to 0.64 cm/min^{-1} when the ferroin concentration was increased. With a standard deviation of $\pm 0.03 \text{ cm min}^{-1}$, this is not a significant decrease. Equation (A.2) predicts that the front velocity should be independent of the ferroin concentration.

Several qualitative observations of the unstirred BZ system were made. The chemical waves in the reaction mixture did not behave like water waves. When one of the blue waves ran into a wall, no reflection was observed; the wave simply terminated. This observation is in agreement with the grass–fire analogy. If a fire encounters a barrier that it cannot penetrate, it will eventually die out, not reflect back to where it had already burned. Also, when two waves collided, there was no interference. Again, the waves annihilated one another.

Superior results can be obtained if the waves are videotaped and the front position is determined from the monitor using a ruler, or better yet, using image analysis software (Pojman and West, 1996).

In order to achieve accurate results from this experiment, convection must be eliminated. An open container allows evaporation, causing temperature gradients and convection. Even if the there is no air–solution interface, concentration and temperature gradients are generated by waves, which can cause convection (see

Chapter 9). To eliminate convection, the reaction can be performed in a gel. A simple approach is to add ultrafine silica gel (CAB-O-SIL, Cabot Corp.). However, reducing the thickness of the solution layer also reduces convection substantially. Such thin layers are difficult to create, because the solution will not wet the Petri dish. Using a surfactant such as TRITON X (Field and Winfree, 1979) or sodium dodecyl sulfate (SDS) can help.

References

Acton, F.S. 1970. *Numerical Methods that Work*. Harper and Row: New York.

Agladze, K.; Dulos, E.; De Kepper, P. 1992. "Turing Patterns in Confined Gel and Gel-Free Media," *J. Phys. Chem. 96*, 2400–2403.

Agladze, K. I.; Krinsky, V. I. 1982. "Multi-Armed Vortices in an Active Medium," *Nature 296*, 424–426.

Alamgir, M.; Epstein, I. R. 1983. "Birhythmicity and Compound Oscillation in Coupled Chemical Oscillators: Chlorite–Bromate–Iodide System," *J. Am. Chem. Soc. 105*, 2500–2502.

Alamgir, M.; Epstein, I. R. 1984. "Experimental Study of Complex Dynamical Behavior in Coupled Chemical Oscillators," *J. Phys. Chem. 88*, 2848–2851.

Alamgir, M.; Epstein, I. R. 1985a. "Complex Dynamical Behavior in a New Chemical Oscillator: The Chlorite–Thiourea Reaction in a CSTR," *Int. J. Chem. Kinet. 17*, 429–439.

Alamgir, M.; Epstein, I. R. 1985b. "New Chlorite Oscillators: Chlorite–Bromide and Chlorite–Thiocyanate in a CSTR," *J. Phys. Chem. 89*, 3611–3614.

Allcock, H. R.; Lampe, F. W. 1981. *Contemporary Polymer Chemistry*. Prentice-Hall: Englewood Cliffs, N.J.

Allnatt, A. R.; Jacobs, P. W. M. 1968. "Theory of Nucleation in Solid State Reactions," *Can. J. Chem. 46*, 111–116.

Andronov, A. A.; Vitt, A. A.; Khaikin, S. E. 1966. *Theory of Oscillations*. Pergamon: New York.

Antar, B. N.; Nuotio-Antar, V. S. 1993. *Fundamentals of Low Gravity Fluid Dynamics and Heat Transfer*. CRC Press: Boca Raton, Fla.

Arkin, A.; Ross, J. 1995. "Statistical Construction of Chemical Reaction Mechanisms from Measured Time-Series," *J. Phys. Chem. 99*, 970–979.

Atkins, P. W. 1995. *Physical Chemistry*, 5th ed.; W. H. Freeman: San Francisco.

Balcom, B. J.; Carpenter, T. A.; Hall, L. D. 1992. "Methacrylic Acid Polymerization. Travelling Waves Observed by Nuclear Magnetic Resonance Imaging," *Macromolecules 25*, 6818–6823.

Bar-Eli, K. 1985. "The Peristaltic Effect on Chemical Oscillations," *J. Phys. Chem. 89*, 2852–2855.

Barkley, D. 1988. "Slow Manifolds and Mixed-Mode Oscillations in the Belousov–Zhabotinskii Reaction," *J. Chem. Phys. 89*, 5547–5559.

Bassett, M. R.; Hudson, J. L. 1987. "The Dynamics of the Electrodissolution of Copper," *Chem. Eng. Commun. 60*, 145–159.

Bassett, M. R.; Hudson, J. L. 1989. "Quasi-Periodicity and Chaos during an Electrochemical Reaction," *J. Phys. Chem. 93*, 2731–2736.

Bauer, G. J.; McCaskill, J. S.; Otten, H. 1989. "Traveling Waves of *in vitro* evolving RNA," *Proc. Natl. Acad. Sci. U.S.A. 86*, 7937–7941.

Bayliss, A.; Matkowsky, B. J. 1987. "Fronts, Relaxation Oscillations and Period Doubling in Solid Fuel Combustion," *J. Comput. Phys. 71*, 147–168.

Bayliss, A.; Matkowsky, B. J. 1990. "Two Routes to Chaos in Condensed Phase Combustion," *SIAM J. Appl. Math. 50*, 437–459.

Bazsa, G.; Epstein, I. R. 1985. "Traveling Waves of Iron(II) Oxidation in Nitric Acid," *J. Phys. Chem. 89*, 3050–3053.

Begishev, V. P.; Volpert, V. A.; Davtyan, S. P.; Malkin, A. Y. 1985. "On some Features of the Anionic Activated ε–Caprolactam Polymerization Process under Wave Propagation Conditions," *Dokl. Phys. Chem. 279*, 1075–1077.

Bellman, R.; Cooke, K. L. 1963. *Differential-Difference Equations*. Academic Press: New York.

Belousov, B. P. "A Periodic Reaction and Its Mechanism," 1958. *Sbornik Referatov po Radiatsionni Meditsine*, 145.

Belousov, B. P. 1981. "A Periodic Reaction and its Mechanism," in *Autowave Processes in Systems with Diffusion* (Grecova, M. T., Ed.). USSR Academy of Sciences, Institute of Applied Physics: Gorky; pp. 176–186.

Belousov, B. P. 1985. "A Periodic Reaction and its Mechanism," in *Oscillations and Traveling Waves in Chemical Systems* (Field, R. J.; Burger, M., Eds.). Wiley: New York; pp. 605–613.

Benini, O.; Cervellati, R.; Fetto, P. 1996. "The BZ Reaction: Experimental and Model Studies in the Physical Chemistry Laboratory," *J. Chem. Ed. 73*, 865–868.

Berding, C.; Haken, H.; Harbich, T. 1983. "A Prepattern Formation Mechanism for the Spiral Type Patterns of the Sunflower Head," *J. Theor. Biol. 104*, 53–57.

Bernasconi, C. 1976. *Reaction Kinetics*. Academic Press: New York.

Bernasconi, C. F., Ed. 1986. *Investigation of Rates and Mechanisms of Reactions. Part II. Investigation of Elementary Reaction Steps in Solution and Fast Reaction Techniques.* Wiley: New York.

Berridge, M. J. 1989. "Cell Signaling through Cytoplasmic Calcium Oscillations," in *Cell Signaling: From Experiments to Theoretical Models* (Goldbeter, A., Ed.). Academic Press: London; pp. 449–459.

Berry, M. V. 1984. "Quantal Phase Factors Accompanying Adiabatic Changes," *Proc. Roy. Soc. London A 392*, 45–57.

Bode, P. M.; Bode, H. R. 1987. "Formation of Pattern in Regenerating Tissue Pieces of *Hydra attenuata*. IV. Three Processes Combine to Determine the Number of Tentacles," *Development 99*, 89–98.

Boga, E.; Kádár; Peintler, G.; Nagypál, I. 1990. "Effect of Magnetic Fields on a Propagating Reaction Front," *Nature 347*, 749–751.

Boissonade, J. 1976. "Theoretical Studies on 'Double Oscillation' in a Chemical Dissipative System," *J. Chim. Phys. 73*, 540–544.

Boissonade, J.; De Kepper, P. 1980. "Transitions from Bistability to Limit Cycle Oscillations. Theoretical Analysis and Experimental Evidence in an Open Chemical System," *J. Phys. Chem. 84*, 501–506.

Boiteux, A.; Hess, B.; Sel'kov, E. E. 1980. "Creative Functions of Instability and Oscillations in Metabolic Systems," *Curr. Topics Cellular Regulation 17*, 171–203.

Boon, J. P.; Dab, D.; Kapral, R.; Lawniczak, A. 1996. "Lattice Gas Automata for Reactive Systems," *Phys. Rep. 273*, 55–148.

Borckmans, P.; DeWit, A.; Dewel, G. 1992. "Competition in Ramped Turing Structures," *Physica A 188*, 137–157.

Boukalouch, M.; Elezgaray, J.; Arneodo, A.; Boissonade, J.; De Kepper, P. 1987. "Oscillatory Instability by Mass Interchange between Two Coupled Steady-State Reactors," *J. Phys. Chem. 91*, 5843–5845.

Bowden, G.; Garbey, M.; Ilyashenko, V. M.; Pojman, J. A.; Solovyov, S.; Taik, A.; Volpert, V. 1997. "The Effect of Convection on a Propagating Front with a Solid Product: Comparison of Theory and Experiments," *J. Phys. Chem. B 101*, 678–686.

Boyce, W. E.; DiPrima, R. C. 1977. *Elementary Differential Equations and Boundary Value Problems*; Wiley: New York.

Bray, W. C. 1921. "A Periodic Reaction in Homogeneous Solution and its Relation to Catalysis," *J. Am. Chem. Soc. 43*, 1262–1267.

Bray, W. C.; Liebhafsky, H. A. 1931. "Reactions Involving Hydrogen Peroxide, Iodine and Iodate Ion. I. Introduction," *J. Phys. Chem. 53*, 38–48.

Brenan, K. E.; Campbell, S. L.; Petzold, L. R. 1996. *Numerical Solution of Initial-Value Problems in Differential-Algebraic Equations*. Society for Industrial and Applied Mathematics: Philadelphia

Briggs, T. S.; Rauscher, W. C. 1973. "An Oscillating Iodine Clock Reaction," *J. Chem. Ed. 50*, 496.

Broeck, C. V. D.; Nicolis, G. 1993. "Noise-Induced Sensitivity to the Initial Conditions in Stochastic Dynamical Systems," *Phys. Rev. E 48*, 4845–4846.

Brusa, M. A.; Perissinotti, L. J.; Colussi, A. J. 1985. "Electron Spin Resonance Kinetic Studies of Malonyl Radical Self-Decay and Oxidation Reactions by Cerium(IV) and Bromate in Acid Aqueous Media. The Role of Free Radicals in the Belousov–Zhabotinskii Oscillator," *J. Phys. Chem. 89*, 1572–1574.

Buchholtz, F.; Schneider, F. W. 1987. "Computer Simulation of T3/T7 Phage Infection Using Lag Times," *Biophys. Chem. 26*, 171–179.

Buchholtz, F.; Golowasch, J.; Marder, E.; Epstein, I. R. 1992. "Mathematical Model of an Identified Stomatogastric Ganglion Neuron," *J. Neurophysiol. 67*, 332–340.

Busse, H. G. 1969. "A Spatial Periodic Homogeneous Chemical Reaction," *J. Phys. Chem. 73*, 750.

Byrne, G. D.; Hindmarsh, A. C. 1987. "Stiff ODE Solvers: A Review of Current and Coming Attractions," *J. Comp. Phys. 70*, 1–62.

Camacho, P.; Lechleiter, J. D. 1993. "Increased Frequency of Calcium Waves in *Xenopus laevis* Oocytes that Express a Calcium-ATPase," *Science 260*, 226–229.

Carlier, M. F.; Melki, R.; Pantaloni, D.; Hill, T. L.; Chen, Y. 1987. "Synchronous Oscillations in Microtubule Polymerization," *Proc. Natl. Acad. Sci. U.S.A. 84*, 5257–5261.

Caroli, B.; Caroli, C.; Roulet, B. 1992. "Instabilities of Planar Solidification Fronts," in *Solids Far From Equilibrium* (Godreche, C., Ed.). Cambridge University Press: Cambridge, U.K.; pp. 155–294.

Castets, V.; Dulos, E.; Boissonade, J.; De Kepper, P. 1990. "Experimental Evidence of a Sustained Standing Turing-Type Nonequilibrium Chemical Pattern," *Phys. Rev. Lett. 64*, 2953–2956.

Celia, M. A.; Gray, W. G. 1992. *Numerical Methods for Differential Equations. Fundamental Concepts for Scientific and Engineering Applications.* Prentice Hall: Englewood Cliffs, N.J.

Cesaro, A.; Benegas, J. C.; Ripoll, D. R. 1986. "Molecular Model of the Cooperative Amylose–Iodine–Triiodide Complex," *J. Phys. Chem. 90*, 2787–2791.

Chance, B.; Pye, E. K.; Ghosh, A. K.; Hess, B., Eds. 1973. *Biological and Biochemical Oscillators.* Academic Press: New York.

Chevalier, T.; Freund, A.; Ross, J. 1991. "The Effects of a Nonlinear Delayed Feedback on a Chemical Reaction," *J. Chem. Phys. 95*, 308–316.

Chopin-Dumas, J. 1978. "Diagramme d'État de la Réaction de Bray," *C.R. Acad. Sci. 287*, 553–556.

Churchill, R. V. 1972. *Operational Mathematics*, 3rd ed. McGraw-Hill: New York.

Citri, O.; Epstein, I. R. 1986. "Mechanism for the Oscillatory Bromate–Iodide Reaction," *J. Am. Chem. Soc. 108*, 357–363.

Citri, O.; Epstein, I. R. 1987. "Dynamical Behavior in the Chlorite–Iodide Reaction: A Simplified Mechanism," *J. Phys. Chem. 91*, 6034–6040.

Citri, O.; Epstein, I. R. 1988. "Mechanistic Study of a Coupled Chemical Oscillator: The Bromate–Chlorite–Iodide Reaction," *J. Phys. Chem. 92*, 1865–1871.

Clarke, B. L. 1976. "Stability of the Bromate–Cerium–Malonic Acid Network. I. Theoretical Formulation," *J. Chem. Phys. 64*, 4165–4178.

Clarke, B. L. 1980. "Stability of Complex Reaction Networks," *Adv. Chem. Phys. 43*, 1–215.

Coffman, K. G.; McCormick, W. D.; Noszticzius, Z.; Simoyi, R. H.; Swinney, H. L. 1987. "Universality, Multiplicity, and the Effect of Iron Impurities in the Belousov–Zhabotinskii Reaction," *J. Chem. Phys. 86*, 119–129.

Colussi, A. J.; Ghibaudi, E.; Yuan, Z.; Noyes, R. M. 1991. "Oscillatory Oxidation of Benzaldehyde by Air. 1. Experimental Observations," *J. Am. Chem. Soc. 112*, 8660–8670.

Cooper, M. S. 1995. "Intercellular Signaling in Neuronal–Glial Networks," *Biosystems 34*, 65–85.

Correia, J. J.; Williams, R. C. 1983. "Mechanisms of Assembly and Disassembly in Microtubules," *Ann. Rev. Biophys. Bioeng. 12*, 211–235.

Cox, B. G. 1994. *Modern Liquid Phase Kinetics.* Oxford University Press: Oxford, U.K.

Crank, J. 1975. *Mathematics of Diffusion*, 2nd. ed.; Clarendon: Oxford, U.K..

Créton, R.; Zivkovic, D.; Speksnijder, J. E.; Dohmen, M. R. 1992. "Manipulation of Cytokinesis Affects Polar Ionic Current around the Eggs of *Lymnaea stagnalis*," *Roux's Arch. Dev. Biol. 201*, 346–353.

Crowley, M.; Field, R. J. 1986. "Electrically Coupled Belousov-Zhabotinskii Oscillators. 1. Experiments and Simulations," *J. Phys. Chem. 90*, 1907–1915.

Crowley, M. F.; Epstein, I. R. 1989. "Experimental and Theoretical Studies of a Coupled Chemical Oscillator: Phase Death, Multistability, and In-Phase and Out-Of-Phase Entrainment," *J. Phys. Chem. 93*, 2496–2502.

Curl, R. L. 1963. "Dispersed Phase Mixing. I. Theory and Effects in Simple Reactors," *AIChE J. 9*, 175–181.

Cussler, E. L. 1984. *Diffusion: Mass Transfer in Fluid Systems*, 2nd ed. Cambridge: London.

D'Ancona, U. 1954. *The Struggle for Existence.* E. J. Brill: Leiden, The Netherlands.

Dab, D.; Lawniczak, A.; Boon, J.-P.; Kapral, R. 1990. "Cellular-Automaton Model for Reactive Systems," *Phys. Rev. Lett.* *64*, 2462–2465.

Dateo, C. E.; Orbán, M.; De Kepper, P.; Epstein, I. R. 1982. "Bistability and Oscillations in the Autocatalytic Chlorite–Iodide Reaction in a Stirred-Flow Reactor," *J. Am. Chem. Soc.* *104*, 504–509.

Davidenko, J. M.; Pertsov, A. V.; Salomonsz, R.; Baxter, W.; Jalife, J. 1992. "Stationary and Drifting Spiral Waves of Excitation in Isolated Cardiac Muscle," *Nature 355*, 349–351.

Davtyan, S. P.; Zhirkov, P. V.; Vol'fson, S. A. 1984. "Problems of Non-isothermal Character in Polymerisation Processes," *Russ. Chem. Rev.* *53*, 150–163.

de Groot, S. R.; Mazur, P. 1984. *Nonequilibrium Thermodynamics.* Dover: New York.

De Kepper, P.; Epstein, I. R. 1982. "A Mechanistic Study of the Oscillations and Bistability in the Briggs–Rauscher Reaction," *J. Am. Chem. Soc.* *104*, 49–55.

De Kepper, P.; Boissonade, J.; Epstein, I. R. 1990. "Chlorite–Iodide Reaction: A Versatile System for the Study of Nonlinear Dynamical Behavior," *J. Phys. Chem.* *94*, 6525–6536.

De Kepper, P.; Epstein, I. R.; Kustin, K. 1981a. "Bistability in the Oxidation of Arsenite by Iodate in a Stirred Flow Reactor," *J. Am. Chem. Soc.* *103*, 6121–6127.

De Kepper, P.; Epstein, I. R.; Kustin, K. 1981b. "A Systematically Designed Homogeneous Oscillating Chemical Reaction. The Arsenite–Iodate–Chlorite System," *J. Am. Chem. Soc.* *103*, 2133–2134.

De Kepper, P.; Epstein, I. R.; Orbán, M.; Kustin, K. 1982. "Batch Oscillations and Spatial Wave Patterns in Chlorite Oscillating Systems," *J. Phys. Chem.* *86*, 170–171.

De Kepper, P.; Rossi, A.; Pacault, A. 1976. "Etude d'une Réaction Chimique Périodique. Diagramme d'État de la Réaction de Belousov–Zhabotinskii," *Compt. Rend. Scéances Acad. Sci., Ser. C. 283*, 371–375.

De Meeus, J.; Sigalla, J. 1966. "Cinétique et Mécanisme de la Reduction du Chlorite par l'Iodure," *J. Chim. Phys. 63*, 453–459.

Decroly, O.; Goldbeter, A. 1982. "Birhythmicity, Chaos, and Other Patterns of Temporal Self-Organization in a Multiply Regulated Biochemical System," *Proc. Natl. Acad. Sci. U.S.A. 79*, 6917–6921.

Degn, H. 1967. "Effect of Bromine Derivatives of Malonic Acid on the Oscillating Reaction of Malonic Acid, Cerium Ions and Bromate," *Nature 213*, 589–590.

Degn, H. 1972. "Oscillating Chemical Reactions in Homogeneous Phase," *J. Chem. Ed. 49*, 302–307.

Degn, H.; Harrison, D. E. F. 1969. "Theory of Oscillations of Respiration Rate in Continuous Culture of *Klebsiella aerogenes*," *J. Theor. Biol. 22*, 238–248.

Ditto, W. L.; Rauseo, S. N.; Spano, M. L. 1990. "Experimental Control of Chaos," *Phys. Rev. Lett. 65*, 3211–3214.

Doedel, E.; Keller, H. B.; Kernévez., J. 1991. "Numerical Analysis and Control of Bifurcation Problems: (I) Bifurcation in Finite Dimensions," *Int. J. Bifurc. Chaos 1*, 493–520.

Dolnik, M.; Epstein, I. R. 1993. "A Coupled Chemical Burster: The Chlorine Dioxide–Iodide Reaction in Two Flow Reactors," *J. Chem. Phys. 98*, 1149–1155.

Dolnik, M.; Schreiber, I.; Marek, M. 1986. "Dynamic Regimes in a Periodically Forced Reaction Cell with Oscillatory Chemical Reaction," *Physica D 21*, 78–92.

Doona, C. J.; Kustin, K.; Orbán, M.; Epstein, I. R. 1991. "Newly Designed Permanganate-Reductant Chemical Oscillators," *J. Am. Chem. Soc. 113*, 7484–7489.

Dushman, S. 1904. "The Rate of the Reaction between Iodic and Hydriodic Acids," *J. Phys. Chem. 8*, 453–482.

Dutt, A. K.; Menzinger, M. 1990. "Stirring and Mixing Effects on Chemical Instabilities: Bistability of the $BrO_3^-/Br^-/Ce^{+3}$ System," *J. Phys. Chem. 94*, 4867–4870.

Duysens, L. N. M.; Amesz, J. 1957. "Fluorescence Spectrophotometry of Reduced Phosphopyridine Nucleotide in Intact Cells in the Near-Ultraviolet and Visible Region," *Biochim. Biophys. Acta 24*, 19–26.

Eager, M. D.; Santos, M.; Dolnik, M.; Zhabotinsky, A. M.; Kustin, K.; Epstein, I. R. 1994. "Dependence of Wave Speed on Acidity and Initial Bromate Concentration in the Belousov–Zhabotinsky Reaction–Diffusion System," *J. Phys. Chem. 98*, 10750–10755.

Edblom, E. C.; Luo, Y.; Orbán, M.; Kustin, K.; Epstein, I. R. 1989. "Kinetics and Mechanism of the Oscillatory Bromate–Sulfite–Ferrocyanide Reaction," *J. Phys. Chem. 93*, 2722–2727.

Edblom, E. C.; Orbán, M.; Epstein, I. R. 1986. "A New Iodate Oscillator: The Landolt Reaction with Ferrocyanide in a CSTR," *J. Am. Chem. Soc. 108*, 2826–2830.

Edelson, D.; Allara, D. L. 1980. "A Computational Analysis of the Alkane Pyrolysis Mechanism: Sensitivity Analysis of Individual Reaction Steps," *Int. J. Chem. Kinet. 12*, 605–621.

Edelson, D.; Thomas, V. M. 1981. "Sensitivity Analysis of Oscillating Reactions. 1. The Period of the Oregonator," *J. Phys. Chem. 85*, 1555–1558.

Edelson, D.; Field, R. J.; Noyes, R. M. 1975. "Mechanistic Details of the Belousov–Zhabotinskii Oscillations," *Int. J. Chem. Kinet. 7*, 417–432.

Edelstein-Keshet, L. 1988. *Mathematical Models in Biology.* McGraw-Hill: New York.

Eggert, J.; Scharnow, B. 1921. "Einige Analogiefälle zur Landolt Reaktion. (Über die Landoltreaktion II.)," *Z. Elektrochem. 27*, 455–470.

Eigen, M. 1963a. "Fast Elementary Steps in Chemical Reaction Mechanism," *Pure and Appl. Chem. 6*, 97–115.

Eigen, M. 1963b. "Ionen-und Ladungsübertragungsreaktionen in Lösungen," *Ber. Bunsenges. Physik. Chem. 67*, 753–762.

Eigen, M.; Schuster, P. 1979. *The Hypercycle. A Principle of Natural Self-Organization.* Springer-Verlag: Berlin.

Eiswirth, M.; Freund, A.; Ross, J. 1991. "Mechanistic Classification of Chemical Oscillators and the Role of Species," *Adv. Chem. Phys. 80*, 127–199.

Eiswirth, M.; Krischer, K.; Ertl, G. 1988. "Transition to Chaos in an Oscillating Surface Reaction," *Surf. Sci. 202*, 565–591.

Eiswirth, R. M.; Krischer, K.; Ertl, G. 1990. "Nonlinear Dynamics in the CO-Oxidation on Pt Single Crystal Surfaces," *Appl. Phys. A 5*, 79–90.

Epstein, I. R. 1987. "Patterns in Time and Space Generated by Chemistry," *Chem. Eng. News, March 30*, 24–36.

Epstein, I. R. 1989. "The Role of Flow Systems in Far-from Equilibrium Dynamics," *J. Chem. Ed. 66*, 191–195.

Epstein, I. R. 1989. "Differential Delay Equations in Chemical Kinetics: Some Simple Linear Model Systems," *J. Chem. Phys. 92*, 1702–1712.

Epstein, I. R.; Golubitsky, M. 1993. "Symmetric Patterns in Linear Arrays of Coupled Cells," *Chaos 3*, 1–5.

Epstein, I. R.; Kustin, K. 1984. "Design of Inorganic Chemical Oscillators," *Structure and Bonding 56*, 1–33.

Epstein, I. R.; Kustin, K. 1985. "A Mechanism for Dynamical Behavior in the Oscillatory Chorite–Iodide Reaction," *J. Phys. Chem. 89*, 2275–2282.

Epstein, I. R.; Luo, Y. 1991. "Differential Delay Equations in Chemical Kinetics. Nonlinear Models: The Cross-Shaped Phase Diagram and the Oregonator," *J. Chem. Phys. 95*, 244–254.

Epstein, I. R.; Dateo, C. E.; De Kepper, P.; Kustin, K.; Orbán, M. 1981. "Bistability in a C.S.T.R.: New Experimental Examples and Mathematical Modeling," in *Nonlinear Phenomena in Chemical Dynamics* (Vidal, C; Pacault, A., Eds.) Springer-Verlag: Berlin; pp. 188–191.

Epstein, I. R.; Lengyel, I.; Kàdàr, S.; Kagan, M.; Yokoyama, M. 1992. "New Systems for Pattern Formation Studies," *Physica A 188*, 26–33.

Ermentrout, G. B.; Kopell, N. 1990. "Oscillator Death in Systems of Coupled Neural Oscillators," *SIAM J. Appl. Math. 50*, 125–146.

Erneux, T.; Nicolis, G. 1993. "Propagating Waves in Discrete Bistable Reaction–Diffusion Systems," *Physica D 67*, 237–244.

Espenson, J. H. 1995. *Chemical Kinetics and Reaction Mechanisms*, 2nd ed. McGraw-Hill: New York.

Falta, J.; Imbihl, R.; Henzler, M. 1990. "Spatial Pattern Formation in a Catalytic Surface Reaction: The Facetting of Pt(110) in CO + O_2," *Phys. Rev. Lett. 64*, 1409–1412.

Farage, V. J.; Janjic, D. 1980. "Effect of Mechanical Stirring on Oscillating Chemical Reactions. Preliminary Communication," *Chimia 34*, 342.

Farage, V. J.; Janjic, D. 1981. "Effects of Mechanical Agitation on Oscillating Chemical Reactions in the Presence of Oxygen or Nitrogen," *Chimia 35*, 289.

Farey, J. 1816. "A letter from Dr. William Richardson to the Countess of Gosford (occasioned by the Perusal of Cuvier's 'Geological Essay') describing the Arrangement of the Strata for 60 Miles on the South and as many on the North of Gosford Castle, in the Armagh County in Ireland. Communicated by Mr. John Farey, Sen.; with some preliminary Remarks and illustrative Notes by him," *Phil. Mag. 47*, 354–364.

Farmer, J. D.; Sidorowich, J. J. 1989. "Exploiting Chaos to Predict the Future and Reduce Noise," in *Evolution, Learning and Cognition* (Lee., Y. C., Ed.). World Scientific: New York; pp. 277–330.

Fechner, A. T. 1828. "Ueber Umkehrungen der Polarität der einfachen Kette," *Schweigg. J. 53*, 61–76.

Feeney, R.; Schmidt, S.; Ortoleva, P. 1981. "Experiments on Electric Field – BZ Chemical Wave Interactions: Annihilation and the Crescent Wave," *Physica D 2*, 536–544.

Feigenbaum, M. J. 1979. "The Universal Metric Properties of Nonlinear Transformations," *J. Stat. Phys. 21*, 669–706.

Feinberg, M. 1980. "Chemical Oscillations, Multiple Equilibria, and Reaction Network Structure," in *Dynamics and Modelling of Reactive Systems* (Stewart, W.; Ray, W. H., Conley, C., Eds.). Academic Press: New York; pp. 59–130.

Feinberg, M.; Horn, F. J. M. 1974. "Dynamics of Open Chemical Systems and the Algebraic Structure of the Underlying Reaction Network," *Chem. Eng. Sci. 29*, 775–787.

Feinberg, M.; Horn, F. J. M. 1977. "Chemical Mechanism Structure and the Coincidence of the Stoichiometric and Kinetic," *Arch. Rat. Mech. Anal. 66*, 83–97.

Ferrone, F. A.; Hofrichter, J.; Eaton, W. A. 1985a. "Kinetics of Sickle Hemoglobin Polymerization I. Studies Using Temperature-Jump and Laser Photolysis Techniques," *J. Mol. Biol. 183*, 591–610.

Ferrone, F. A.; Hofrichter, J.; Eaton, W. A. 1985b. "Kinetics of Sickle Hemoglobin Polymerization II. A Double Nucleation Mechanism," *J. Mol. Biol. 183*, 611–631.

Field, R. J. 1989. "The Language of Dynamics," *J. Chem. Ed. 66*, 188–189.

Field, R. J.; Burger, M. 1985. *Oscillations and Traveling Waves in Chemical Systems*. Wiley: New York.

Field, R. J.; Försterling, H. D. 1986. "On the Oxybromine Chemistry Rate Constants with Cerium Ions in the Field–Körös–Noyes Mechanism of the Belousov–Zhabotinskii Reaction," *J. Phys. Chem. 90*, 5400–5407.

Field, R. J.; Noyes, R. M. 1972. "Explanation of Spatial Band Propagation in the Belousov Reaction," *Nature (London)* 237, 390–392.

Field, R. J.; Noyes, R.M. 1974a. "Oscillations in Chemical Systems. V. Quantitative Explanation of Band Migration in the Belousov–Zhabotinskii Reaction," *J. Am. Chem. Soc.* 96, 2001–2006.

Field, R. J.; Noyes, R. 1974b. "Oscillations in Chemical Systems. IV. Limit Cycle Behavior in a Model of a Real Chemical Reaction," *J. Chem. Phys.* 60, 1877–1884.

Field, R. J.; Noyes, R. M. 1977. "Oscillations in Chemical Systems. 18. Mechanisms of Chemical Oscillators: Conceptual Bases," *Acc. Chem. Res.* 10, 214–221.

Field, R. J.; Schneider, F. W. 1989. "Oscillating Chemical Reactions and Nonlinear Dynamics," *J. Chem. Ed.* 66, 195–204.

Field, R. J.; Winfree, A. T. 1979. "Travelling Waves of Chemical Activity in the Zaikin–Zhabotinskii–Winfree Reagent," *J. Chem. Ed.* 56, 754.

Field, R. J.; Körös, E.; Noyes, R. 1972. "Oscillations in Chemical Systems. II. Thorough Analysis of Temporal Oscillation in the Bromate–Cerium–Malonic Acid System," *J. Am. Chem. Soc.* 94, 8649–8664.

Fife, P. 1984. "Propagator–Controller Systems and Chemical Patterns," in *Non-equilibrium Dynamics in Chemical Systems*; (Vidal, C.; Pacault, A., Eds.). Springer-Verlag: Berlin; pp. 76–88.

Fink, T.; Imbihl, R.; Ertl, G. 1989. "Excitation of Chemical Waves in a Surface Reaction by a Laser-Induced Thermal Desorption: CO Oxidation on Pt(100)," *J. Chem. Phys.* 91, 5002–5010.

Fisher, R. A. 1937. "The Wave of Advance of Advantageous Genes," *Ann. Eugen.* 7, 355–369.

Fitzhugh, R. 1961. "Impulses and Physiological States in Models of Nerve Membrane," *Biophys. J.* 1, 445–466.

Foerster, P.; Müller, S. C.; Hess, B. 1988. "Curvature and Propagation Velocity of Chemical Waves," *Science* 241, 685–687.

Försterling, H.-D.; Noszticzius, Z. 1989. "An Additional Negative Feedback Loop in the Classical Belousov–Zhabotinsky Reaction: Malonyl Radical as a Second Control Intermediate," *J. Phys. Chem.* 93, 2740–2748.

Försterling, H.-D.; Stuk, L. 1992. "Effects of Ce^{4+}/Sulfato Complex Formation in the Belousov–Zhabotinskii Reaction: ESR Studies of Malonyl Radical Formation," *J. Phys. Chem.* 96, 3067–3072.

Försterling, H.-D.; Murányi, S.; Noszticzius, Z. 1990. "Evidence of Malonyl Controlled Oscillations in the Belousov–Zhabotinsky Reaction (Malonic Acid–Bromate–Cerium System)," *J. Phys. Chem.* 94, 2915–2921.

Franck, U. F. 1985. "Spontaneous Temporal and Spatial Phenomena in Physicochemical Systems," in *Temporal Order* (Rensing, L.; Jaeger, N., Eds.). Springer: Berlin; pp. 2–12.

Furukawa, H. 1992. "Phase Separation by Directional Quenching and Morphological Transition," *Physica A* 180, 128–155.

Gaal, O.; Vereczkey, L.; Medgyesi, G. A. 1980. *Electrophoresis in the Separation of Biological Macromolecules*. Wiley: New York.

Ganapathisubramanian, N.; Noyes, R. M. 1982a. "A Discrepancy between Experimental and Computational Evidence for Chemical Chaos," *J. Chem. Phys.* 76, 1770–1774.

Ganapathisubramanian, N.; Noyes, R. M. 1982b. "Additional Complexities during Oxidation of Malonic Acid in the Belousov–Zhabotinsky Oscillating Reaction," *J. Phys. Chem.* 86, 5158–5162.

Gardiner, C. W. 1990. *Handbook of Stochastic Methods for Physics, Chemistry and the Natural Sciences*, 2nd ed. Springer-Verlag: Berlin.

Garfinkel, A.; Spano, M. L.; Ditto, W. L.; Weiss, J. N. 1992. "Controlling Cardiac Chaos," *Science 257*, 1230–1235.

Gáspár, V.; Showalter, K. 1990. "A Simple Model for the Oscillatory Iodate Oxidation of Sulfite and Ferrocyanide," *J. Phys. Chem. 94*, 4973–4979.

Gáspár, V.; Maselko, J.; Showalter, K. 1991. "Transverse Coupling of Chemical Waves," *Chaos 1*, 435–444.

Gatti, R. A.; Robinson, W. A.; Dinare, A. S.; Nesbit, M.; McCullogh, J. J.; Ballow, J. J.; Good, R. A. 1973. "Cyclic Leukocytosis in Chronic Myologenous Leukemia," *Blood 41*, 771–782.

Gear, C. W. 1971. *Numerical Initial Value Problems in Ordinary Differential Equations.* Prentice-Hall: Englewood Cliffs, N.J.

Gerhardt, M.; Schuster, H.; Tyson, J. J. 1990. "A Cellular Automaton Model of Excitable Media Including Curvature and Dispersion," *Science 247*, 1563–1566.

Gerisch, G. 1968. "Cell Aggregation and Differentiation in *Dictyostelium*," in *Current Topics in Developmental Biology* (Moscona, A.; Monroy, A., Eds.). Academic Press: New York; pp. 157–197.

Gershuni, G. Z.; Zhukhovitsky, E. M. 1976. *Convective Stability of Incompressible Fluids* (translated from Russian). Israel Program for Scientific Translations: Jerusalem; Springfield, Va.

Ghosh, A. K.; Chance, B. 1964. "Oscillations of Glycolytic Intermediates in Yeast Cells," *Biochem. Biophys. Res. Commun. 16*, 174–181.

Gillespie, D. T. 1977. "Exact Stochastic Simulation of Coupled Chemical Reactions," *J. Phys. Chem. 81*, 2340–2361.

Glass, L.; Mackey, M. C. 1979. "Pathological Conditions Resulting from Instabilities in Physiological Control Systems," *Ann. N.Y. Acad. Sci. 316*, 214–235.

Glass, L.; Mackey, M. C. 1988. *From Clocks to Chaos: The Rhythms of Life.* Princeton University Press: Princeton, N.J.

Glasser, L. 1989. "Order, Chaos and All That!," *J. Chem. Ed. 66*, 997–1001.

Gleick, J. 1987. *Chaos: Making a New Science.* Viking: New York.

Goldbeter, A. 1996. *Biochemical Oscillations and Cellular Rhythms: The Molecular Bases of Periodic and Chaotic Behaviour.* Cambridge University Press: Cambridge, U.K.

Golowasch, J. 1990. Characterization of a Stomatogastric Ganglion. A Biophysical and a Mathematical Description. Ph.D. Thesis, Brandeis University: Waltham, Mass.

Golowasch, J.; Marder, E. 1992. "Ionic Currents of the Lateral Pyloric Neuron of the Stomatogastric Ganglion of the Crab," *J. Neurophysiol. 67*, 318–331.

Golubitsky, M. 1985. *Singularities and Groups in Bifurcation Theory.* Springer-Verlag: New York.

Gorelova, N. A.; Bures, J. 1983. "Spiral Waves of Spreading Depression in the Isolated Chicken Retina," *J. Neurobiol. 14*, 353–363.

Gray, P.; Scott, S. 1986. "A New Model for Oscillatory Behaviour in Closed Systems: The Autocatalator," *Ber. Bunsenges. Phys. Chem. 90*, 985–996.

Gray, P.; Scott, S. K. 1983. "Autocatalytic Reactions in the Isothermal, Continuous Stirred Tank Reactor: Isolas and Other Forms of Multistability," *Chem. Eng. Sci. 38*, 29–43.

Gray, P.; Scott, S. K. 1990. *Chemical Oscillations and Instabilities.* Clarendon: Oxford, U.K.

Graziani, K. R.; Hudson, J. L.; Schmitz, R. A. 1976. "The Belousov–Zhabotinskii Reaction in a Continuous Flow Reactor," *Chem. Eng. J. 12*, 9–12.

Grebogi, C.; Ott, E.; Yorke, J. A. 1982. "Chaotic Attractors in Crisis," *Phys. Rev. Lett. 48*, 1507–1510.

Grutzner, J. B.; Patrick, E. A.; Pellechia, P. J.; Vera, M. 1988. "The Continuously Rotated Cellular Reactor," *J. Am. Chem. Soc. 110*, 726–728.

Guckenheimer, J.; Holmes, P. 1983. *Nonlinear Oscillations, Dynamical Systems, and Bifurcations of Vector Fields*. Springer-Verlag: New York.

Gunton, J. D.; San Miguel, M.; Sahni, P. S. 1983. "The Dynamics of First-Order Phase Transitions," in *Phase Transitions and Critical Phenomena* (Domb, C.; Lebowitz, J. L., Eds.). Academic Press: London; pp. 267–466.

Györgyi, L.; Field, R. J. 1989. "A Periodicity Resulting from Two-Cycle Coupling in the Belousov–Zhabotinskii Reaction. III. Analysis of a Model of the Effect of Spatial Inhomogeneities at the Input Ports of a Continuous-Flow Stirred Tank Reactor," *J. Chem. Phys. 91*, 6131–6141.

Györgyi, L.; Field, R. J. 1991. "Simple Models of Deterministic Chaos in the Belousov–Zhabotinsky Reaction," *J. Phys. Chem. 95*, 6594–6602.

Györgyi, L.; Field, R. J. 1992. "A Three-Variable Model of Deterministic Chaos in the Belousov–Zhabotinsky Reaction," *Nature 355*, 808–810.

Györgyi, L.; Field, R. J.; Noszticzius, Z.; McCormick, W. D.; Swinney, H. L. 1992. "Confirmation of High Flow Rate Chaos in the Belousov–Zhabotinsky Reaction," *J. Phys. Chem. 96*, 1228–1233.

Hairer, E.; Norsett, S. P.; Wanner, G. 1996. *Solving Ordinary Differential Equations* (2 vols.), 2nd ed. Springer-Verlag: New York.

Hale, J. K. 1979. "Nonlinear Oscillations in Equations with Delays," *Lec. Appl. Math. 17*, 157–185.

Hamilton, C. A.; Kautz, R. L.; Lloyd, F. L.; Steiner, R. L.; Field, B. F. 1987. "The NBS Josephson Array Voltage Standard," *IEEE Trans. Instrum. Meas. (U.S.A.) 36*, 258–261.

Hammel, S. M.; Yorke, J. A.; Grebogi, C. 1987. "Do Numerical Orbits of Complex Processes Represent True Orbits?," *J. Complexity 3*, 136–145.

Hammes, G. G. 1982. *Enzyme Catalysis and Regulation*. Academic Press: New York.

Hanna, A.; Saul, A.; Showalter, K. 1982. "Detailed Studies of Propagating Fronts in the Iodate Oxidation of Arsenous Acid," *J. Am. Chem. Soc. 104*, 3838–3844.

Hannay, J. H. 1985. "Angle Variable Holonomy in Adiabatic Excursion of an Integrable Hamiltonian," *J. Phys. A 18*, 221–230.

Hansen, E. W.; Ruoff, P. 1989, "Determination of Enolization Rates and Overall Stoichiometry from ^1H NMR Records of the Methylmalonic Acid BZ Reaction," *J. Phys. Chem. 93*, 2696–2704.

Hardy, G. H.; Wright, E. M., Eds. 1989. *An Introduction to the Theory of Numbers*. Clarendon: Oxford, U.K..

Hardy, J.; de Pazzis, O.; Pomeau, Y. 1976. "Molecular Dynamics of a Classical Lattice Gas: Transport Properties and Time Correlation Functions," *Phys Rev. A13*, 1949–1961.

Hasal, P.; Nevorl, V.; Schreiber, I.; Sevciková, H.; Snita, D.; Marek, M. 1997. "Waves in Ionic Reaction–Diffusion–Migration Systems," in *Nonlinear Dynamics, Chaotic and Complex Systems* (Infeld, E.; Zelazny, R.; Galkowski, A., Eds.). Canbridge University Press: Cambridge, U.K.; pp. 72–98.

Hauser, M. J. B.; Simoyi, R. H. 1994a. "Inhomogeneous Precipitation Patterns in a Chemical Wave," *Phys. Lett. A 191*, 31–38.

Hauser, M. J. B.; Simoyi, R. H. 1994b. "Inhomogeneous Precipitation Patterns in a Chemical Wave. Effect of Thermocapillary Convection," *Chem. Phys. Lett. 227*, 593–600.

Hayes, N. D. 1950. "Roots of the Transcendental Equations Associated with a Certain Difference-Differential Equation," *J. London Math. Soc. 25*, 226–232.

He, X.; Kustin, K.; Nagypál, I.; Peintler, G. 1994. "A Family of Magnetic Field Dependent Chemical Waves," *Inorg. Chem. 33*, 2077–2078.

Higgins, J. 1967. "The Theory of Oscillating Reactions," *Ind. Eng. Chem. 59*, 18–62.

Hindmarsh, A. C. 1974. *GEAR—Ordinary Differential Equation System Solver. UCID-30001 Rev. 3*. Lawrence Livermore Laboratory: Livermore, Calif.

Hocker, C. G. 1994. "Applying Bifurcation Theory to Enzyme Kinetics," in *Numerical Computer Methods* (Johnson, M. L.; Brand, L., Eds.). Academic Press: New York; pp. 781–816.

Hocker, C. G.; Epstein, I. R. 1989. "Analysis of a Four-Variable Model of Coupled Chemical Oscillators," *J. Chem. Phys. 90*, 3071–3080.

Hocker, C. G.; Epstein, I. R.; Kustin, K.; Tornheim, K. 1994. "Glycolytic pH Oscillations in a Flow Reactor," *Biophys. Chem. 51*, 21–35.

Hodges, A.1983. *Alan Turing: The Enigma*. Simon and Schuster: New York.

Hodgkin, A. L.; Huxley, A. F. 1952. "A Quantitative Description of Membrane Current and its Application to Conduction and Excitation in Nerve," *J. Physiol. (London) 117*, 500–544.

Hofrichter, J. 1986. "Kinetics of Sickle Hemoglobin Polymerization. III. Nucleation Rates Determined from Stochastic Fluctuations in Polymerization Progress Curves," *J. Mol. Biol. 189*, 553–571.

Hollinger, H. B.; Zenzen, M. J. 1991. "Thermodynamic Irreversibility I. What is it?," *J. Chem. Ed. 68*, 31–34.

Hudson, J. L.; Hart, M.; Marinko, D. 1979. "An Experimental Study of Multiple Peak Periodic and Nonperiodic Oscillations in the Belousov–Zhabotinskii-Reaction," *J. Chem. Phys. 71*, 1601–1606.

Hunt, E. R. 1991. "Stabilizing High-Period Orbits in a Chaotic System: The Diode Resonator," *Phys. Rev. Lett. 67*, 1953–1955.

Hwang, J.-T.; Dougherty, E. P.; Rabitz, S.; Rabitz, H. 1978. "The Green's Function Method of Sensitivity Analysis in Chemical Kinetics," *J. Chem. Phys. 69*, 5180–5191.

Hynne, F.; Sørensen, P. G. 1987. "Quenching of Chemical Oscillations," *J. Phys. Chem. 91*, 6573–6577.

Imbihl, R.; Reynolds, A. E.; Kaletta, D. 1991. "Model for the Formation of a Microscopic Turing Structure: The Faceting of Pt(110) during Catalytic Oxidation of CO," *Phys. Rev. Lett. 67*, 275–278.

Ives, D. J. G.; Janz, G. J. 1961. *Reference Electrodes: Theory and Practice*. Academic Press: New York.

Jacobs, S. S.; Epstein, I. R. 1976. "Effects of Chloride Ion on Oscillations in the Bromate–Cerium–Malonic Acid System," *J. Am. Chem. Soc. 98*, 1721–1724.

Jahnke, W.; Winfree, A. T. 1991. "Recipes for Belousov–Zhabotinsky Reagents," *J. Chem. Ed. 68*, 320–324.

Jahnke, W.; Henze, C.; Winfree, A. T. 1988. "Chemical Vortex Dynamics in Three-Dimensional Excitable Media," *Nature 336*, 662–665.

Jahnke, W.; Skaggs, W. E.; Winfree, A. T. 1989. "Chemical Vortex Dynamics in the Belousov–Zhabotinsky Reaction and in the Two-Variable Oregonator Model," *J. Phys. Chem. 93*, 740–749.

Jakubith, S.; Rotermund, H. H.; Engel, W.; von Oertzen, A.; Ertl, G. 1990. "Spatiotemporal Concentration Patterns in a Surface Reaction: Propagating and Standing Waves, Rotating Spirals, and Turbulence," *Phys. Rev. Lett. 65*, 3013–3016.

Jalife, J.; Antzelevitch, C. 1979. "Phase Resetting and Annihilation of Pacemaker Activity in Cardiac Tissue," *Science 206*, 695–697.

Johnson, B. R.; Scott, S. K. 1990. "Period Doubling and Chaos during the Oscillatory Ignition of the CO + O_2 Reaction," *J .Chem. Soc., Faraday Trans. 86*, 3701–3705.

Johnson, B. R.; Griffiths, J. F.; Scott, S. K. 1991. "Bistability, Oscillations and Chaos in CO + O_2 Combustion," *Chaos 1*, 387–395.

Johnson, K. A.; Borisy, G. G. 1977. "Kinetic Analysis of Microtubule Self-Assembly *in Vitro*," *J. Mol. Biol. 117*, 1–31.

Jordan, P. C. 1979. *Chemical Kinetics and Transport*. Plenum Press: New York.

Kádár, S.; Lengyel, I.; Epstein, I. R. 1995. "Modeling of Transient Turing-Type Patterns in the Closed Chlorine–Dioxide–Iodine–Malonic Acid–Starch Reaction System," *J. Phys. Chem. 99*, 4054–4058.

Kagan, M. L.; Kepler, T. B.; Epstein, I. R. 1991. "Geometric Phase Shifts in Chemical Oscillators," *Nature 349*, 506–508.

Kaner, R. J.; Epstein, I. R. 1978. "Induction and Inhibition of Chemical Oscillations by Iodide Ion in the Belousov–Zhabotinskii Reaction," *J. Am. Chem. Soc. 100*, 4073–4079.

Kapral, R.; Lawniczak, A.; Masiar, P. 1992. "Reactive Dynamics in a Multi-Species Lattice-Gas Automaton," *J. Chem. Phys. 96*, 2762–2776.

Kasperek, G. J.; Bruice, T. C. 1971. "Observations on an Oscillating Reaction. The Reaction of Potassium Bromate, Ceric Sulfate, and a Dicarboxylic Acid," *Inorg. Chem. 10*, 382–386.

Kawczynski, A. L.; Gorecki, J. 1992. "Molecular Dynamics Simulations of Sustained Oscillations in a Thermochemical System," *J. Phys. Chem. 96*, 1060–1067.

Keane, T. R. 1972. "Single-Phase Polymerization Reactors," *Chemical Reaction Engineering: Proceedings of the Fifth European/Second International Symposium on Chemical Reaction Engineering* (Fortuin, J. M. H., Ed.). Elsevier:Amsterdam; pp. A7-1–A7-9.

Keener, J. P.; Tyson, J. J. 1986. "Spiral Waves in the Belousov–Zhabotinsky Reaction," *Physica 21D*, 307–324.

Kepler, T. B.; Kagan, M. L. 1991. "Geometric Phase Shifts under Adiabatic Parameter Changes in Classical Dissipative Systems," *Phys. Rev. Lett. 66*, 847–849.

Keresztessy, A.; Nagy, I. P.; Bazsa, G.; Pojman, J. A. 1995. "Traveling Waves in the Iodate–Sulfite and Bromate–Sulfite Systems," *J. Phys. Chem. 99*, 5379–5384.

Kern, D. M.; Kim, C. H. 1965. "Iodine Catalysis in the Chlorite–Iodide Reaction," *J. Am. Chem. Soc. 87*, 5309–5313.

Khan, A. M.; Pojman, J. A. 1996. "The Use of Frontal Polymerization in Polymer Synthesis," *Trends Polym. Sci. (Cambridge, U.K.) 4*, 253–257.

Kim, C.; Teng, H.; Tucker, C. L.; White, S. R. 1995. "The Continuous Curing Process for Thermoset Polymer Composites. Part 1: Modeling and Demonstration," *J. Comp. Mater. 29*, 1222–1253.

Kondepudi, D. K.; Sabanayagam, C. 1994. "Secondary Nucleation that Leads to Chiral Symmetry Breaking in Stirred Crystallization," *Chem. Phys. Lett. 217*, 364–368.

Kondepudi, D. K.; Kaufman, R. J.; Singh, N. 1990. "Chiral Symmetry Breaking in Sodium Chlorate Crystallization," *Science 250*, 975–976.

Korotkov, V. N.; Chekanov, Y. A.; Rozenberg, B. A. 1993. "The Simultaneous Process of Filament Winding and Curing for Polymer Composites," *Comp. Sci. Tech. 47*, 383–388.

Kuhnert, L.; Agladze, K. I.; Krinsky, V. I. 1989. "Image Processing Using Light-Sensitive Chemical Waves," *Nature 337*, 244–245.

Kumpinsky, E.; Epstein, I. R. 1985. "A Model for Stirring Effects on Transitions in Bistable Chemical Systems," *J. Chem. Phys. 82*, 53–57.

Lang, S. 1986. *Introduction to Linear Algebra*, 2nd ed. Springer-Verlag: New York.

Laplante, J. P.; Erneux, T. 1992. "Propagation Failure in Arrays of Coupled Bistable Chemical Reactors," *J. Phys. Chem. 96*, 4931–4934.

Lechleiter, J.; Girard, S.; Peralta, E.; Clapham, D. 1991. "Spiral Calcium Wave Propagation and Annihilation in *Xenopus laevis* Oocytes," *Science 252*, 123–126.

Lechleiter, J. D.; Clapham, D. E. 1992. "Molecular Mechanisms of Intracellular Calcium Excitablility in *X. laevis* Oocytes," *Cell 69*, 283–294.

Lee, K. J.; McCormick, W. D.; Ouyang, Q.; Swinney, H. L. 1993. "Pattern Formation by Interacting Chemical Fronts," *Science 261*, 192–194.

Lee, K. J.; McCormick, W. D.; Swinney, H. L.; Noszticzius, Z. 1992. "Turing Patterns Visualized by Index of Refraction Variations," *J. Chem. Phys. 95*, 4048–4049.

Legros, J. C.; Dupont, O.; Queeckers, P.; Vaerenbergh, S. V. 1990. "Thermodynamic Instabilities and Capillary Flows," *Progress in Astronautics and Aeronautics 130*, 207–239.

Lehninger, A. L. 1993. *Principles of Biochemistry*, 2nd ed. Worth: New York.

Lemarchand, A.; Ben Aim, R. I.; Nicolis, G. 1989. "Experimental Study of Sensitivity of Thermal Explosion," *Chem. Phys. Lett. 162*, 92–98.

Lengyel, I.; Epstein, I. R. 1991. "Modeling of Turing Structures in the Chlorite–Iodide–Malonic Acid–Starch Reaction System," *Science 251*, 650–652.

Lengyel, I.; Epstein, I. R. 1992. "A Chemical Approach to Designing Turing Patterns in Reaction–Diffusion Systems," *Proc. Natl. Acad. Sci. U.S.A. 89*, 3977–3979.

Lengyel, I.; Epstein, I. R. 1995. "The Chemistry Behind the First Experimental Chemical Examples of Turing Patterns," in *Chemical Waves and Patterns* (Kapral, R.; Showalter, K., Eds.). Kluwer: Dordrecht; pp. 297–322.

Lengyel, I.; Kádár, S.; Epstein, I. R. 1992a. "Quasi-Two-Dimensional Turing Patterns in an Imposed Gradient," *Phys. Rev. Lett. 69*, 2729–2732.

Lengyel, I.; Kádár, S.; Epstein, I. R. 1993. "Transient Turing Structures in a Gradient-Free Closed System," *Science 259*, 493–495.

Lengyel, I.; Li, J.; Epstein, I. R. 1992b "Dynamical Study of the Chlorine Dioxide–Iodide Open System Oscillator," *J. Phys. Chem. 96*, 7032–7037.

Lengyel, I.; Li, J.; Kustin, K.; Epstein, I. R. 1996. "Rate Constants for Reactions between Iodine- and Chlorine-Containing Species: A Detailed Mechanism of the Chlorine Dioxide/Chlorite–Iodide Reaction," *J. Am. Chem. Soc. 118*, 3708–3719.

Lengyel, I.; Rábai, G.; Epstein, I. R. 1990a. "Batch Oscillation in the Reaction of Chlorine Dioxide with Iodine and Malonic Acid," *J. Am. Chem. Soc. 112*, 4606–4607.

Lengyel, I.; Rábai, G.; Epstein, I. R. 1990b. "Experimental and Modeling Study of Oscillations in the Chlorine Dioxide–Iodine–Malonic Acid Reaction," *J. Am. Chem. Soc. 112*, 9104–9110.

Levitan, I. B.; Kaczmarek, L. K. 1997. *The Neuron: Cell and Molecular Biology*, 2nd ed. Oxford University Press: New York.

Liebhafsky, H. A.; Roe, G. M. 1979. "The Detailed Mechanism of the Dushman Reaction Explored by Computer," *Int. J. Chem. Kinet. 11*, 693–703.

Lohmann, K. J. 1992. "How Sea Turtles Navigate," *Sci. Amer. 266*, 100–107.

López-Tomàs, L.; Sagués, F. 1991. "New Features of Stirring Sensitivities of the Belousov–Zhabotinskii Reaction," *J. Phys. Chem. 95*, 701–705.

Lorenz, E. N. 1964. "The Problem of Deducing the Climate from the Governing Equations," *Tellus 16*, 1–11.

Lorenz, E. N. 1973. "Deterministic Nonperiodic Flow," *J. Atmos. Sci. 20*, 130–141.

Lotka, A. J. 1910. "Contribution to the Theory of Periodic Reactions," *J. Phys. Chem. 14*, 271–274.

Lotka, A. J. 1920a. "Analytical Note on Certain Rhythmic Relations in Organic Systems," *Proc. Natl. Acad. Sci. U.S.A. 6*, 410–415.

Lotka, A. J. 1920b. "Undamped Oscillations Derived from the Law of Mass Action," *J. Am. Chem. Soc. 42*, 1595–1599.

Lotka, A. J. 1925. *Elements of Physical Biology*. Williams and Wilkins: Baltimore, Md.

Luo, Y.; Epstein, I. R. 1986. "Stirring and Premixing Effects in the Oscillatory Chlorite–Iodide Reaction," *J. Chem. Phys. 85*, 5733–5740.

Luo, Y.; Epstein, I. R. 1990. "Feedback Analysis of Mechanisms for Chemical Oscillators," *Adv. Chem. Phys. 79*, 269–299.

Luo, Y.; Epstein, I. R. 1991. "A General Model for pH Oscillators," *J. Am. Chem. Soc. 113*, 1518–1522.

Luo, Y.; Orbán, M.; Kustin, K.; Epstein, I. R. 1989. "Mechanistic Study of Oscillations and Bistability in the Cu(II)-Catalyzed Reaction between H_2O_2 and KSCN," *J. Am. Chem. Soc. 111*, 4541–4548.

Macdonald, N. 1989. *Biological Delay Systems: Linear Stability Theory*. Cambridge University Press: Cambridge, U.K.

Mackey, M. C.; Glass, L. 1977. "Oscillation and Chaos in Physiological Control Systems," *Science 197*, 287–289.

Malthus, T. R. 1798. *An Essay on the Principle of Population*. Oxford University Press: London.

Mandelkow, E.; Mandelkow, E.-M.; Hotani, H.; Hess, B.; Müller, S. C. 1989. "Spatial Patterns from Oscillating Microtubules," *Science 246*, 1291–1293.

Mandelkow, E.-M.; Lange, G.; Jagla, A.; Spann, U.; Mandelkow, E. 1988. "Dynamics of the Microtubule Oscillator: Role of Nucleotides and Tubulin–MAP Interactions," *EMBO J. 7*, 357–365.

Marcus, R. A. 1993. "Electron Transfer Reactions in Chemistry. Theory and Experiment," *Rev. Mod. Phys. 65*, 599–610.

Marek, M.; Schreiber, I. 1991. *Chaotic Behavior of Deterministic Dissipative Systems*. Cambridge University Press: Cambridge, U.K..

Marek, M.; Svobodova, E. 1975. "Nonlinear Phenomena in Oscillatory Systems of Homogeneous Reactions. Experimental Observations," *Biophys. Chem. 3*, 263–273.

Margolis, S. B. 1983. "An Asymptotic Theory of Condensed Two-Phase Flame Propagation," *SIAM J. Appl. Math. 43*, 331–369.

Markus, M.; Hess, B. 1990. "Isotropic Cellular Automaton for Modelling Excitable Media," *Nature 347*, 56–58.

Maruyama, M. 1963. "The Second Cybernetics: Deviation-Amplifying Mutual Causal Processes," *Am. Sci. 51*, 164–179.

Maselko, J.; Epstein, I. R. 1984. "Dynamical Behavior of Coupled Chemical Oscillators: Chlorite–Thiosulfate–Iodide–Iodine," *J. Phys. Chem. 88*, 5305–5308.

Maselko, J.; Showalter, K. 1989. "Chemical Waves on Spherical Surfaces," *Nature 339*, 609–611.

Maselko, J.; Showalter, K. 1991. "Chemical Waves in Inhomogeneous Excitable Media," *Physica D 49*, 21–32.

Maselko, J.; Swinney, H. L. 1986. "Complex Periodic Oscillations and Farey Arithmetic in the Belousov–Zhabotinskii Reaction," *J. Chem. Phys. 85*, 6430–6441.

Maselko, J.; Swinney, H. L. 1987. "A Farey Triangle in the Belousov–Zhabotinskii Reaction," *Phys. Lett. A 119*, 403–406.

Maselko, J.; Alamgir, M.; Epstein, I. R. 1986. "Bifurcation Analysis of a System of Coupled Oscillators: Bromate–Chlorite–Iodide," *Physica 19D*, 153–161.

Maselko, J.; Reckley, J. S.; Showalter, K. 1989. "Regular and Irregular Spatial Patterns in an Immobilized-Catalyst BZ Reaction," *J. Phys. Chem. 93*, 2774–2780.

Masere, J.; Vasquez, D. A.; Edwards, B. F.; Wilder, J. W.; Showalter, K. 1994. "Nonaxisymmetric and Axisymmetric Convection in Propagating Reaction–Diffusion Fronts," *J. Phys. Chem. 98*, 6505–6508.

Matkowsky, B. J.; Sivashinsky, G. I. 1978. "Propagation of a Pulsating Reaction Front in Solid Fuel Combustion," *SIAM J. Appl. Math. 35*, 465–478.

May, R. M. 1974. *Stability and Complexity in Model Ecosystems*; 2nd ed. Princeton University Press: Princeton, N.J.

McCaskill, J. S.; Bauer, G. J. 1993. "Images of Evolution: Origin of Spontaneous RNA Replication Waves," *Proc. Natl. Acad. Sci. U.S.A. 90*, 4191–4195.

Melichercik, M.; Mrakavová, M.; Nagy, A.; Olexová, A.; Treindl, L. 1992. "Permanganate Oscillators with Keto Dicarboxylic Acid," *J. Phys. Chem. 96*, 8367–8368.

Melka, R. F.; Olsen, G.; Beavers, L.; Draeger, J. A. 1992. "The Kinetics of Oscillating Reactions," *J. Chem. Ed. 69*, 596–598.

Menzinger, M.; Dutt, A. K. 1990. "The Myth of the Well-Stirred CSTR in Chemical Instability Experiments: The Chlorite/Iodide Reaction," *J. Phys. Chem. 94*, 4510–4514.

Meron, E.; Procaccia, I. 1987. "Gluing Bifurcations in Critical Flows: The Route to Chaos in Parametrically Excited Surface Waves," *Phys. Rev. A 35*, 4008–4011.

Metcalfe, G.; Ottino, J. M. 1994. "Autocatalytic Processes in Mixing Flows," *Phys. Rev. Lett. 72*, 2875–2878.

Metropolis, N.; Stein, M. L.; Stein, P. R. 1973. "On Finite Limit Sets for Transformations on the Unit Interval," *J. Comb. Theory 15A*, 25–44.

Meyer, T.; Stryer, L. 1988. "Molecular Model for Receptor-Stimulated Calcium Spiking," *Proc. Natl. Acad. Sci. U.S.A. 85*, 5051–5055.

Millero, F. J. 1972. "Partial Molal Volumes of Electrolytes in Aqueous Solutions," in *Water and Aqueous Solutions: Structure, Thermodynamics and Transport Processes* (Horne, R. A., Ed.). Wiley: New York; pp. 519–595.

Min, K. W.; Ray, W. H. 1974. "On the Mathematical Modeling of Emulsion Polymerization Reactors," *J. Macro. Sci. (Revs.) 11*, 177–255.

Mirollo, R. E.; Strogatz, S. H. 1990. "Synchronization of Pulse-Coupled Biological Oscillator," *SIAM J. Appl. Math. 50*, 1645–1662.

Möckel, P. 1977. "Photochemisch induzierte dissipative Strukturen," *Naturwissenschaften 64*, 224.

Morgan, J. S. 1916. "The Periodic Evolution of Carbon Monoxide," *J. Chem. Soc. Trans. 109*, 274–283.

Müller, S. C.; Plesser, T.; Hess, B. 1985. "The Structure of the Core of the Spiral Wave in the Belousov–Zhabotinsky Reagent," *Science 230*, 661–663.

Mullins, W. W.; Sekerka, R. F. 1964. "Stability of a Planar Interface During Solidification of a Dilute Binary Alloy," *J. Appl. Phys. 35*, 444–451.

Münster, A. F.; Hasal, F.; Snita, D.; Marek, M. 1994. "Charge Distribution and Electric Field Effects on Spatiotemporal Patterns," *Phys. Rev. E 50*, 546–550.

Murdoch, K. A. 1992. "The Amylose–Iodine Complex," *Carbohydr. Res. 233*, 161–174.

Murray, A.; Hunt, T. 1993. *The Cell Cycle: An Introduction*. Freeman: New York.

Murray, J. D. 1993. *Mathematical Biology*. 2nd corr. ed. Springer-Verlag: Berlin.

Nagorcka, B. N.; Mooney, J. R. 1982. "The Role of a Reaction–Diffusion System in the Formation of Hair Fibres," *J. Theor. Biol. 98*, 575–607.

Nagumo, J. S.; Arimoto, S.; Yoshizawa, S. 1962. "An Active Pulse Transmission Line Simulating Nerve Axon," *Proc. IRE 50*, 2061–2070.

Nagy, I. P.; Pojman, J. A. 1993. "Multicomponent Convection Induced by Fronts in the Chlorate–Sulfite Reaction," *J. Phys. Chem. 97*, 3443–3449.

Nagy, I. P.; Pojman, J. A. 1996. "Suppressing Convective Instabilities in Propagating Fronts by Tube Rotation," *J. Phys. Chem. 100*, 3299–3304.

Nagy, I. P.; Keresztessy, A.; Pojman, J. A. 1995a. "Periodic Convection in the Bromate–Sulfite Reaction: A 'Jumping Wave'," *J. Phys. Chem. 99*, 5385–5388.

Nagy, I. P.; Keresztessy, A.; Pojman, J. A.; Bazsa, G.; Noszticzius, Z. 1994. "Chemical Waves in the Iodide–Nitric Acid System," *J. Phys. Chem. 98*, 6030–6037.

Nagy, I. P.; Sike, L.; Pojman, J. A. 1995b. "Thermochromic Composite Prepared via a Propagating Polymerization Front," *J. Am. Chem. Soc. 117*, 3611–3612.

Nagypál, I.; Epstein, I. R. 1986. "Fluctuations and Stirring Rate Effects in the Chlorite–Thiosulfate Reaction," *J. Phys. Chem. 90*, 6285–6292.

Nagypál, I.; Epstein, I. R. 1988. "Stochastic Behavior and Stirring Rate Effects in the Chlorite–Iodide Reaction," *J. Chem. Phys. 89*, 6925–6928.

Nagypál, I.; Bazsa, G.; Epstein, I. R. 1986. "Gravity Induced Anisotropies in Chemical Waves," *J. Am. Chem. Soc. 108*, 3635–3640.

Neta, P.; Huie, R. E.; Ross, A. B. 1988. "Rate Constants for Reactions of Inorganic Radicals in Aqueous Solutions," *J. Phys. Chem. Ref. Data 17*, 1027–1284.

Nicogossian, A. E.; Huntoon, C. L.; Pool, S. L. 1994. *Space Physiology and Medicine*, 3rd ed. Williams and Wilkins: Baltimore, Md.

Nicolis, C.; Nicolis, G. 1993. "Finite Time Behavior of Small Errors in Deterministic Chaos and Lyapunov Exponents," *Intl. J. Bifur. Chaos 3*, 1339–1342.

Nicolis, G.; Baras, F. 1987. "Intrinsic Randomness and Spontaneous Symmetry-Breaking in Explosive Systems," *J. Stat. Phys. 48*, 1071–1090.

Nicolis, G.; Gaspard, P. 1994. "Toward a Probabilistic Approach to Complex Systems," *Chaos, Solitions & Fractals 4*, 41–57.

Nicolis, G.; Prigogine, I. 1977. *Self-Organization in Nonequilibrium Systems*. Wiley: New York.

Nicolis, G.; Prigogine, I. 1989. *Exploring Complexity*. W. H. Freeman: New York.

Norrish, R. G. W.; Smith, R. R. 1942. "Catalyzed Polymerization of Methyl Methacrylate in the Liquid Phase," *Nature 150*, 336–337.

Noszt*i*czius, Z.; Bódiss, J. 1979. "A Heterogeneous Chemical Oscillator. The Belousov–Zhabotinskii-Type Reaction of Oxalic Acid," *J. Am. Chem. Soc. 101*, 3177–3182.

Noszt*i*czius, Z.; Farkas, H.; Schelly, Z. A. 1984a. "Explodator: A New Skeleton Mechanism for the Halate Driven Chemical Oscillators," *J. Chem. Phys. 80*, 6062–6070.

Noszt*i*czius, Z.; Horsthemke, W.; McCormick, W. D.; Swinney, H. L. 1990. "Stirring Effects in the BZ Reaction with Oxalic Acid–Acetone Mixed Substrate in a Batch Reactor and in a CSTR," in *Spatial Inhomogeneities and Transient Behavior in Chemical Kinetics* (Gray, P.; Nicolis, G.; Baras, F.; Borckmans, P.; Scott, S. K., Eds.) Manchester University Press: Manchester, U.K. pp. 647–652.

Noszt*i*czius, Z.; Horsthemke, W.; McCormick, W. D.; Swinney, H. L.; Tam, W. Y. 1987. "Sustained Chemical Waves in an Annular Gel Reactor: A Chemical Pinwheel," *Nature 329*, 619–620.

Noszt*i*czius, Z.; Noszt*i*czius, E.; Schelly, Z. A. 1982. "On the Use of Ion-Selective Electrodes for Monitoring Oscillating Reactions. 1. Potential Response for Silver Halide Membrane Electrodes to Hypohalous Acids," *J. Am. Chem. Soc. 104*, 6194–6199.

Noszt*i*czius, Z.; Noszt*i*czius, E.; Schelly, Z. A. 1983. "On the Use of Ion-Selective Electrodes for Monitoring Oscillating Reactions. 2. Potential Response of Bromide- and Iodide-Selective Electrodes in Slow Corrosive Processes. Disproportionation of Bromous and Iodous Acid. A Lotka–Volterra Model for the Halate Driven Oscillators," *J. Phys. Chem. 87*, 510–524.

Noszt*i*czius, Z.; Ouyang, Q.; McCormick, W. D.; Swinney, H. L. 1992. "Effect of Turing Pattern Indicators of CIMA Oscillators," *J. Phys. Chem. 96*, 6302–6307.

Noszt*i*czius, Z.; Stirling, P.; Wittmann, M. 1985. "Measurement of Bromine Removal Rate in the Oscillatory BZ Reaction of Oxalic Acid. Transition from Limit Cycle Oscillations to Excitability via Saddle-Node Infinite Period Bifurcation," *J. Phys. Chem. 89*, 4914–4921.

Noszticzius, Z.; Wittman, M.; Stirling, P. 1984b. "A New Bromide-Selective Electrode for Monitoring Oscillating Reactions," in *4th Symposium on Ion-Selective Electrodes, Mátafüred*. Elsevier: Amsterdam; pp. 579–589.

Noyes, R. M. 1980. "A Generalized Mechanism for Bromate-Driven Oscillators," *J. Am. Chem. Soc. 102*, 4644–4649.

Noyes, R. M. 1989. "Some Models of Chemical Oscillators," *J. Chem. Ed. 66*, 190–191.

Noyes, R. M.; Field, R. J. 1974. "Oscillatory Chemical Reactions," *Annu. Rev. Phys. Chem. 25*, 95–119.

Noyes, R. M.; Field, R. J.; Körös, E. 1972. "Oscillations in Chemical Systems. I. Detailed Mechanism in a System Showing Temporal Oscillations," *J. Am. Chem. Soc. 94*, 1394–1395.

Noyes, R. M.; Field, R. J.; Thompson, R. C 1971.. "Mechanism of Reaction of Br(V) with Weak, One-Electron Reducing Agents," *J. Am. Chem. Soc. 93*, 7315–7316.

Nozakura, T.; Ikeuchi, S. 1984. "Formation of Dissipative Structures in Galaxies," *Astrophys. J. 279*, 40–52.

Ochiai, E.-I.; Menzinger, M. 1990. "Chemical Instabilities: A Spectroscopic Study of Spatial Inhomogeneities in the ClO_2^-/I^- Reaction in a Continuously Stirred Tank Reactor," *J. Phys. Chem. 94*, 8866–8868.

Odian, G. 1991. *Principles of Polymerization*, 3rd ed. Wiley: New York.

Okinaka, J.; Tran-Cong, Q. 1995. "Directional Phase Separation of a Binary Polymer Mixture Driven by a Temperature Gradient," *Physica D 84*, 23–30.

Olsen, L. F.; Degn, H. 1977. "Chaos in an Enzyme Reaction," *Nature 267*, 177–178.

Orbán, M. 1990. "Copper(II)-Catalyzed Oscillatory Chemical Reactions," *React. Kinet. Catal. Lett. 42*, 333–338.

Orbán, M.; Epstein, I. R. 1983. "Inorganic Bromate Oscillators. Bromate–Chlorite–Reductant," *J. Phys. Chem. 87*, 3212–3219.

Orbán, M.; Epstein, I. R. 1989a. "Chemical Oscillators in Group VIA: The Cu(II)-Catalyzed Reaction between Thiosulfate and Peroxodisulfate Ions," *J. Am. Chem. Soc. 111*, 2891–2896.

Orbán, M.; Epstein, I. R. 1989b. "The Minimal Permanganate Oscillator: The Guyard Reaction in a CSTR," *J. Am. Chem. Soc. 111*, 8543–8544.

Orbán, M.; Epstein, I. R. 1990. "Minimal Permanganate Oscillator and Some Derivatives: Oscillatory Oxidation of $S_2O_3^{2-}$, SO_3^{2-} and S^{2-} by Permanganate in a CSTR," *J. Am. Chem. Soc. 112*, 1812–1817.

Orbán, M.; Körös, E. 1979. "Chemical Oscillations during the Uncatalyzed Reaction of Aromatic Compounds with Bromate. 2. A Plausible Skeleton Mechanism," *J. Phys. Chem. 83*, 3056–3057.

Orbán, M.; Dateo, C.; De Kepper, P.; Epstein, I. R. 1982a. "Chlorite Oscillators: New Experimental Examples, Tristability and Preliminary Classification," *J. Am. Chem. Soc. 104*, 5911–5918.

Orbán, M.; De Kepper, P.; Epstein, I. R. 1982b. "Minimal Bromate Oscillator: Bromate–Bromide–Catalyst," *J. Am. Chem. Soc. 104*, 2657–2658.

Orbán, M.; De Kepper, P.; Epstein, I. R.; Kustin, K. 1981. "New Family of Homogeneous Chemical Oscillators: Chlorite–Iodate–Substrate," *Nature 292*, 816–818.

Orbán, M.; Epstein, I. R. 1982. "Complex Periodic and Aperiodic Oscillation in the Chlorite–Thiosulfate Reaction," *J. Phys. Chem. 86*, 3907–3910.

Orbán, M.; Körös, E.; Noyes, R. M. 1978. "Chemical Oscillations during the Uncatalyzed Reaction of Aromatic Compounds with Bromate. Part I. Search for Chemical Oscillators," *J. Phys. Chem. 82*, 1672–1674.

Orbán, M.; Lengyel, I.; Epstein, I. R. 1991. "A Transition Metal Oscillator: Oscillatory Oxidation of Manganese(II) by Periodate in a CSTR," *J. Am. Chem. Soc. 113*, 1978–1982.

Ostwald, W. 1899. "Periodisch veraenderliche Reaktionsgeschwindigkeiten," *Phys. Zeitsch. 8*, 87–88.

Ott, E.; Grebogi, C.; Yorke, J. A. 1990. "Controlling Chaos," *Phys. Rev. Lett. 64*, 1196–1199.

Ouyang, Q.; Boissonade, J.; Roux, J. C.; De Kepper, P. 1989. "Sustained Reaction–Diffusion Structures in an Open Reactor," *Phys. Lett. A 134*, 282–286.

Ouyang, Q.; Castets, V.; Boissonade, J.; Roux, J. C.; Kepper, P. D.; Swinney, H. L. 1991. "Sustained Patterns in Chlorite–Iodide Reactions in a One-Dimensional Reactor," *J. Chem. Phys. 95*, 351–360.

Ouyang, Q.; Swinney, H. L.; Roux, J. C.; Kepper, P. D.; Boissonade, J. 1992. "Recovery of Short-Lived Chemical Species in a Couette Flow Reactor," *AIChE J. 38*, 502–510.

Ouyang, Q.; Tam, W. Y.; DeKepper, P.; McCormick, W. D.; Noszticzius, Z.; Swinney, H. L. 1987. "Bubble-Free Belousov–Zhabotinskii-Type Reactions," *J. Phys. Chem. 91*, 2181–2184.

Pacault, A.; Hanusse, P.; De Kepper, P.; Vidal, C.; Boissonade, J. 1976. "Phenomena in Homogeneous Chemical Systems far from Equilibrium," *Acc. Chem. Res. 9*, 438–445.

Pagola, A.; Vidal, C. 1987. "Wave Profile and Speed near the Core of a Target Pattern in the Belousov–Zhabotinsky Reaction," *J. Phys. Chem. 91*, 501–503.

Parker, T. S. 1989. *Practical Numerical Algorithms for Chaotic Systems*. Spinger-Verlag: New York.

Peard, M.; Cullis, C. 1951. "A Periodic Chemical Reaction," *Trans. Faraday Soc. 47*, 616–630.

Pearson, J. E. 1993. "Complex Patterns in a Simple System," *Science 261*, 189–192.

Peintler, G. 1995. *Zita, A Comprehensive Program Package for Fitting Parameters of Chemical Reaction Mechanisms, 4.0*. Attila József University: Szeged, Hungary.

Peng, B.; Petrov, V.; Showalter, K. 1991. "Controlling Chemical Chaos," *J. Phys. Chem. 95*, 4957–4959.

Perraud, J. J.; Agladze, K.; Dulos, E.; De Kepper, P. 1992. "Stationary Turing Patterns versus Time-Dependent Structures in the Chlorite–Iodide Malonic Acid Reaction," *Physica A 188*, 1–16.

Petrov, V.; Gáspár, V.; Masere, J.; Showalter, K. 1993. "Controlling Chaos in the Belousov–Zhabotinsky Reaction," *Nature 361*, 240–243.

Petrov, V.; Scott, S. K.; Showalter, K. 1992. "Mixed-Mode Oscillations in Chemical Systems," *J. Chem. Phys. 97*, 6191–6198.

Pinsky, M. A. 1995. *The EMF Book: What You Should Know about Electromagnetic Fields, Electromagnetic Radiation, and Your Health*. Warner Books: New York.

Plant, R. E. 1981. "Bifurcation and Resonance in a Model for Bursting Nerve Cells," *J. Math. Biol. 11*, 15–32.

Plesser, T.; Müller, S. C.; Hess, B. 1990. "Spiral Wave Dynamics as a Function of Proton Concentration in the Ferroin-Catalyzed Belousov–Zhabotinsky Reaction," *J. Phys. Chem. 94*, 7501–7507.

Pojman, J. A. 1990. "A Simple Demonstration of Convective Effects on Reaction–Diffusion Systems: A Burning Cigarette," *J. Chem. Ed. 67*, 792–794.

Pojman, J. A. 1991. "Traveling Fronts of Methacrylic Acid Polymerization," *J. Am. Chem. Soc. 113*, 6284–6286.

Pojman, J. A.; West, W. W. 1996. "A Unified Physical Chemistry Lab Based on Oscillating Reactions and Traveling Fronts," *J. Chem. Ed. 73*, 35.

Pojman, J. A.; Craven, R.; Khan, A.; West, W. 1992a. "Convective Instabilities in Traveling Fronts of Addition Polymerization," *J. Phys. Chem. 96*, 7466–7472.

Pojman, J. A.; Craven, R.; Leard, D. 1994. "Oscillating Reactions and Traveling Waves in the Physical Chemistry Lab," *J. Chem. Ed. 71*, 84–90.

Pojman, J. A.; Curtis, G.; Ilyashenko, V. M. 1996a. "Frontal Polymerization in Solution," *J. Am. Chem. Soc. 115*, 3783–3784.

Pojman, J. A.; Dedeaux, H.; Fortenberry, D. 1992b. "Stirring Effects in the Mn-Catalyzed Belousov–Zhabotinskii Reaction with a Mixed Hypophosphite/Acetone Substrate in a Batch Reactor," *J. Phys. Chem. 96*, 7331–7333.

Pojman, J. A.; Epstein, I. R.; Karni, Y.; Bar-Ziv, E. 1991a. "Stochastic Coalescence–Redispersion Model for Molecular Diffusion and Chemical Reactions. 2. Chemical Waves," *J. Phys. Chem. 95*, 3017–3021.

Pojman, J. A.; Epstein, I. R.; McManus, T.; Showalter, K. 1991b. "Convective Effects on Chemical Waves. 2. Simple Convection in the Iodate–Arsenous Acid System," *J. Phys. Chem. 95*, 1299–1306.

Pojman, J. A.; Epstein, I. R.; Nagy, I. 1991c. "Convective Effects on Chemical Waves. 3. Multicomponent Convection in the Iron(II)–Nitric Acid System," *J. Phys. Chem. 95*, 1306–1311.

Pojman, J. A.; Ilyashenko, V. M.; Khan, A. M. 1995a. "Spin Mode Instabilities in Propagating Fronts of Polymerization," *Physica D 84*, 260–268.

Pojman, J. A.; Ilyashenko, V. M.; Khan, A. M. 1996b. "Free-Radical Frontal Polymerization: Self-Propagating Thermal Reaction Waves," *J. Chem. Soc. Faraday Trans. 92*, 2825–2837.

Pojman, J. A.; Khan, A. M.; Mathias, L. J. 1997a. "Frontal Polymerization in Microgravity: Results from the Conquest I Sounding Rocket Flight," *Microg. Sci. Tech., X*, 36–40.

Pojman, J. A.; Komlósi, A.; Nagy, I. P. 1996c. "Double-Diffusive Convection in Traveling Waves in the Iodate–Sulfite System Explained," *J. Phys. Chem. 100*, 16209–16212.

Pojman, J. A.; Leard, D. C.; West, W. 1992c. "The Periodic Polymerization of Acrylonitrile in the Cerium-Catalyzed Belousov–Zhabotinskii Reaction," *J. Am. Chem. Soc. 114*, 8298–8299.

Pojman, J. A.; Nagy, I. P.; Salter, C. 1993. "Traveling Fronts of Addition Polymerization with a Solid Monomer," *J. Am. Chem. Soc. 115*, 11044–11045.

Pojman, J. A.; West, W. W.; Simmons, J. 1997b. "Propagating Fronts of Polymerization in the Physical Chemistry Laboratory," *J. Chem. Ed. 74*, 727–730.

Pojman, J. A.; Willis, J.; Fortenberry, D.; Ilyashenko, V.; Khan, A. 1995b. "Factors Affecting Propagating Fronts of Addition Polymerization: Velocity, Front Curvature, Temperature Profile, Conversion and Molecular Weight Distribution," *J. Polym. Sci. Part A: Polym. Chem. 33*, 643–652.

Pojman, J. A.; Willis, J. R.; Khan, A. M.; West, W. W. 1996d. "The True Molecular Weight Distributions of Acrylate Polymers Formed in Propagating Fronts," *J. Polym. Sci. Part A: Polym. Chem. 34*, 991–995.

Pomeau, Y.; Roux, J. C.; Rossi, A.; Bachelart, S.; Vidal, C. 1981. "Intermittent Behaviour in the Belousov–Zhabotinsky Reaction," *J. Phys. Lett. 42*, L271–L273.

Póta, G.; Lengyel, I.; Bazsa, G. 1991. "Traveling Waves in the Acidic Nitrate–Iron(II) Reaction: Analytical Description of the Wave Velocity," *J. Phys. Chem. 95*, 4381–4386.

Prigogine, I. 1955. *Thermodynamics of Irreversible Processes*. Wiley: New York.

Prigogine, I. 1980. *From Being to Becoming*. W. H. Freeman: New York.

Prigogine, I.; Lefever, R. 1968. "Symmetry Breaking Instabilities in Dissipative Systems. II," *J. Chem. Phys. 48*, 1695–1700.

Puhl, A.; Nicolis, G. 1987. "Normal Form Analysis of Multiple Bifurcations in Incompletely Mixed Chemical Reactors," *J. Chem. Phys.* 87, 1070–1078.

Rábai, G.; Beck, M. T. 1987. "Kinetics and Mechanism of the Autocatalytic Reaction between Iodine and Chlorite Ion," *Inorg. Chem.* 26, 1195–1199.

Rábai, G.; Beck, M. T. 1988. "Exotic Chemical Phenomena and their Chemical Explanation in the Iodate–Sulfite–Thiosulfate System," *J. Phys. Chem.* 92, 2804–2807.

Rábai, G.; Epstein, I. R. 1989. "Oxidation of Hydroxylamine by Periodate in a CSTR: A New pH Oscillator," *J. Phys. Chem.* 93, 7556–7559.

Rábai, G.; Epstein, I. R. 1990. "Large Amplitude pH Oscillation in the Oxidation of Hydroxylamine by Iodate in a Continuous-Flow Stirred Tank Reactor," *J. Phys. Chem.* 94, 6361–6365.

Rábai, G.; Epstein, I. R. 1992. "pH Oscillations in a Semibatch Reactor," *J. Am. Chem. Soc.* 114, 1529–1530.

Rábai, G.; Bazsa, G.; Beck, M. T. 1979. "Design of Reaction Systems Exhibiting Overshoot–Undershoot Kinetics," *J. Am. Chem. Soc.* 101, 6746–6748.

Rábai, G.; Beck, M. T.; Kustin, K.; Epstein, I. R. 1989. "Sustained and Damped pH Oscillation in the Periodate–Thiosulfate Reaction in a CSTR," *J. Phys. Chem.* 93, 2853–2858

Rábai, G.; Kustin, K.; Epstein, I. R. 1989b. "A Systematically Designed pH Oscillator: The Hydrogen Peroxide–Sulfite–Ferrocyanide Reaction in a Continuous Stirred Tank Reactor," *J. Am. Chem. Soc.* 111, 3870–3874.

Rábai, G.; Nagy, Z.; Beck, M. T. 1987. "Quantitative Description of the Oscillatory Behavior of the Iodate–Sulfite–Thiourea System in CSTR," *React. Kinet. Catal. Lett.* 33, 23–29.

Rábai, G.; Orbán, M.; Epstein, I. R. 1990. "Design of pH-Regulated Oscillators," *Acc. Chem. Res.* 23, 258–263.

Rastogi, R. P.; Das, I.; Singh, A. R. 1984. "A New Iodate Driven Nonperiodic Oscillatory Reaction in a Continuously Stirred Tank Reactor," *J. Phys. Chem.* 88, 5132–5134.

Rawlings, J. B.; Ray, W. H. 1987. "Stability of Continuous Emulsion Polymerization Reactors: A Detailed Model Analysis," *Chem. Eng. Sci.* 42, 2767–2777.

Rayleigh, J. W. 1899. *Scientific Papers, ii.* Cambridge University Press: Cambridge, U.K.

Reynolds, W. L. 1958. "The Reaction between Potassium Ferrocyanide and Iodine in Aqueous Solution," *J. Am. Chem. Soc.* 80, 1830–1835.

Rice, F.; Reiff, O. 1927. "The Thermal Decomposition of Hydrogen Peroxide," *J. Phys. Chem.* 31, 1352–1356.

Richetti, P. 1987. Études Théoriques et Numériques des Dynamiques Chaotiques de la Réaction de Belousov–Zhabotinskii. Ph.D. Thesis, University of Bordeaux, France.

Richter, P. H.; Ross, J. 1980. "Oscillations and Efficiency in Glycolysis," *Biophys. Chem.* 12, 285–297.

Rinzel, J. 1981. "Models in Neurobiology," in *Nonlinear Phenomena in Physics and Biology* (Enns, R. H.; Jones, B. L.; Miura, R. M.; Ragnekar, S. S., Eds.. Plenum: New York; pp. 345–367.

Rinzel, J. 1987. *Lecture Notes in Biomathematics.* Springer: New York.

Rinzel, J.; Lee, Y. S. 1987. "Dissection of a Model for Neuronal Parabolic Bursting," *J. Math. Biol.* 25, 653–675.

Roebuck, J. R. 1902. "The Rate of the Reaction between Arsenous Acid and Iodine in Acidic Solution. The Rate of the Reverse Reaction and the Equilibrium between Them," *J. Phys. Chem.* 6, 365–398.

Roelofs, M. G.; Jensen, J. H. 1987. "EPR Oscillations During Oxidation of Benzaldehyde," *J. Phys. Chem.* 91, 3380–3382.

Röhrlich, B.; Parisi, J.; Peinke, J.; Rössler, O. E. 1986. "A Simple Morphogenetic Reaction–Diffusion Model Describing Nonlinear Transport Phenomena in Semiconductors," *Z. Phys. B: Cond. Matter 65*, 259–266.

Rosenthal, J. 1991. "A Spectroelectrochemical Demonstration of a Modified Belousov–Zhabotinski Reaction," *J. Chem. Ed. 68*, 794–795.

Rössler, O. E. 1976. "Chaotic Behavior in Simple Reaction Systems," *Z. Naturforsch. 31A*, 259–264.

Roux, J. C.; De Kepper, P.; Boissonade, J. 1983a. "Experimental Evidence of Nucleation Induced Transition in a Bistable Chemical System," *Phys. Lett. 97A*, 168–170.

Roux, J. C.; Rossi, A.; Bachelart, S.; Vidal, C. 1980. "Representation of a Strange Attractor from an Experimental Study of Chemical Turbulence," *Phys. Lett. A77*, 391–393.

Roux, J. C.; Simoyi, R. H.; Swinney, H. L. 1983b. "Observation of a Strange Attractor," *Physica 8D*, 257–266.

Rovinsky, A. B. 1987. "Turing Bifurcation and Stationary Patterns in the Ferroin Catalyzed Belousov–Zhabotinsky Reaction," *J. Phys. Chem. 91*, 4606–4613.

Roy, R.; Murphy, T. W.; Maier, T. D.; Gillis, Z.; Hunt, E. R. 1992. "Dynamical Control of a Chaotic Laser: Experimental Stabilization of a Globally Coupled System," *Phys. Rev. Lett. 68*, 1259–1262.

Ruff, I.; Friedrich, V. J.; Csillag, K. 1972. "Transfer Diffusion. III. Kinetics and Mechanism of the Triiodide–Iodide Exchange Reaction," *J. Phys. Chem. 76*, 162–165.

Rundle, R. E. 1943. "The Configuration of Starch and the Starch–Iodine Complex: I. The Dichroism of Flows of Starch–Iodine Solutions," *J. Am. Chem. Soc. 65*, 544–557.

Ruoff, P. 1984. "Phase Response Relationships of the Closed Bromide-Perturbed Belousov–Zhabotinsky Reaction. Evidence of Bromide Control of the Free Oscillating State without Use of a Bromide-Detecting Device," *J. Phys. Chem. 88*, 2851–2857.

Sagués, F.; Ramirez-Piscina, L.; Sancho, J. M. 1990. "Stochastic Dynamics of the Chlorite–Iodide Reaction," *J. Chem. Phys. 92*, 4786–4792.

Samardzija, N.; Greller, L. D.; Wasserman, E. 1989. "Nonlinear Chemical Kinetic Schemes Derived from Mechanical and Electrical Dynamics Systems," *J. Chem. Phys. 90*, 2296–2304.

Saunders, P. T., Ed. 1992. *Collected Works of A.M. Turing. Morphogenesis.* North-Holland Publishing: Amsterdam.

Savostyanov, V. S.; Kritskaya, D. A.; Ponomarev, A. N.; Pomogailo, A. D. 1994. "Thermally Initiated Frontal Polymerization of Transition Metal Nitrate Acrylamide Complexes," *J. Poly. Sci. Part A: Poly. Chem. 32*, 1201–1212.

Schell, M.; Ross, J. 1986. "Effects of Time Delay in Rate Processes," *J. Chem. Phys. 85*, 6489–6503.

Schmidt, S.; Ortoleva, P. 1977. "A New Chemical Wave Equation for Ionic Systems," *J. Chem. Phys. 67*, 3771–3776.

Schmidt, S.; Ortoleva, P. 1979. "Multiple Chemical Waves Induced by Applied Electric Field," *J. Chem. Phys. 71*, 1010–1015.

Schmidt, S.; Ortoleva, P. 1981. "Electric Field Effects on Propagating BZ Waves: Predictions of an Oregonator and New Pulse Supporting Models," *J. Chem. Phys. 74*, 4488–4500.

Schmitz, R. A.; Graziani, K. R.; Hudson, J. L. 1977. "Experimental Evidence of Chaotic States in the Belousov–Zhabotinskii Reaction," *J. Chem. Phys. 67*, 3040–3044.

Schneider, F. W. 1985. "Periodic Perturbations of Chemical Oscillators: Experiments," *Ann. Rev. Phys. Chem. 36*, 347–379.

Schork, F. J.; Ray, W. H. 1987. "The Dynamics of the Continuous Emulsion Polymerization of Methylmethacrylate," *J. Appl. Poly. Sci. (Chem.) 34*, 1259–1276.

Scott, S. K. 1991. *Chemical Chaos*. Clarendon Press: Oxford.

Sevcik, P.; Adamčiková, L. 1982. "Oscillating Heterogeneous Reaction with Oxalic Acid," *Collection Czechoslovak Chem. Commun. 47*, 891–898.

Sevcik, P.; Adamčiková, I. 1988. "Stirring Rate Effects in the Belousov–Zhabotinskii Reaction," *Chem. Phys. Lett. 146*, 419–421.

Sevcik, P.; Adamčiková, I. 1989. "Stirring Rate Effect in the Closed, Batch Belousov–Zhabotinsky System with Oxalic Acid," *J. Chem. Phys. 91*, 1012–1014.

Ševčíková, H.; Marek, M. 1983. "Chemical Waves in Electric Field," *Physica D 9*, 140–156.

Ševčíková, H.; Marek, M. 1986. "Chemical Waves in Electric Field—Modelling," *Physica D 21*, 61–77.

Seydel, R. 1994. *Practical Bifurcation and Stability Analysis.*, 2nd ed. Springer-Verlag: New York.

Shakhashiri, B. Z. 1985. *Chemical Demonstrations: A Handbook for Teachers. Volume 2.* University of Wisconsin Press: Madison.

Shapere, A.; Wilczek, F., Ed. 1989. *Geometric Phases in Physics.* World Scientific: Singapore.

Sharbaugh, A. H., III; Sharbaugh, A. H., Jr. 1989. "An Experimental Study of the Liesegang Phenomenon and Crystal Growth in Silica Gels," *J. Chem. Ed. 66*, 589–594.

Sharma, K. R.; Noyes, R. M. 1975. "Oscillations in Chemical Systems. VII. Effects of Light and of Oxygen on the Bray–Liebhafsky Reaction," *J. Am. Chem. Soc. 91*, 202–204.

Sharp, A. 1994. Single Neuron and Small Network Dynamics Explored with the Dynamic Clamp. Ph.D. Thesis, Brandeis University, Waltham, Mass.

Sharpe, F. R.; Lotka, A. J. 1923. "Contribution to the Analysis of Malaria Epidemiology. IV. Incubation Lab," *Am. J. Hygiene Suppl. 3*, 96–110.

Shaw, D. H.; Pritchard, H. O. 1968. "The Existence of Homogeneous Oscillating Reactions," *J. Phys. Chem. 72*, 2692–2693.

Shcherbak, S. B. 1983. "Unstable Combustion of Samples of Gas-Free Compositions in the Forms of Rods of Square and Circular Cross Section," *Combust. Explos. Shock Waves 19*, 542–545.

Shkadinsky, K. G.; Khaikin, B. I.; Merzhanov, A. G. 1971. "Propagation of Pulsating Exothermic Reaction Front in the Condensed Phase," *Combust. Explos. Shock Waves 1*, 15–22.

Showalter, K. 1981. "Trigger Waves in the Acidic Bromate Oxidation of Ferroin," *J. Phys. Chem. 85*, 440.

Showalter, K. 1995. "Bringing out the Order in Chaos," *Chem. in Britain 31*, 202–205.

Showalter, K.; Noyes, R. M.; Turner, H. 1979. "Detailed Studies of Trigger Wave Initiation and Detection," *J. Am. Chem. Soc. 101*, 7463–7469.

Siegert, F.; Weijer, C. 1989. "Digital Image Processing of Optical Density Wave Propagation in *Dictyostelium discoideum* and Analysis of the Effects of Caffeine and Ammonia," *J. Cell Sci. 93*, 325–335.

Simoyi, R. H.; Wolf, A.; Swinney, H. L. 1982. "One-Dimensional Dynamics in a Multicomponent Chemical Reaction," *Phys. Rev. Lett. 49*, 245–248.

Sivashinsky, G. I. 1981. "On Spinning Propagation of Combustion," *SIAM J. Appl. Math. 40*, 432–438.

Smith, K. W.; Noyes, R. M. 1983. "Gas Evolution Oscillators. 3. A Computational Model of the Morgan Reaction," *J. Phys. Chem. 87*, 1520–1524.

Soltzberg, L. J. 1989. "Self-Organization in Chemistry—The Larger Context," *J. Chem. Ed. 66*, 187.

Sørensen, P. G. 1974. "Physical Chemistry of Oscillatory Phenomena—General Discussion," *Faraday Symp. Chem. Soc., 9*, 88–89.

Sørensen, P. G.; Hynne, F. 1989. "Amplitudes and Phases of Small-Amplitude Belousov-Zhabotinskii Oscillations Derived from Quenching Experiments," *J. Phys. Chem. 93*, 5467–5474.

Sparrow, C. 1982. *The Lorenz Equations: Bifurcations, Chaos and Strange Attractors.* Springer-Verlag: New York.

Steinbock, O.; Kettunen, P.; Showalter, K. 1995a. "Anisotropy and Spiral Organizing Centers in Patterned Excitable Media," *Science 269*, 1857–1860.

Steinbock, O.; Tóth, A.; Showalter, K. 1995b. "Navigating Complex Labyrinths: Optimal Paths from Chemical Waves," *Science 267*, 868–871.

Stewart, I. 1989. *Does God Play Dice? The Mathematics of Chaos.* Blackwell: Oxford, U.K.

Strizhak, P.; Menzinger, M. 1996. "Nonlinear Dynamics of the BZ Reaction: A Simple Experiment that Illustrates Limit Cycles, Chaos, Bifurcations and Noise," *J. Chem. Ed. 73*, 868–873.

Strobl, G. 1996. *The Physics of Polymers.* Springer: New York, 1996.

Strogatz, S. H. 1994. *Nonlinear Dynamics and Chaos.* Addison-Wesley: Reading, Mass.

Strunina, A. G.; Dvoryankin, A. V.; Merzhanov, A. G. 1983. "Unstable Regimes of Thermite System Combustion," *Combust. Explos. Shock Waves 19*, 158–163.

Stryer, L. 1995. *Biochemistry*, 4th ed. Freeman: New York.

Stuchl, I.; Marek, M. 1982. "Dissipative Structures in Coupled Cells: Experiments," *J. Chem. Phys. 77*, 2956–2963.

Su, S.; Menzinger, M.; Armststrong, R. L.; Cross, A.; Lemaire, C. 1994. "Magnetic Resonance Imaging of Kinematic Wave and Pacemaker Dynamics in the Belousov-Zhabotinsky Reaction," *J. Phys. Chem. 98*, 2494–2498.

Summerlin, L. R.; Borgford, C. L.; Ealy, J. B. 1988. *Chemical Demonstrations. Volume 1.* American Chemical Society: Washington, D.C.

Swanson, P. D.; Ottino, J. M. 1990. "A Comparative Computational and Experimental Study of Chaotic Mixing of Viscous Fluids," *J. Fluid Mech. 213*, 227–249.

Swartz, C. J. 1969. "On Chemical Kinetics," *J. Chem. Ed. 46*, 308–309.

Swinney, H. L. 1983. "Observations of Order and Chaos in Nonlinear Systems," *Physica 7D*, 3–15.

Szabo, A. 1988. "Fluctuations in the Polymerization of Sickle Hemoglobin. A Simple Analytic Model," *J. Mol. Biol. 199*, 539–542.

Szalay, J.; Nagy, I. P.; Barkai, I.; Zsuga, M. 1996. "Conductive Composites Prepared via a Propagating Polymerization Front," *Die Ang. Makr. Chem. 236*, 97–109.

Takens, F. 1981. "Detecting Strange Attractors in Turbulence," *Lec. Notes Math. 898*, 366–381.

Talanquer, V. 1994. "A Microcomputer Simulation of the Liesegang Phenomena," *J. Chem. Ed. 71*, 58–62.

Tam, W. Y.; Swinney, H. L. 1987. "Mass Transport in Turbulent Couette–Taylor Flow," *Phys. Rev. A36*, 1374–1381.

Tam, W. Y.; Horsthemke, W.; Noszticzius, Z.; Swinney, H. L. 1988a. "Sustained Spiral Waves in a Continuously Fed Unstirred Chemical Reactor," *J. Chem. Phys. 88*, 3395–3396.

Tam, W. Y.; Vastano, J. A.; Swinney, H. L.; Horsthemke, W. 1988b. "Regular and Chaotic Spatiotemporal Patterns," *Phys. Rev. Lett. 61*, 2163–2166.

Taylor, G. 1950. "The Instability of Liquid Surfaces when Accelerated in a Direction Perpendicular to their Planes. I.," *Proc. Roy. Soc. (London), Ser. A 202*, 192–196.

Taylor, G. I. 1954. "Diffusion and Mass Transport in Tubes," *Proc. Phys. Soc. 67b*, 57–69.

Teymour, F.; Ray, W. H. 1992a. "The Dynamic Behavior of Continuous Polymerization Reactors—V. Experimental Investigation of Limit-Cycle Behavior for Vinyl Acetate Polymerization," *Chem. Eng. Sci. 47*, 4121–4132.

Teymour, F.; Ray, W. H. 1992b. "The Dynamic Behavior of Continuous Polymerization Reactors—VI. Complex Dynamics in Full-Scale Reactors," *Chem. Eng. Sci. 47*, 4133–4140.

Tran-Cong, Q.; Harada, A. 1996. "Reaction-Induced Ordering Phenomena in Binary Polymer Mixtures," *Phys. Rev. Lett. 76*, 1162–1165.

Tritton, D. J. 1988. *Physical Fluid Dynamics.* Oxford University Press: Oxford, U.K.

Trommsdorff, E.; Köhle, H.; Lagally, P. 1948. "Zur Polymerisation des Methacrylsäuremethylesters," *Makromol. Chem. 1*, 169–198.

Turányi, T.; Tomlin, A. S.; Pilling, M. J. 1993. "On the Error of the Quasi-Steady Approximation," *J. Phys. Chem. 97*, 163–172.

Turing, A. M. 1952. "The Chemical Basis of Morphogenesis," *Philos. Trans. Roy. Soc. London, Ser. B 237*, 37–72.

Turner, J. S. 1965. "The Coupled Turbulent Transports of Salt and Heat Across a Sharp Density Interface," *Int. J. Heat Mass Transfer 8*, 759–767.

Turner, J. S. 1979. *Buoyancy Effects in Fluids.* Cambridge University Press: Cambridge, U.K.

Turner, J. S. 1985. "Multicomponent Convection," *Annu. Rev. Fluid Mech. 7*, 11–44.

Turner, J. S.; Roux, J. C.; McCormick, W. D.; Swinney, H. L. 1981. "Alternating Periodic and Chaotic Regimes in a Chemical Reaction—Experiment and Theory," *Phys. Lett. A85*, 9–12.

Tyson, J. 1973. "Some Further Studies of Nonlinear Oscillations in Chemical Systems," *J. Chem. Phys. 58*, 3919–3930.

Tyson, J. 1985. "A Quantitative Account of Oscillations, Bistability, and Traveling Waves in the Belousov–Zhabotinskii Reaction," in *Oscillations and Traveling Waves in Chemical Systems.* (Field, R. J.; Burger, M., Eds.). Wiley: New York; pp. 93–144.

Tyson, J. J. 1975. "Classification of Instabilities in Chemical Reaction Systems," *J. Chem. Phys. 62*, 1010–1015.

Tyson, J. J.; Fife, P. C. 1980. "Target Patterns in a Realistic Model of the Belousov–Zhabotinskii Reaction," *J. Chem. Phys. 73*, 2224–2237.

Tyson, J. J.; Keener, J. P. 1988. "Singular Perturbation Theory of Traveling Waves in Excitable Media," *Physica D 32*, 327–361.

Tyson, J. J.; Novak, B.; Odell, G.; Chen, K.; Thron, C. D. 1996. "Chemical Kinetic Understanding of Cell-Cycle Regulation," *Trends Biochem. Sci. 21*, 89–96.

Tzalmona, A.; Armstrong, R. L.; Menzinger, M.; Cross, A.; Lemaïre, C. 1990. "Detection of Chemical Waves by Magnetic Resonance Imaging," *Chem. Phys. Lett. 174*, 199–202.

Tzalmona, A.; Armstrong, R. L.; Menzinger, M.; Cross, A.; Lemaire, C. 1992. "Measurement of the Velocity of Chemical Waves by Magnetic Imaging," *Chem. Phys. Lett. 188*, 457–461.

Ulbrecht, J. J.; Pearson, G. K. 1985. *Mixing of Liquids by Mechanical Agitation.* Gordon and Breach: London.

Vajda, S.; Valko, P.; Turányi, T. 1985. "Principal Component Analysis of Kinetic Models," *Int. J. Chem. Kinet. 17*, 55–81.

Vance, W.; Ross, J. 1988. "Experiments on Bifurcation of Periodic Structures into Tori for a Periodically Forced Chemical Oscillator," *J. Chem. Phys. 88*, 5536–5546.

Váradi, Z.; Beck, M. T. 1973. "Inhibition of a Homogeneous Periodic Reaction by Radical Scavengers," *J. Chem. Soc., Chem. Commun.* 30–31.

Vemaulpalli, G. K. 1990. "A Discourse on the Drinking Bird," *J. Chem. Ed. 67*, 457–458.

Venkataraman, B.; Sørensen, P. G. 1991. "ESR Studies of the Oscillations of the Malonyl Radical in the Belousov–Zhabotinsky Reaction in a CSTR," *J. Phys. Chem. 95*, 5707–5712.

Vidal, C.; Roux, J. C.; Rossi, A. 1980. "Quantitative Measurements of Intermediate Species in Sustained Belousov–Zhabotinsky Oscillations," *J. Am. Chem. Soc. 102*, 1241–1245.

Villermaux, J. 1983. "Mixing in Chemical Reactors," *ACS Symp. Ser. 226*, 135–186.

Villermaux, J. 1991. "Mixing Effects on Complex Chemical Reactions in a Stirred Reactor," *Rev. Chem. Eng. 7*, 51–108.

Vlachos, D. G. 1995. "Instabilities in Homogeneous Nonisothermal Reactors: Comparison of Deterministic and Monte Carlo Simulations," *J. Chem. Phys. 102*, 1781–1790.

Volpert, V. A.; Mergabova, I. N.; Davtyan, S. P.; Begishev, V. P. 1986. "Propagation of the Caprolactam Polymerization Wave," *Combust. Explos. Shock Waves 21*, 443–447.

Volpert, V. A.; Volpert, V. A.; Garbey, M.; Pojman, J. A. 1996. "Stability of Reaction Fronts," in *Gas-Phase Chemical Reaction Systems: Experiments and Models 100 Years after Max Bodenstein* (Wolfrum, J.; Volpp, H.-R.; Rannacher, R.; Warnatz, J., Eds.). Springer-Verlag: Berlin, Heidelberg; pp. 309–317.

Volterra, V. 1926. "Fluctuations in the Abundance of a Species Considered Mathematically," *Nature 118*, 558–560.

von Bünau, G.; Eigen, M. 1962. "Zur Kinetik der Jod-Sulfit Reaktion," *Z. Phys. Chem. 32*, 27–50.

Weimer, J. J.; Smith, W. L. 1994. "An Oscillating Reaction as a Demonstration of Principles Applied in Chemistry and Chemical Engineering," *J. Chem. Ed. 71*, 325–327.

Weiner, J.; Schneider, F. W.; Bar-Eli, K. 1989. "Delayed-Output-Controlled Chemical Oscillations," *J. Phys. Chem. 93*, 2704–2711.

Weiss, G. H. 1967. "First Passage Time Problems in Chemical Physics," *Adv. Chem. Phys. 13*, 1–18.

Welsh, B. J.; Gomatam, J. 1990. "Diversity of Three-Dimensional Chemical Waves," *Physica D 43*, 304–317.

White, S. R.; Kim, C. 1993. "A Simultaneous Lay-Up and in situ Cure Process for Thick Composites," *J. Reinforced Plastics and Comp. 12*, 520–535.

Whitney, H. 1935. "Differential Manifolds in Euclidean Space," *Proc. Natl. Acad. Sci. U.S.A. 21*, 462–464.

Wiesenfeld, K.; Hadley, P. 1989. "Attractor Crowding in Oscillator Arrays," *Phys. Rev. Lett. 62*, 1335–1338.

Wiesenfeld, K.; Bracikowski, C.; James, G.; Roy, R. 1990. "Observation of Antiphase States in a Multimode Laser," *Phys. Rev. Lett. 65*, 1749–1752.

Winfree, A. 1972. "Spiral Waves of Chemical Activity," *Science 175*, 634–635.

Winfree, A. T. 1980. *The Geometry of Biological Time*. Springer: New York.

Winfree, A. T. 1983. "Sudden Cardiac Death: A Problem in Topology," *Sci. Am. 248(5)*, 144–160.

Winfree, A. T. 1984. "The Prehistory of the Belousov–Zhabotinsky Oscillator," *J. Chem. Ed. 61*, 661–663.

Winfree, A. T. 1986. "Benzodiazepines Set the Clock," *Nature 321*, 114–115.

Winfree, A. T. 1991. "Varieties of Spiral Wave Behavior: An Experimentalist's Approach to the Theory of Excitable Media," *Chaos 1*, 303–334.

Winfree, A. T.; Strogatz, S. H. 1983a. "Singular Filaments Organize Chemical Waves in Three Dimensions. I. Geometrically Simple Waves," *Physica 8D*, 35–49.

Winfree, A. T.; Strogatz, S. H. 1983b. "Singular Filaments Organize Chemical Waves in Three Dimensions. II. Twisted Waves," *Physica 9D*, 65–80.

Winfree, A. T.; Strogatz, S. H. 1983c. "Singular Filaments Organize Chemical Waves in Three Dimensions. III. Knotted Waves," *Physica 9D*, 333–345.

Winfree, A. T.; Strogatz, S. H. 1984. "Singular Filaments Organize Chemical Waves in Three Dimensions. IV. Wave Taxonomy," *Physica 13D*, 221–233.

Winston, D.; Arora, M.; Maselko, J.; Gáspar, V.; Showalter, K. 1991. "Cross-Membrane Coupling of Chemical Spatiotemporal Patterns," *Nature 351*, 132–135.

Wolf, A.; Swift, J. B.; Swinney, H. L.; Vastano, J. A. 1985. "Determining Lyapunov Exponents from a Time Series," *Physica D 16*, 285–317.

Yamaguchi, T.; Kuhnert, L.; Nagy-Ungvarai, Z.; Müller, S. C.; Hess, B. 1991. "Gel Systems for the Belousov–Zhabotinskii Reaction," *J. Phys. Chem. 95*, 5831–5837.

Yokota, K.; Yamazaki, I. 1977. "Analysis and Computer Simulation of Reduced Nicotinamide Dinucleotide Catalyzed by Horseradish Peroxidase," *Biochemistry 16*, 1913–1920.

Zaikin, A. N.; Zhabotinskii, A. M. 1970. "Concentration Wave Propagation in Two-Dimensional Liquid-Phase Self-Oscillating System," *Nature 225*, 535–537.

Zaikin, A. N.; Zhabotinsky, A. M. 1973. "A Study of a Self-Oscillatory Chemical Reaction: II. Influence of Periodic External Force," in *Biological and Biochemical Oscillators* (Chance, B.; Pye, E. K.; Ghosh, A. K.; Hess, B., Eds.). Academic Press: New York; pp. 81–88.

Zeldovich, Y. B.; Barenblatt, G. I.; Librovich, V. B.; Makhviladze, G. M. 1985. *The Mathematical Theory of Combustion and Explosions.* Consultants Bureau: New York.

Zhabotinsky, A. M. 1964a. "Periodic Kinetics of Oxidation of Malonic Acid in Solution (Study of the Belousov Reaction Kinetics)," *Biofizika 9*, 306–311.

Zhabotinsky, A. M. 1964b. "Periodic Liquid-Phase Oxidation Reactions," *Dokl. Akad. Nauk SSSR 157*, 392–395.

Zhabotinsky, A. M.; Buchholtz, F.; Kiyatkin, A. B.; Epstein, I. R. 1994. "Oscillations and Waves in Metal-Ion-Catalyzed Bromate Oscillating Reactions in Highly Oxidized States," *J. Phys. Chem. 97*, 7578–7584.

Zhabotinsky, A. M.; Eager, M. D.; Epstein, I. R. 1993. "Refraction and Reflection of Chemical Waves," *Phys. Rev. Lett. 71*, 1526–1529.

Zhabotinsky, A. M.; Müller, S. C.; Hess, B. 1990. "Interaction of Chemical Waves in a Thin Layer of Microheterogeneous Gel with a Transversal Chemical Gradient," *Chem. Phys. Lett. 172*, 445–448.

Zhabotinsky, A. M.; Müller, S. C.; Hess, B. 1991. "Pattern Formation in a Two-Dimensional Reaction–Diffusion System with a Transversal Chemical Gradient," *Physica D 49*, 47–51.

Zimm, B. H.; Bragg, J. K. 1959. "Theory of the Phase Transition between Helix and Random Coil in Polypeptide Chains," *J. Chem. Phys. 32*, 526–535.

Zimmerman, E. C.; Ross, J. 1984. "Light Induced Bistability in $S_2O_6F_2 \leftrightarrow SO_3F$: Theory and Experiment," *J. Chem. Phys. 80*, 720–729.

Zimmerman, E. C.; Schell, M.; Ross, J. 1984. "Stabilization of Unstable States and Oscillatory Phenomena in an Illuminated Thermochemical System: Theory and Experiment," *J. Chem. Phys. 81*, 1327–1335.

Zwietering, T. N. 1984. "A Backmixing Model Describing Micromixing in Single-Phase Continuous-Flow Systems," *Chem. Eng. Sci. 39*, 1765–1778.

Zykov, V. S. 1980. "Analytical Evaluation of the Dependence of the Speed of an Excitation Wave in a Two-Dimensional Excitable Medium on the Curvature of its Front," *Biophysics 25*, 906–911.

Index